# STEWARDS OF THE LAND

# STEWARDS OF THE LAND

## The American Farm School and Greece in the Twentieth Century

Brenda L. Marder

Mercer University Press

Macon, Georgia

Stewards of the land : the American Farm School and Greece in the twentieth century

© 2004 Mercer University Press
1400 Coleman Avenue
Macon, Georgia 31207

∞The paper used in this publication meets the minimum requirements of American National Standard for Information Sciences—Permanence of Paper for Printed Library Materials, ANSI Z39.48-1992.

Library of Congress Cataloging-in-Publication Data

Marder, Brenda L.
 Stewards of the land : the American Farm School and Greece in the twentieth century / Brenda L. Marder.— 1st ed.
    p. cm.
Book 1 was originally published: Boulder : East European Quarterly, 1979.
Includes bibliographical references and indexes.
 ISBN 0-86554-844-7 (alk. paper)—ISBN 0-86554-849-8 (pbk. : alk. paper)
1. American Farm School (Greece)—History. 2. Agriculture—Study and teaching (Higher)—Greece—History. 3. Greece—History. I. Title.
 S539.G8A455 2004
 630'.71'1495--dc22
2003026713

For Cynthia, John, David, and Everett with whom
I shared my life in Greece

# CONTENTS FOR BOOK 1

# CONTENTS FOR BOOK 2

# Foreword to Book One

## Peter Bien

Viewing the destructive first half of the twentieth century with its many wars, we are tempted to consider human nature inherently corrupt, or at least to conclude that those years were a time of disintegrating values. How reassuring, then, to discover in reading book 1 of Brenda Marder's centenary history that from 1904 to 1949, on a tiny plot of ground in Macedonia, the human spirit functioned as it truly should: creatively.

Yet the American Farm School, part of whose encouraging story is here related, was not a Prince Charming who triumphed effortlessly over numerous ogres. Nor did it succeed because of disembodied idealism. Its idealism was of the applied, practical, realistic kind, where chores in henhouse, cowbarn and pigpen played a central role. Whatever the precise cause of its success (I suspect a mutually fructifying collaboration between work and idealism—purpose energizing action, action keeping purpose adaptable), the result is clear. The school survived in spite of the disintegrative forces that ought to have destroyed it many times over.

Most spectacularly, it survived politics. Brenda Marder's account unavoidably becomes the history of Balkan turmoil, seen from a vantage point often ignored in newspaper headlines or governmental pronouncements, and authenticated by documents from the school's archives. Readers gain in two ways: they learn about Balkan affairs while at the same time they are shown how forces such as the farm school, which acted quietly over this period to improve international relations between Greece and the United States, accomplish political ends through non-political means.

Less spectacularly, perhaps, but no less importantly, the school survived religion. The great energy that formed and sustained it was

American Protestant evangelicalism (with important additions from British and American Quakerism). I do not suggest that the farm school ever repudiated this initial energy, but it did escape the dangers of dogmatism, inflexibility, and religious imperialism lurking as potentials in its missionary heritage. Just as the school's political energies survived because of a refusal to be political, so the original religious energies survived because of a refusal to allow official religiosity to kill them. Certainly the school has never aspired to purvey superior dogma, although it most definitely aspired to purvey superior milk, seed, and cattle.

Clearly, this idealism—practical, adaptable, capable of changing shape without losing integrity, attentive to individuals, and believing that small is beautiful—was able not only to survive, but to flourish, even in the first half of the twentieth century with its disintegrating values.

Official blurbs from the school office will emphasize, quite understandably, what the institution has introduced into Greek agriculture over the years: the first course in farm machinery maintenance and repair, the first pasteurized milk operation, improved breeding stock, and so forth. All this is important. But I choose to emphasize once again the school's effect on individual people, who are so often forgotten when we invoke abstractions such as agriculture or education. A botanical metaphor is appropriate here. We know that cross-fertilization can be more interestingly creative than self-fertilization, and furthermore (I quote Darwin himself) that "the good effects of cross-fertilization are transmitted by plants to the next generation." Decade after decade, the American Farm School has stimulated cross-fertilization between Americans and Greeks. The northern and southern peoples have so much to offer each other, since they possess such different yet complementary virtues. But cultural exchange organized in governmental offices, not to mention tourism, is shallow and may even be harmfully deceptive. The farm school has allowed cross-fertilization to occur in natural, leisurely ways, randomly, in a context of purposeful work—beneath a tractor oil sump or on top of a wagon carrying in the harvest. And, as all of us who have enjoyed such contacts know, the good effects are easily transmitted to the next generation.

If only by a jot, these good effects will increase the chance that the human spirit may continue to bud creatively in the future, despite all the hatred, violence, and dogmatism of the half century during which the American Farm School quietly carried out its constructive mission of peace, economic development, agricultural education, and international understanding.

Let us read book 1 of its history with thanks, hoping—and expecting—that the story presented here will be but prologue to an equally reassuring tale written at the twentieth century's end.

Hanover, New Hampshire
January 29, 1979

Thessaloniki, Greece
October 30, 2000

Riparius, New York
July 23, 2002

# AUTHOR'S NOTE

Book 1 of *Stewards of the Land: The American Farm School and Greece in the Twentieth Century* was originally published as *Stewards of the Land: The American Farm School and Modern Greece* (Boulder: East European Quarterly/ Columbia University Press, 1979). Book 1 as reprinted in this volume has been edited only for inaccuracies that have been called to my attention over the years, some infelicitous sentence structure, and certain verbal tense constructions that render today's reading anachronistic. No fresh research was conducted and no new interpretations or conclusions were added for this republication of book one.

The reader should be aware that events in the course of the last three decades since book 1 was researched and written have changed the map of the Balkans. Macedonia is viewed here in its nineteenth-and early twentieth-century form, first as a Turkish province, and later as it was divided between competing Balkan states. Consequently, the discussion pertaining to Macedonia, valid at the time of the research and writing of book 1, could not have taken cognizance of the new nation called the Former Yugoslav Republic of Macedonia (FYROM), created in 1991.

# PREFACE

Book 1 of this history of the American Farm School,[1] covering the period 1904–1949, is an effort to explain the spirit and survival of this unique agricultural and vocational institution amid some of the most chaotic and violent events in twentieth-century European history. To accomplish this I have integrated the development of the school into the broader themes of Macedonian, Ottoman, Balkan, and modern Greek history and the two world wars. Each of these subjects is examined to determine in what measure it affected the internal development of the school. By the same token, this book describes the ways in which the farm school was able to play a positive role in the lives of the Greek people as they struggled against the turbulence created by these events. Also, this narrative traces the Hellenization of the farm school as it evolved, conceptually, from an American outpost in the Balkans into an institution principally Greek in character.

To tell this story I have examined all available documents, testing one against the other to arrive at a balanced interpretation. Yet I must advise the reader that this account has been more than a detached study: it has been a labor of love. I spent the academic year 1966–1967 at the American Farm School with my husband and three children and taught English at the Quaker Girls School. I came to know and love the youngsters. At that time I heard tales about the individuals who guided the school through the first five decades of its life. The farm school is much the tale of four gigantic personalities: the founder, John Henry House; his wife, Susan Adeline House; their son, Charles Lucius House, director from 1929 to 1955; and his wife, Ann Kellogg House. Their determination and courage in the Balkan wilderness sustained the school through some of the larger events in European history. Indeed, the

---

[1] The official name of the school is the Thessalonika Agricultural and Industrial Institute. The name American Farm School was first given to it by a number of British servicemen who became acquainted with the school while serving on the Macedonian front in World War I. (Francis Crow, British Consul General, Graduation Day Speech, 20 June 1925, AFS Archives). I preferred throughout the book to use the name American Farm School, since it is by this title that the school is popularly known in Greece and abroad.

forceful character of these four individuals goes a long way in explaining the survival of the American Farm School through this turbulent era. They were supported by a host of equally inspired individuals. Foremost among them is Ruth Eleanor House, daughter of the founder, who, along with others mentioned in this book, dedicated her life to the rural youth of Greece.

The American historian Richard Hofstadter wrote in the *Progressive Historians* that "memory is the thread of personal identity. Men who have achieved any civic existence at all must, to sustain it, have some kind of history." And this is the reason I have written this book, that the history of the American Farm School might remain intact, and that many of those who have had their "civic existence" upon its campus might be sustained within history.

Since my book tells more how the spirit of the school was forged than how it was absorbed by the students, I hope that some day a sociologist will undertake to study the ways in which the perceptions of the youngsters have been influenced by their farm-school experience.

Book 1 of this narrative ties the school to a series of dramatic events that terminate with the end of the Greek civil war in 1949. The reconstruction of postwar Greece and the institution's role under the dynamic leadership of Bruce M. Lansdale are studied in book 2.

The book was only half of my twofold project. In 1971, I began to gather documents from Europe and the United States for the establishment of the farm school archives. The archives now house a wealth of primary source materials. (See the footnotes and selected bibliography for a description of holdings.) Ann Kellogg House—in her tireless search for materials, the depositing of her own personal papers in the archives, and her enthusiasm—contributed enormously to the project. Her remarkable memoirs, which she made available to me for the preparation of this book, were started in 1971 when she was eighty-four years old. Her cooperation, constant encouragement, and trust in me were certainly one of the main factors in the completion of the book and the progress of the archives.

I am grateful to many other people who helped me in the course of writing. To Nike Myer, who took over the job of cataloguing the archival materials, I owe many thanks. Great credit must be given to Joice Nankivell Loch, whose biography of the founder, John Henry House, *A Life for the Balkans*, covering the early period (1902–1917), is the only published work on this subject.[2] Because she deals thoroughly with the details of the elder Houses' biography, I have written only a synopsis of their lives, to set their personalities into the perspective from which I view their story. The main objective of book 1 of this history is to present a detailed account of the farm school under the directorship of Charles Lucius House, for his story has not yet been told and the source materials are rich and untapped for this period.

I wish to thank Apostolos Vacalopoulos, professor of history, emeritus, of the University of Thessaloniki, who in a lengthy interview put into clear focus for me the many contradictions of the Macedonian Struggle. I am indebted to Vasilis Dimitriadis, director of the Macedonian Archives, Thessaloniki, who patiently researched, and translated from Turkish, tax entries and the original 1902 American Farm School land purchase. I thank Louisa Laourda of the Institute for Balkan Studies for graciously directing me to interviewees. To Tad (Elizabeth) and Bruce Lansdale, who extended to me every hospitality and freedom at the school, I wish to express my special debt.

Herbert P. Lansdale, Jr., read my manuscript almost in its entirety. I have leaned on him not only as a primary source of information, since he lived in Greece in an official capacity throughout many of these years, but also for his wise and careful counsel. S. Victor Papacosma, professor of Balkan and Modern Greek History at Kent State University, generously took time from his own research in Greece to read the entire manuscript and offered detailed suggestions on its historical and stylistic aspects. My daughter Cynthia typed portions of the first draft and edited the penultimate one. My husband, Everett, encouraged and advised me at every stage.

It is to Professor Peter Bien of Dartmouth College that I owe inexpressible gratitude. His interest in the manuscript and efforts in its behalf were most instrumental in bringing it to publication.

Brenda L. Marder
Palio Psychiko
Athens, Greece
1978

---

[2] *A Life in the Balkans: The Story of John Henry House of the American Farm School, Thessaloniki, Greece* was written under her maiden name J. M. Nankivell.

# Book 1

## 1904–1949

# INTRODUCTION

The American Farm School, a secondary level agricultural and industrial institute, is situated just outside the port city of Thessaloniki in Greek Macedonia. The school's location has placed it in the eye of an historic maelstrom: from the moment of its founding in 1904 it has witnessed and been affected by a series of dramatic events. Macedonia has been the scene of violent clashes among antagonistic Balkan nationalities, familiarly known as the "Macedonian Struggle"; the setting for the 1908 Young Turk Revolt; the main battlefield for the Macedonian front in the First World War; and the region for the resettlement of more than a half-million Greek refugees from Asia Minor after 1922. Occupied by the Axis from 1941 to 1944, this heartland in the Balkans further suffered from the tragic Greek civil war, which was not resolved until 1949. By virtue of its geographic position, the American Farm School was shaped in one way or another by all of these events. Even the founding of this institution by three American missionaries was a direct consequence of the Turkish domination of Macedonia and the Macedonian Struggle.

Considering the magnitude and close succession of these events, the mere survival of the farm school is itself an epic tale. The fact that its original spirit has not been warped or the dynamism of its life sapped is remarkable. When Bruce Lansdale, who became director of the school in 1955, was asked to explain its survival he answered, "The affection the people have for it."[1] The affection is, no doubt, a real factor, yet the school's survival is due to a complexity of reasons.

---

[1] Bruce Lansdale, interview with author, Thessaloniki, 7 September 1966, written, AFS Archives

A commentator has suggested that one of the main factors in the school's survival has been the continuity of policy maintained by its first three directors.[2] On my initial visit in 1966 I strolled with Director Lansdale past the gardens in front of James Hall where the bronze busts of Dr. John Henry House (founder-director, 1904–1929) and his son, Charles (director, 1929–1955) are displayed. Pointing to the nearest bust, I asked if that were the first director. "Yes," answered Lansdale. "He is the father, over there is the son, so I must be the—" Lansdale's voice trailed off leaving the religious innuendo to be carried by a blast of the Vardaris, a violent wind that funnels through Northern Greece, sweeping down from the northern border. He strode double-time toward his office, his coat jacket two sizes too large and a few decades out of style, unbuttoned and flapping in the wind.

However irreverent and hyperbolic, the intended reference is apt, for each of the three "persons" is somehow constructed from the same substance. Each is, of course, a product of his own generation. Dr. House, born before the American Civil War, would have been incapable of Lansdale's reference to the Holy Ghost, for he was a pious missionary, abstemious regarding drink and tobacco, dressed in suit and tie even when he stood knee-deep in Macedonian mud behind a plough. He conceived of the school as "an American outpost in the Balkans," and it was first and foremost a method of bringing the boys to God as John Henry House conceived of him. Charles, his son, was a man closer to our time. Born in 1887, he became a civil engineer and worked at that profession for eight years before coming to the school in 1917. Although a person of steadfast faith, he was liberal in outlook and never formally joined any church. Any analysis of the documents and of Charlie's personality would reveal that he regarded his selection as director to be a "call" in the religious sense. Yet his main mission as he conceived it was to teach the boys to be good, productive farmers, to bring into their lives the principles of honesty and cooperation, to reinforce their Christianity through many means. He was not a missionary by strict definition but more a humanitarian, an educator. He was not adverse to taking a drink, and his smoking remained a bone of contention between him and his father, even

---

[2] Helen Polychronis, "Charles House," 1936, AFS Archives.

though he confined his "vices" to off-campus locations. He never regarded the school as an American outpost, but took the first steps toward integrating it into Greek life, the process that I call Hellenization.

The differences in personality between father and son were not basic, but created by the span of a generation. They were both strong, stubborn men and their dedication to the school was total even when devotion meant personal sacrifice and danger. Both believed that they were answering a call, both believed in the same educational and moral principles, and both had a deep respect for the Balkan farmer. Bruce Lansdale's personality, in most respects, is a contemporary extension of his predecessors'. Most significant in the common personality of the three men is their feeling for humanity, a characteristic that transcends any other they may share. The survival of the school, too, depended on the ability of the directors to keep themselves apolitical in the sense that they have never allowed their personal or political beliefs to color their dealings with the multitude of governments under whose rule the region was placed.

The school's spirit is just as intrinsic to its history as is its survival. The spirit still operative on campus is the result of a common perception. It was articulated by the founder, amplified by his son, and perpetuated by Bruce Lansdale. When the spirit was caught by the Greek staff and students, it was adapted to Greek values and became, through this diffusion, a bicultural phenomenon.

# CHAPTER 1

# THE BALKAN MAELSTROM

The people in the Balkans at the time of John Henry House's arrival in 1872, although in a state of transition, were still for the most part isolated from European developments. As centuries of Ottoman domination wore on, the Balkans had become a breeding ground for violence and hatred. By the turn of the nineteenth century, the Ottoman Empire in the Balkans had become a dying entity. The Greek revolution of 1821 had succeeded in freeing the Peloponnesos and Central Greece from Turkish rule. Bulgaria, which was House's first missionary station, began emerging from Ottoman rule half a century later. Still, in 1900 multitudes of Greeks and other nationalities dispersed throughout the Ottoman Empire remained trapped under Turkish rule. With conscious spite the Turks set these various Balkan nationalities at each other's throats over disputed areas that would gradually become free as the Ottomans lost their grip.

At the time of the founding of the American Farm School in 1904, the territory extending from the Black Sea as far as the Adriatic coast of today's Albania was European Turkey. This area comprised what is known as Eastern and Western Thrace, Epiros, Bosnia-Herzegovina (nominally under Ottoman rule but controlled by Austria after 1878), Albania, and Eastern and Western Macedonia. The peoples of this territory, mostly Christian and Muslim, were of Greek, Bulgarian, Serbian, or Albanian extraction, with a sprinkling of Vlachs (traced to Romanian-Roman origin) and a minority of Jews. Serbia, Greece, Bulgaria, and Albania (which gained statehood only in

1913) claimed nationals throughout much of the area.[1] Most of these peoples were attempting to forge nation states in the nineteenth and twentieth century and looked toward territorial expansion through annexations of lands that they claimed were predominantly inhabited by their own ethnic group.

In 1894 the Houses moved to Macedonia, the area of greatest friction. The question of what and where Macedonia is remains vaguely defined because it has never, since ancient times, been a concrete racial, linguistic, or political entity. For the sake of some designation for this discussion, we can call it a central area in the Balkans bounded on the east by the Rila and Rhodopi Mountains, on the north by the hills north of Skopje and the Shar Mountains, on the south by the Aegean coast and Mount Olympus stretching to the Pindos range, and on the west by Lake Ochrid and the Prespes Lakes.[2]

The last Macedonian state was conquered by the Romans in 146 BCE. The *Encyclopedia Britannica* expresses the issue precisely: "Macedonia is rather a political problem than a geographical entity."[3] (In the late twentieth century a nation called Macedonia was formed but it contained only a small portion of the old Ottoman province. The remaining portion of the former province is currently split between Bulgaria, Greece, and Albania.)

How did the notion of Macedonia as a political problem start? This dramatic chapter in the "Eastern Question" (the decay of the Ottoman Empire) began when the Turks—under pressure from the Russians and with a desire to play off one Balkan group against the other—freed the Bulgarian Orthodox Church from the Greek Patriarch in 1870. In 1882 the Greek Patriarch in Constantinople, resenting this action which reduced his power, excommunicated the Bulgarian Exarch (chief of the autonomous Bulgarian Church). The

---

[1] Charles Lucius House, "The New Greece and the American Farm School," n.d., AFS Archives. For a survey that excludes Greece but gives an excellent account of the rest of the Balkans, see Robert Lee Wolff, *The Balkans in Our Time* (New York: W.W. Norton & Co., 1967).

[2] For a discussion of the boundaries and national character of Macedonia see Elizabeth Barker, *Macedonia: Its Place in Balkan Power Politics* (London: Oxford University Press, 1950); L. S. Stavrianos, *The Balkans Since 1453* (New York: Dryden, 1958) 519; George Zotiades, *The Macedonian Controversy* (Thessaloniki: Institute for Balkan Studies, 1961) 13.

[3] *Encyclopedia Britannica* (1974) 16:561.

establishment of the Bulgarian Church and the resulting schism between the Greeks and the Bulgarians inaugurated the struggle between these groups for control of Macedonia, since both had significant populations there and an historical claim.[4] The Serbians, too, entered the struggle, which led to a bloody campaign to achieve control over this unfortunate Turkish province. So mixed was the ethnic dispersion of population that the French chose the word "Macedoine" to mean a salad full of diverse fruits.

The Bulgarians at first were the most active participants in the struggle, mounting in the last decade of the nineteenth century a fierce campaign to compel the inhabitants of Macedonia to declare themselves under the ecclesiastical control of the Bulgarian Exarch and thereby becoming ipso facto Bulgarians. The chief arm of the Bulgarian extremists was an organization called the Internal Macedonian Revolutionary Organization, familiarly known by its acronym, IMRO. Its radical members were Bulgarian "patriots" who aimed to lead Macedonia to autonomy and later to annex it to Bulgaria. By 1900 the Macedonian Struggle became full-blown when IMRO started to launch bloody attacks with armed guerrillas known as *komitadjis*. These bands became notorious for their crimes and outrages within the villages and countryside of disputed areas. They roamed into Macedonia, using terror and murder as a principal tool, burning whole villages and the farm crops of those who refused to accept Bulgarian nationality.[5]

In response to the Bulgarian revolutionary movement, the Greeks also formed squads of armed men and the Hellenic Council and the clergy of the diocese of Thessaloniki took it upon themselves to secure weapons for the Greek rebels. Their purpose was to win over villagers and protect them from intimidation by the opposing forces. An elaborate organization of Greek noncombatants also worked within the villages to keep defections at a minimum.

The most vicious of all the combatants during the period of the 1870s were the *Bashi-bazouks,* a special branch of irregulars attached to the Turkish army. Wild looking creatures wrapped in sashes packed

---

[4] Edward S. Forster, *A Short History of Modern Greece* (London: Methuen & Co., 1960) 38.

[5] Douglas Dakin, "British Sources Concerning the Greek Struggle in Macedonia, 1901–1909," *Balkan Studies* 2 1/2 (1961): 73.

with daggers and pistols, they sported large turbans, huge mustaches, and baggy trousers. Their terrain was the countryside and mountains, where they kept up a reign of terror. They were always ready to attack any Christians suspected of plotting an uprising. They kept the villages in such a state of fear that it was in many cases the *Bashi-bazouks* who were responsible for goading the locals into an uprising. They would then sweep in and respond to the revolt with a massacre. When the foreign powers called the Turks to task for these outrages, the Turks responded that the army was not involved in the excesses, and that they had no control over the *Bashi-bazouks*.[6]

Another important factor in the conflict was the interest of the Great Powers. Russia, for instance, always wanting to expand toward the Mediterranean, tended to support the Balkan peoples in their struggle against the Turks. Austria was most interested in acquiring access to the Aegean and feared the emergence of a strong Slavic state, such as a greater Bulgaria, that might absorb the landmass of Macedonia and be subject to Russian influence. The British, for their part, supported the Ottomans as a buffer to keep the Russians away from Constantinople and the straits. Under the provisions of an international agreement with the Great Powers, the Ottoman authorities, supervised by Austrian and Russian officials, conducted a population census in 1903 shortly before the first boys were gathered into the farm school. According to this census, Macedonia contained 1,720,007 Muslims, 648,962 Greeks, 577,734 Bulgarians, and 167,601 Serbs.[7]

Throughout these struggles in the latter part of the nineteenth century, the Turks, realizing their declining power and fearing the national tendencies and ethnic stirring of the Balkan peoples, employed methods calculated to divide the Christian subjects still further. In this way, they hoped to impede any coalition that Christians might form as a base for a united rebellion to dislodge the Ottoman authorities. Thus, fully aware of the divisive effects of religious differences, they gave Protestant missionaries rather free range throughout their empire. They furnished missionaries with the sultan's firman, a flashy, large document with the sultan's cipher in

---

[6] Nankivell, *A Life for the Balkans,* 41.

[7] Stavrianos, *The Balkans*, 517. Also, Zotiades, *Controversy*, 16. The Muslim population figure is considered to be inflated.

the center. This document commanded all officials to give hospitality to the missionaries and their animals.[8] As another means of dividing their Christian subjects, the Ottoman authorities encouraged them to compete with each other in laying claim to disputed areas. Since the language of a people was sometimes supposed to be a determining factor in ethnic identity, the Ottomans granted rights to the ethnic groups to operate schools. Naturally, educational institutions stressed the language, religion, history, and ambitions of their sponsors.

The schools were patterned after the European system, offering a classical education. The graduates of secondary schools went on to universities in the Balkan capitals or in Western Europe, siphoning from the rural areas young men with ability. As the population of the Balkan states was overwhelmingly peasant, such an educational structure was hardly suited to solve the problems of the great masses of people. No vocational or agricultural school existed to educate a Christian peasant who wished to better his life while remaining on the farm. Furthermore, school recruitment was discriminatory, since students were sought only from the professional, military, and bureaucratic classes.[9] Although a certain pride and genuine willingness to improve the living conditions of their own countrymen was developed in the students at these schools, their characters were not molded to the ideals of service, cooperation, and leadership; cooperation with alien nationalities and a spirit of tolerance were not features of nationalistic schools.

By the turn of the century, the role of education in the Balkans, especially in what was to become Greek Macedonia, was increasingly political. As such, it threatened to destroy social and traditional kinship patterns as the Macedonian Struggle intensified. Nationality papers, consisting mainly of baptismal records, had no meaning in those days; thus if a boy attended a school, he generally adopted the ethnic orientation of that school. A father might have several sons attending different institutions, each son thereby acquiring a different ethnic identity.

At this time, the Greeks, forming a national organization to halt the spread of Slavism in Macedonia, enlisted the aid of the Patriarch of Constantinople and the fledgling modern Greek state. By 1902, the

---

[8] Nankivell, *A Life for the Balkans*, 41.
[9] Charles Lucius House, "New Greece," AFS Archives.

year of the land purchase for what was to become the American Farm
School, there were over 1,000 Greek schools in Macedonia serving
over 70,000 students.[10] Not one of these was devoted to teaching
practical skills or agriculture. The fact that Greeks in Macedonia were
engaged principally in commerce and lived in the more urban areas
explains to some extent the exclusion of technical courses.

Similarly, the Bulgarian Exarchate, pressing for influence in
Macedonia, founded 592 schools serving about 30,000 pupils. The
Serbs, too, opened a number of their own schools. It is instructive to
note once again that no technical schools existed even though Serbs
and Slavs were widely engaged in agriculture. The Greek historian
Apostolos Vacalopoulos claims that in the second half of the
nineteenth century many workers came from Bulgaria to Monastir, a
major town in western Macedonia, to work in the fields. The
Bulgarians had some fame as farmers, while most Greeks remained
merchants until the end of the Balkan Wars.[11]

In stressing the political role of the schools and the critical
influence they exerted on youth, Douglas Dakin writes: "It was
precisely in those regions where Greek schools were numerous that
Hellenism was strongest and it was precisely in those regions where
exarchist and patriarchist school populations were evenly matched
that the struggle between the two churches was the fiercest. Both the
weakness and the strength of Hellenism in Macedonia corresponded
precisely to the position of Greek education."[12]

The educational system within Bulgaria gave large impetus to the
revolutionary movement in Macedonia, for the leaders of the
movement were products of that system. The missionaries felt that,
for the most part, this system turned out semi-educated ne'er-do-wells
who despised trade and manual labor. The economy of the country
was not suited to absorb them—as a result, embittered and idle, they

---

[10] Douglas Dakin, *The Unification of Greece, 1770–1923* (London: Ernest Benn, Ltd, 1972) 160. Also, Stephanos Papadopoulos, "Ecoles et associations grècques dans la Macédoine du Nord durant le dernier siecle de la domination turque," *Balkan Studies* 3 1/2 (1962): 397–442.

[11] Apostolos Vacalopoulos, interview with author, Thessaloniki, 19 June 1972, written, AFS Archives.

[12] Dakin, *Unification*, 161.

became revolutionaries. With the passing of time, the exarchist schools in Macedonia fell into the hands of the revolutionaries.

The Greeks, by and large, both in Greece itself and in Macedonia, did not produce revolutionary individuals. The greater portion of Greek students had some expectations after graduation such as business interests belonging to their families or relatives abroad who were willing to help them start some enterprise.[13] Historians and sociologists have often remarked on the Greek propensity for trade, government, or business. Traian Stoianovich comments as follows on the Greek work ethic: "Since that time [the eighteenth century] the Greek aversion toward cultivating the land has led to an increasingly persistent flight from the farm to the shop."[14]

Dr. John Henry House began working, after 1872, in the emerging state of Bulgaria. He had originally intended to direct his mission to the Turks but, because proselytizing among Muslims was forbidden by Ottoman law, he confined his activities to the Orthodox population.[15] He felt that under the stifling effects of four centuries of Ottoman domination Bulgarian Christians had fallen into what he considered demoralized ways. He wrote: "We came first with the desire to go to the Muslims and we found our brother Christians fallen in the mud and it was necessary to pick them up first and then we could go together to help our Muslim brothers."[16] But his attitude was one of understanding, not condemnation. He respected the Bulgarians and was quick to recognize that the problems of the people were sociological and economic as well as religious. Moreover, by the time he came to European Turkey a number of Bulgarians had already been converted to Protestantism by earlier groups of missionaries, giving House and his companions sympathetic and established communities in which to operate. Any animosity that he encountered from hostile groups was somewhat counterbalanced by the warmth and sense of

---

[13] Ibid.

[14] Traian Stoianovich, *A Study in Balkan Civilization* (New York: Alfred A. Knopf, 1967) 171. For an excellent account of industry and agriculture in the Ottoman Empire see Bernard Lewis, *The Emergence of Modern Turkey* (London: Oxford University Press, 1968) 31–36.

[15] William St. Clair, *That Greece Might Still Be Free: The Philhellenes in the War of Independence* (London: Oxford University Press, 1972) 196.

[16] John Henry House, "Stories About Our Life in Eski Zaghra, Turkey," 1936, AFS Archives.

success generated by the already converted communities. Of all the peoples of the Balkans, the Bulgarians were the most receptive to the American Protestant missionaries in the pre-World War I period.[17]

The four hundred years of Turkish dominance reinforced certain tendencies in the Balkan work ethic. Ottoman rule left little scope for private enterprise and gave scant promise that hard labor would contribute to a farmer's wellbeing. Thrift, a cornerstone of Dr. House's philosophy, was a value that was never fostered among the Christian inhabitants of the Ottoman Empire. What profit was there in being thrifty when the foreign authorities would take it all away? As a result of the onerous tax burden they shouldered, the more prosperous peasants abandoned fertile fields on the plain and moved to scrawny patches on the mountainside to avoid the tax collectors. By working their fertile fields they were taxed to the point of diminishing return.

One of the most corrosive aspects of the nationalist schools in Macedonia was their method of recruiting students. Because they wanted pupils for political purposes and were in keen competition to attract the youngsters, they offered not only free tuition but enticing benefits. In a society where labor was scorned, free education without any financial responsibility on the part of the student or his family reinforced the ethic that labor to gain desired ends was an unnecessary value. Nothing could run more contrary to the philosophy of John Henry House, founder of the American Farm School.

---

[17] W. W. Hall, *Puritans in the Balkans, The American Board of Mission in Bulgaria: A Study in Purpose and Procedure* (Sofia: Cultura Printing House, 1938) 45–46.

# CHAPTER 2

# THE FOUNDER: JOHN HENRY HOUSE

Born in 1845, John Henry House was the product of the Protestant work ethic. He was the grandson of a pioneer who in 1822 left Massachusetts to reconstruct his life in the Western Reserve (Ohio). Young John Henry was raised in Painesville, Ohio, where his father ran a general store and tended to his farm in nearby Leroy. This simple, rural, almost pioneer life developed in him a respect for labor, a toughness, and that element that Americans like to think of as Yankee ingenuity. He recalled in later life that "as a boy I was surrounded by tools and had a farm near with the opportunity of seeing things grow, taking care of horses, cows, etc. I could make kites fly and the active part of my youth gave me satisfaction. Children love to make things."[1]

He graduated from Western Reserve University in 1868 and received his bachelor of divinity degree from Union Theological Seminary in 1871. In 1872 he married Susan Adeline Beers. That same year the newlyweds left for their life in the Balkans, a life that would stretch into more than sixty years of service.

Of medium height, he carried himself so erect that he gave the illusion of being taller. Dignified and patriarchal in appearance, with penetrating blue eyes, his chin was rimmed by a trim beard, which in his declining years turned white. He had a tolerance for all peoples: nationality or color was never a measure by which he judged the worth of a person. His commitment to his mission, to his fellow human being, regardless of his own safety and comfort, was the mainstay of

---

[1] John Henry House, "The Ideal Education," n.d., AFS Archives.

his life, and this commitment, which was total, kept him always on balance, always on course. Whatever his political opinions, he never allowed them to influence his actions where the welfare of his mission, or later, the farm school, was involved. His ability to keep his distance from the factional and political animosities that swirled around him in the Balkans is, in fact, one of the fundamental reasons for the survival of the American Farm School through all of these turbulent years. In spite of the misery, poverty, violence, antagonism, jealousies, duplicities, and pettiness to which he was witness during the Balkan upheavals, he remained remarkably gentle.

The words of William St. Clair writing about an entirely different personality, Lord Byron, apply curiously enough to Dr. John Henry House: he "was never tempted to either cynicism or to withdrawal... [H]e never despaired and he never despised. These were rare and precious gifts."[2]

His students looked to him with great respect and admiration, for even as young boys they realized that he represented something quite exceptional. One of his former students referred to him and his son Charlie as "truly civilized men."[3] Against the severe backdrop of violence and hatred, the qualities of this man probably stood out in inordinate relief. He, in turn, respected the farm boys and especially the Balkan peasants, with whom he patiently discussed problems and techniques that created an atmosphere of mutual respect.

A man never idle, he was always at work with his hands, roaming the farm school, properly albeit incongruously dressed in suit jacket, suit pants, vest hung with a Phi Beta Kappa key, white collar and soft hat, as he grafted trees, sawed wood, or demonstrated the uses of a modern plough. He possessed a sense of tact and persuasiveness, for the success of his mission and later that of the school depended on his

---

[2] William St. Clair, *That Greece Might Still Be Free: The Philhellenes in the War of Independence* (London: Oxford University Press, 1972) 151.

[3] Alexander Andreou, interview with author, Thessaloniki, 8 June 1972, written, AFS Archives. The character analysis of Dr. House is a composite of several interviews by the author: Ann Kellogg House, Athens, 25 May 1971; Demitris Hadjis, Thessaloniki, 8 September 1972; Alexandros Andreou, Thessaloniki, 8 June and 6 September 1972; Patroclis Krisopoulos, Ekali, 4 October 1972; Ioannis Kanelopoulos, Thessaloniki, 14 September 1973; Edward Howell, Thessaloniki, 7 September 1972; Polly Beers Dillard, Ekali, 18 April 1972. All interviews are written and located in the AFS Archives.

ability to attract contributors, to please the American Board of Commissioners for Foreign Missions (ABCFM), to convince the school's board of trustees, to arouse his parishioners, and to walk the tightrope among the Turks, Bulgarians, and Greeks. Yet despite all this he was a soft-spoken man and, in many situations, a shy man. Only in the pulpit did his voice rise above the normal threshold; there he would give forth thunderously with admonitions about God and Satan, Good and Evil, Heaven and Hell. At such moments his blue eyes would be ablaze with an inner fire.

Like De Gaulle and Adenauer, who came into their full powers at an advanced age, and Goethe, who was capable of riding a horse for six hours at the age of sixty-four, Dr. House in his later years had the energy of a man in his prime. At the age of sixty, when he founded the American Farm School, his hair was gray, not yet white, and he looked much younger than his years.

He was a man of his time but also a visionary. With a natural bent for the land, a deep religious faith, a turn-of-the-century missionary's concern for "Christianizing" and saving souls, he had a dream: that of founding an educational institution for Balkan youths that would be unique. It would have as its goal Harvard President Eliot's precept of the "whole man." He also had the turn-of-the century belief in progress. He was convinced that improvement of the Balkan peasants' material conditions would contribute to the maintenance of peace in southeastern Europe. His intention was to get at what he considered the major problem in the Balkans. As he put it, "...the key to the solution in this whole Balkan problem lies in striking at the roots of the trouble, through the kind of education that will replace the bitter social and religious antagonism, the jealousy and misunderstandings that have helped to make this region the storm center of Europe, by the Christian ideals of brotherhood and service."[4] He thought that, by training youngsters to be not only modern farmers but also leaders in their rural communities, he could replace poverty with economic stability especially on the small farm. His desire was to travel outside the normal missionary range, which was usually confined to preaching, and to deal with the basic material and community needs of the peasant. As a missionary he realized, as

---

[4] John Henry House, "An American Outpost in the Balkans," n.d., AFS Archives.

few of his colleagues did, that it was important to educate as well as to spiritualize human beings mired in poverty, condemned to it through superstition and the use of ancient methods.

He had an interesting turn of mind in that it was both idealistic and practical: idealistic in the sense that he had dreams and schemes, practical in the sense that he was fully capable of translating those dreams into reality. Spiritually inclined, he saw the hand of God in everything. Polly Beers Dillard, a niece of Susan Adeline House's, remembers, "If a thing was to be, God would give Father House a sign. Any narrow escapes from danger, and they were innumerable, were effected by God. With the Houses, everything, but everything, was God."[5] In one anecdote, a pious visitor was touring the farm school fields. He commented to House, "Your crops are splendid. You must pray a lot." "Yes, we pray a lot, but we also spray," replied the minister.

The main thrust of his life was his "mission" in the traditional sense of the missionary's duty—that is, to bring the word of the Protestant God to the unenlightened and thus to lead them into the way of salvation. House's intention was always basically religious. The farm school, indeed each one of his practical and educational endeavors, was aimed primarily at spiritualizing the students. His philosophy of education was an intrinsic part of his mission: his institutions served as a vehicle to carry the Word. In this respect he fitted perfectly into his time and into the missionary mold.

Without a doubt, he was a man of action possessing a hard-driving, stubborn temperament that always looked for real solutions to a problem. Once an idea took hold of him he would not let it go until he saw it through to realization. This accounts for Ray Stannard Baker's reference to Dr. House "as a man on fire": "I have never been greatly interested in missions or missionary work—and am not—but I was deeply attracted by Dr. House. He was a man on fire if I ever knew one. He had faith and courage and boundless energy. I was astonished, I remember at that time, by the variety and daring of his activities and was impressed by his tact, his wisdom in meeting the difficulties that confronted him."[6]

---

[5] Polly Beers Dillard, interview with author, Ekali, 18 April 1972, AFS Archives.

[6] Ray Stannard Baker, talk delivered at an American Farm School dinner, Boston, 14 October 1931, AFS Archives.

In his activities at the farm school it became clear that his scope was wide and his approach to education novel. Certainly he was too conservative theologically to break the missionary mold, but he was radical enough in his sociological ideas and educational theories to crack it considerably, for he was one of the first missionaries to grasp the value of practical Christianity.

Joice Loch, Dr. House's biographer, understood his unique contribution. She writes that from the time of Saint Paul many people had heard the cry to come over into Macedonia and had responded. But, she stresses, "I think it is safe to say that Father House was probably the first man who not only heard the cry from the people but from the stony soil itself. He realized early in his mission life that help to the people must mean help to the land if it were to be effective. He was one of the first to preach practical Christianity."[7] His first post in European Turkey, where he came with his new bride in 1872, was located in Eski Zaghra (today Stara Zagora), an area that was later to become Bulgaria.

In one respect, at least, his young wife—"Mother House," as she became known to students and all close friends—seemed hardly a suitable candidate for a life in the Balkans. Born into very comfortable means in the heart of New York City, she had little in her background enabling her to steel herself against the rigors, dangers, and rural character of European Turkey. Unfortunately there is nothing in the American Farm School archives to give a clue as to how this naive young woman arrived at maturity, endured extreme privation, bore seven children and raised them amid such primitive circumstances. A great void exists in her biography, for we leave her in Joice Loch's book (as well as in the archival material) departing for Bulgaria while, according to the new Mrs. House, her own grandmother "expressed herself to some purpose, and said that she was scandalized at the idea of my being a missionary's wife. She said no girl should be married before she could make bread and cut out and sew a shirt."[8] We take up the thread of her life again a few weeks later and find her a fully mature individual having arrived in the Balkans, a monument to self-control and courage. Sadly, no documents exist enabling us to trace the development of this remarkable character.

---

[7] Joice Nankivell Loch, "Two Unforgettable People," 1952, AFS Archives.

[8] Nankivell, *A Life for the Balkans,* 33.

She was born Susan Adeline Beers, 14 September 1850 to a prominent, aristocratic family. Both her father and grandfather were Protestant ministers. It was in the home of her great-great-grandfather that a group of ministers planned the founding of Yale College. Her uncle, John Bigelow, was ambassador to France. The environment of her childhood was in direct contrast to that of her husband. Born on East Fourteenth Street, which was at that time very fashionable, she was treated to the sophistication of Gramercy Park, where she had relatives. As pastoral as pre-civil war New York appears to us now, it figures as poor trial ground for the primitive conditions awaiting her among Turks and Bulgarians.

It was perhaps this social and economic milieu that contributed to her being called a snob by her son Charlie. However, her snobbishness was never directed to the poor or the students. Her sense of superiority was most likely leveled at those people passing through the Balkans who evinced no interest in the Houses' life work. Yet she covered her snobbishness well, for her home in Thessaloniki later became an open, hospitable center for all travelers.

We do know that as a child she was petrified of the dark. As her mother died when she was a baby, she was sent to live with an aunt. The aunt, who was very strict, determined to break the little girl of this phobia and thus forced her to mount the stairs in the dark, lighting her lamp only when she reached her room. This merely reinforced her fears, so the phobia lingered into adulthood. Yet during her early life in the Balkans, when her own children were small, she was able to remain at home with tight self-control, caring for her children while her husband went off on extensive missionary trips at a time when the region was so torn by strife that both his life and hers were constantly in danger.[9]

Her heart opened to the Bulgarians as quickly as did her husband's: "As we became more versed in the ways of the country we found much to fascinate us. When callers came I dispensed hospitality in the manner of the country, serving our guests with a spoon of jam followed by a glass of water. As it is offered you say, 'May it be sweet to you.'"[10]

---

[9] Ann Kellogg House, tape recorded notes to the author, 1 September 1972, AFS Archives.

[10] Nankivell, *A Life for the Balkans*, 57.

The couple's first child, John Henry House, Jr., was born in 1873, barely a year after their arrival on station. Undaunted by the birth of their baby, the mother went almost immediately on missionary trips with Father House (he, too, was addressed with paternal nomenclature by students and intimates), the baby in its mother's arms, while at night the infant slept "in a cradle slung from a low ceiling," village style.[11]

Both Dr. House and Mrs. House lived to see seven children grow up and produce grandchildren and great-grandchildren. As the children grew up they went one by one to America for their education; thus the family was always separated. During their parents' lifetime the whole family managed to get together only three times.

To keep in touch, Mother House wrote the children weekly letters until she was in her eighties. When her fingers became so crippled by arthritis that she found handwriting unbearable, she learned to use a typewriter in the last decade of her life. Then she would peck away using one or two fingers making five copies (two children, Ruth and Charlie, were working at the farm school after 1917), adding to each a few words of personal advice. It is from this prodigious correspondence that much material was gathered for this history. She wrote with a definite sense of historical conscience, many of her letters to her children being an effort to articulate and inscribe for posterity the forces that shaped her and her husband's lives.

A person of tremendous energy and creativity, she always had ideas and was continually planning. For this her son Charlie would tease her by calling her "the general." But it is true that a great many projects carried out during their missionary days and at the school were initiated by her suggestions and plans. Father House, with his practical turn of mind, would implement her ideas.[12] Some former students have implied that perhaps she was the brilliant force at the school. But the facts indicate that husband and wife were a complementary pair, each with individual abilities and strengths. They both contributed so equally to the school that the institution as it stood (and as it stands today) would be inconceivable without the contribution of both personalities.

---

[11] Ibid., 59.

[12] Ann Kellog House, notes, 1 September 1972, AFS Archives.

Not only a dreamer, and a planner, she was very practical and thus was also the prodder and the organizer. She also had a strong sense of propriety, knowing intuitively when it was right to speak. Of an age and social class in which women were reluctant to speak on profound matters, her comprehensive mind was all the more unique. She wrote and spoke particularly on morals and politics: "It is impossible for me to believe that our country will go back to rum and the saloon again. I have not heard of any advantage offered as any excuse for the return of this curse. Oh, it seems as if I could not bear to think of it. Those who love God and their fellow men must surely continue to fight this terrible evil."[13]

There is little doubt about her acute intelligence. The inventory of her personal library on file at the American Farm School archives gives testimony to the breadth of her interests. Politically oriented, she remained current and keen regarding issues in the United States. The fact that Charlie and his brother became Democrats, while she considered herself a Republican (Father House was Republican in sentiment but eschewed family political debates), caused heated arguments. She and Charlie would both enjoy a passionate political discussion, and apparently she was not lacking in rhetorical ability.

Along with her aversion to liquor, she harbored strong reservations about card playing. Since she loved solitaire and other card games, she salved her conscience by using mock cards with substitute symbols.[14]

Her early photos as a young wife give the impression of a woman of striking confidence and poise. Her face was strong but not hard, and her dark eyes in each picture glow with a keen intelligence while the set of her chin proclaims a personality both unfearing and extremely positive. Even the pictures taken a few years before her death at ninety-six, which show her having aged into a frail wisp of a woman, still reflect her characteristic purposefulness.

Former students all registered a deep and abiding affection for this unusual person. She was always positive, always optimistic, and always cheerful to them. "No matter what messy thing we had created," a former student reported, "she would say, 'Oh, my, how

---

[13] Susan Adeline House to her children, Thessaloniki, 9 December 1932, AFS Archives.

[14] Ann Kellogg House, notes, 1 September 1972, AFS Archives.

very nice!'"[15] Her door was open to every boy. The classes at the
school were small in the early years, so they could be run with family
intimacy. Her world outlook, her life-style, her attitude toward the
boys, and the House family's approach to one another were main
factors in the formation of the character of the boys who attended.
Her role as the mother, both to her own family and as mother-in-
residence to the students, did much to set the tenor of that institution.
Given the conditions of the time, the hostile influences that shaped
Macedonian village life, few boys had ever encountered such human
relationships where kindness, gentleness, and cooperation were
constant and overriding values. No one was degraded; everyone was
encouraged. Even the animals were accorded a sense of dignity.
Former students have stressed that they learned their most enduring
lessons from the example set by Dr. and Mrs. House. This, in fact,
was the real meaning of the school: its spirit was personified by the
composite of these two personalities.

At their first post in Eski Zaghra, Dr. House served in the
beginning as pastor to an already established Protestant community.
He began immediately to travel into the villages in an effort to
become acquainted with the people. Using simple phrases, practicing
the few words he knew, he gradually expanded his fluency. It is a mark
of the man's intelligence that within six months he had taught
himself enough Bulgarian to be able to carry on a reasonable
conversation in that language. After one year he was preaching his
sermons in Bulgarian and would write them out beforehand in his
newly mastered tongue. Perfected in such a short time, his writing
ability proved a superb tool, for later he edited two newspapers in the
Bulgarian language, reaching many more people than he ever could
through speech alone.

The Houses' home in Eski Zaghra was always open to everyone,
and peasants came to sit at his table. His language capability, his
accessibility to the common man, put him into direct communication
with the indigenous population. In this way he was unique. Ann
Kellogg House, wife of Charlie House, remarks that their open-door
approach to the villagers was their own way of reaching the people, a
way not commonly shared by their fellow missionaries.

---

[15] Alexandros Andreou, interview with author, 6 September 1972, written, AFS
Archives.

Although he was in easy communication with the peasants—apparently they spoke openly to him—they nonetheless held him in some awe. Often when he walked through bandit-infested mountains, he would be closely trailed by groups of peasants who felt that if they went along with him they would be safe from the bandits. His patriarchal, bearded face, his lack of fear, and his manner with men all combined to give him in the minds of the superstitious folk the aura of a protecting saint.[16]

Even if all missionary hearts and doors did not open as widely and freely to the Bulgarians as did those of John Henry and Susan Adeline House, credit has been given in general to the Protestant missionaries for their positive influence in the emerging Bulgarian state and even in the area of the Ottoman Empire that was to become the modern state of Turkey after World War I. Robert Lee Wolff notes that Western ideas penetrated Bulgaria to a large extent through Protestant missionary schools. During the 1850s, some twenty years before the arrival of Dr. House in this region, Congregational and Methodist groups had founded schools there. "These institutions emphasized educational and social service rather than proselytizing. These kind, intelligent, and selfless men and women proved an inspiration to many Bulgarians."[17] Wickham T. Stead, editor of the *Review of Reviews*, corroborates and exaggerates this view of American missionaries. In citing the flow of Western ideas into the Ottoman Empire, he claims, "It is not too much to say that the only infusion of the ideas of western civilization into the eastern races has come not from Great Britain or Germany, but from America." He stresses that although American colleges (classical high schools) in the region were all religious schools, the main emphasis was enlightenment and education, not proselytizing. Stead explains:

Great Britain, absorbed in diplomatic naval and military affairs, has spent untold millions of dollars in propping up the political system established in the East. The American government on the other hand spends nothing and has

---

[16] Ann Kellogg House, interview with author, Athens, 25 May 1971, written, AFS Archives.
[17] Robert Lee Wolff, *The Balkans in Our Time* (New York: W.W. Norton & Co., 1967) 82. House was a Congregationalist.

accomplished nothing. But private American citizens subscribing out of their own pockets sums that in 50 years might have equaled the amount spent to build one modern ironclad, have left in every province of the Ottoman Empire the imprint of their intelligence and character.[18]

House's contacts with his fellow missionaries throughout the Ottoman Empire had a profound influence on his intellectual growth. Of his colleagues he himself wrote, "There were a number of very fine missionaries there; some of these men were among the best linguists in the missionary world and of great experience in the affairs of the Near East, and it was impossible for a young missionary not to be inspired."[19] Foremost among these was Dr. Cyrus Hamlin, the founder of Robert College, whose book *Forty Years in Turkey* has served as a guide and inspiration for others. The greatest linguist in the missionary circle was Dr. Elias Riggs. Dr. House credits the latter with being able to converse in eighteen languages. As a young missionary, House served on the translating committee for all of the Bible work in the Near East that worked with Greek, Armenian, Bulgarian, and Turkish as well as other languages and sharpened his language skills by assisting the brilliant Riggs in proofreading. Yet almost from the very beginning of his service one distinguishes certain ideas taking root in this young missionary that move him intellectually out of his time and place him well ahead of most of his colleagues. Actually the genesis of his developing philosophy took place in Samakov (now in Bulgaria).

In 1874, a vote by the ABCFM reassigned him, despite his reluctance, to the theological school at Samakov. He had expected to devote his life to evangelical work, so his transfer to a theological school came as a disappointment, for he greatly preferred preaching to teaching. The school was on the junior high school level with a course in theology added to a curriculum of general studies to train preachers and leaders for village work. Dr. House served first as teacher; later, in 1887, he became director.

---

[18] Wickham T. Stead, "The Americanizing of Turkey," ca. 1910, AFS Archives.
[19] John Henry House, "From My Association with the Mission at Constantinople," n.d., AFS Archives.

With this introduction to an educational institution tailored to serve Bulgarian youth, House noticed immediately a striking sociological trend that was endemic throughout this area: the village boys who came to the theological school, and for that matter the other boys he was observing in the whole secondary state school organization outside of the missionary system, had learned to despise village life. He became convinced that the mode of education had to be changed to indoctrinate these boys with the ethic that it was not degrading for educated people to work with their hands. The motto of his later creation, the American Farm School, "to work is to pray," is an indication of how the religious and practical were intertwined in Dr. House's philosophy. Once he had analyzed the sociological and educational problems, he made real attempts to reorient the philosophy of the Bulgarian youth toward a more productive course. The early steps in constructing a sounder educational core at Samakov were the seminal elements in the institution he was to found at Thessaloniki, Macedonia, thirty years later. "The story of this work in Samakov," he wrote, "was the first step on the way to what I considered my ideal of missionary education."[20] He began to work out a plan that, ideally, "should train the children along things of the heart, the head, and the hands, that is the whole man."[21] He felt that the best place for such a school was out on the land, but as the Samakov school was situated in town he would have to adapt his plan to an urban environment. Although he could not start an agricultural program at that time, he could introduce some kind of industrial courses into the missionary school system.

In the post-World War II era and even as early as the 1920s and 1930s, when state agricultural schools began to flourish and vocational schools became part of the public school system of the United States, John Henry House's notions seem hardly revolutionary. Yet he was an innovator and a pioneer in this field, for in the last few decades of the nineteenth century such modifications in the school structure were considered outside of the scope of missionary endeavor, the sole purpose of which was to improve the spiritual lot of the peasants. Evangelical mentality had not yet decided that there was a link between material and spiritual wellbeing.

---

[20] John Henry House, "The Ideal Education."
[21] Ibid.

American missionary educational philosophy was only a reflection of a widespread attitude in the United States. As Richard Hofstadter explains, "Old line educators reared on the traditions of the classical colleges often could not really accept the legitimacy of agricultural and mechanical education."[22] This view prevailed around 1862 when the passage of the Morrill Act had to be engineered by the federal government as there was no public interest in anything as "educational" for farmers as an agricultural school.

As the American missionary board was not sympathetic toward House's plans, any expenses incurred by his innovative programs were to be defrayed by donations given to him and earmarked by the donor for this purpose. It was plain that the mission's money was not to be "wasted" on such digressive activities. This narrow definition of missionary range would be a hindrance thirty years later when Dr. House approached the ABCFM once again for funds to help found the American Farm School.

Dr. House's industrial program at Samakov was launched with the donation of a very small printing press. The boys, studying the movement of this small object's gears and levers, could marvel at the fact that they could themselves reproduce the printed word. In 1877, money was donated by Mother House's family to build an extension to the school to house the industrial arts program. A course in carpentry was instituted by House himself, who confessed that since he was an amateur in this particular skill, he had to work hard to keep one step ahead of his students. This stint of carpentry teaching was consonant with one of the chief tenets of his philosophy.

> The teacher must be out of doors working with the students. That is where one can get hold of the hearts of students and learn to know them. Perhaps this is the most important thing.... [A] chance to become acquainted naturally teaches that I am not bigger than they and that I can learn, too, that we are working together. If a teacher does this, he gains the respect and friendship of the students and gets at their hearts. Law and commands do not do much good, but

---

[22] Richard Hofstadter, *Anti-Intellectualism in American Life* (New York: Alfred A. Knopf, 1970) 28.

when one has gotten the hearts of boys they will do anything.[23]

To reinforce the work ethic, Dr. House insisted that poor boys, who were supported by donations, should work two hours a day to reduce their tuition costs. It was no mean triumph for this missionary to see a boy, with his sleeves rolled up, pushing a wheelbarrow through the streets to defray his tuition costs, a boy who just a few months ago would have disdained such a menial task. At one point, in order to raise funds for the expansion of the school, Dr. House made an appeal to the Bulgarians. As most supporting funds came from the United States, this appeal to the indigenous population was an innovation. It was his opinion that such requests promote a feeling of community responsibility. The successful response indicated that the Bulgarians had indeed begun to feel a responsibility for the school and that they endorsed the aims and methods that had been redefined by John Henry House.[24]

His accomplishments in Bulgaria were substantial. When he went to Samakov there were between fifteen and twenty students living in a rented house. When he departed the number of students had reached about seventy-five and the campus now consisted of two large school buildings, an industrial department, and a church.

While he was in Bulgaria three very strong personality factors came into clear focus, traits that were to remain constant during his later work in Macedonia. These were unshakable religious faith, an unswerving sense of mission, and an abiding feeling for humanity.

As for the first, there is no doubt that unshakable religious faith was the wellspring of his motivation and also the basic ingredient that made up the personality of his wife, his son Charlie, and Charlie's wife, Ann Kellogg House. This strong religious persuasion was paramount in his philosophy of education: he wanted to found a school, "out on the land where, surrounded by nature, one has a chance to refer naturally to God. The first emphasis is that education must be religious in spirit."[25] His religious faith and its connection to his mission were expressed in the following letter to his daughter-in-

---

[23] John Henry House, "The Ideal Education."

[24] Nankivell, *A Life for the Balkans*, 97.

[25] John Henry House, "The Ideal Education."

law, Ann: "If one is called of God to do work—then he is sure of God's ever accompanying Presence in doing it. This to me has been the wonderful and sustaining power of my life. Nothing else could take the place of that. Mrs. House and I in a true sense went out, not knowing whither we went, but we were assured that we were going at the call of our Master, and that was enough for us to know."[26]

Mother House, too, possessed this compelling sense of mission: she was as determined as her husband to bring the word of their God to this primitive and hostile corner of Europe. It was their mission; they had been "called"; nothing could explain their capacity to endure hardship and danger other than this total dedication.

Protestants were feared by Bulgarian peasants, who were superstitious of anyone outside their own closed society. The following extract from the Houses' diary, although hurriedly and roughly written, gives a graphic sampling of the danger and hardship that were everyday occurrences in their early lives in European Turkey:

> The next day we planned to spend in the village of Omarchovo in Bulgaria and visit a friendly priest. On arrival we found he was holding service in another village, so we talked with his wife. However, we soon discovered that a crowd of people had surrounded the house. Not wishing to be the cause of any injury, we withdrew to the market place where our wagons stood. Mrs. House got into one of them, while I [Dr. House] entered a coffee shop to arrange for a place to spend the night. While this was going on, Mrs. House became the object of curiosity to a crowd with sticks. The head man arranged our stopping place and here we were kindly received by an old lady who provided some roast lamb. Soon, however, we were turned out of the house where we had been so kindly received, and the whole village was in an uproar. Evening was approaching and we with the baby were in the street! However, as I carried the sultan's firman which commanded all authorities to give us lodging and food, the head man with much difficulty found a woman who was willing

---

[26] John Henry House to Ann Kellogg, Thessaloniki, 12 December 1922, AFS Archives.

to take us to her little house. The next day we set out, but our Turkish drivers were frightened because of the treatment we had received and would not take us to Kayaludere, but stopped at a village five or six miles away, and no persuasion would induce them to take us to the village for which we started. Perhaps it was well that we did not get there, for even as it was, the friends whom we were to visit got a good beating because we had gotten within five or six miles.[27]

Experiences such as these merely strengthened the resolve of the Houses and underscored Dr. House's notion that a spirit of brotherhood needed to be spread among these hostile people. By the time of their arrival in Macedonia, in 1894, they had learned how to adjust to the worst situations.

The third and very strong element in Dr. House's character—touched on earlier as being a trait common to the first three directors of the farm school—was a feeling for humanity. A strong religious fervor coupled with an equally strong sense of mission, but without a feeling for human beings, would never have produced an institution such as the American Farm School. This humane factor was evidenced early in Dr. House's mission when he went to Constantinople for a short while during the Russo-Turkish War, which started in 1877 and ended the next year. It was there that he visited jails that were overflowing with political prisoners who were in a pitiful state. Russia was aiding Bulgaria in its fight for independence from the Turks; therefore, anyone suspected of siding with the Russians was hanged, banished, or imprisoned. The prisoners included other nationalities besides Russian and Bulgarian. In an effort to relieve the suffering of these prisoners, Dr. House collected bundles of clothing that he gave to an Armenian porter along with funds to bribe the Turkish guards, thus allowing the clothes to be distributed to

---

[27] "Glimpses of the American Farm School," *The Sower: Sixty Year Issue*, 1964–1965. *The Sower*, launched in 1926, was an occasional house medium for the farm school. During the period covered in book 1, the *Sower* began as nothing more than a pamphlet printed on a large sheet of paper and folded into four sides. The publication kept farm school supporters abreast of the progress and news of the school.

the prisoners. He then organized teams of schoolteachers to see that the clothes were distributed to the most needy within the jails.

Fearing that the preliminary terms of the Peace Treaty of San Stefano would not contain provisions for the release of the prisoners and political exiles, he approached Prince Reuss, the German ambassador, to propose that the Russian commissioners insert in the treaty an article that would provide for the freeing of all political prisoners and the political exiles. The article, at Dr. House's insistence, was inserted and prisoners in as remote a reach as the Sahara desert were freed largely as a result of that insertion. Another such intervention, again through the German ambassador as intermediary, had freed some of the prisoners before the treaty was even signed.[28]

Ann House was later to define this feeling for humanity more precisely as a "spirit of unselfish helpfulness" that she valued as being more important to the development of the American Farm School's staff members than any "intellectual or practical qualifications" they might possess. She wrote, "My association with the members of our staff...convince[s] me that the spirit of the House tradition which has its roots in our Lord, 'Whatsoever ye would that man should do to you, do ye even so to them,' is still a living reality."[29]

It was finally the Macedonian Struggle that forced the Houses out of Bulgaria and sent them into Macedonia. Around 1894 conditions close to the Bulgarian-Macedonian border became chaotic and acutely dangerous. Bulgaria, now autonomous, had partisans who desired to attach Macedonia to Bulgaria; they formed the guerrilla bands that we noted earlier were called *komitadjis*. These bands, which disrupted life along the border regions, were joined by ordinary thugs and bandits who saw a chance to expand their own operations, heightening the danger and chaos. Border lawlessness became rampant. Authorities, both Bulgarian and Turkish, found it nearly impossible to control these guerrilla elements.

As a result, missionary work was bogged down, mobility becoming more and more of a problem. It was the responsibility of the Samakov mission station to work the Razlog and Struma districts, which meant that the missionaries were constantly exposed along these border

---

[28] Nankivell, *A Life for the Balkans*, 74.
[29] Ann Kellogg House to Edward Howell, 12 March 1972, AFS Archives.

areas to the *komitadjis* who, calling themselves the "Macedonian government," enforced their will at knife-point. As a solution to these difficulties, the ABCFM decided to open a station in Thessaloniki, Macedonia, for now three railroad lines from that city had opened, allowing the missionaries to reach their territories safely by avoiding travel in the mountain areas where the brigands roamed freely. Since Dr. House knew Turkish and Bulgarian, widely used languages in this territory, he was chosen along with Reverend Edward B. Haskell to minister to the evangelical community in a wide radius around Thessaloniki. He left behind him in Samakov twenty years of service and brought with him a wealth of experience and personal growth as he stepped over into Macedonia, not only with a dream but by now with a well thought out plan. He had before him thirty-two more years before the fire would go out.

# CHAPTER 3

## THE KIDNAPPING OF ELLEN STONE AND THE FOUNDING OF THE AMERICAN FARM SCHOOL

Almost a decade passed from the time Dr. House first arrived in Thessaloniki (his station headquarters) until the school was founded. If life had been dangerous for the missionaries in Bulgaria, in Macedonia it was perilous. Nothing dramatized more vividly the political unrest and its attendant danger than the kidnapping of Ellen Stone and Katerina Stefanova Tsilka. Indeed, the founding of the American Farm School was tied to this internationally publicized event in two ironic ways.

On 3 September 1901, Tsilka and Stone, both Protestant missionaries working in Macedonia, were captured by a band of Bulgarian revolutionaries, members of the IMRO. Mrs. Tsilka, a Bulgarian by birth but educated in America, was married to Reverend Gregory Tsilka and at the time of her capture was four and one-half months pregnant. Miss Ellen M. Stone, a native of Massachusetts, had been a missionary for twenty-three years prior to this event and had carried out much of her mission in the European portion of the Ottoman Empire.

Accompanied by colleagues who were likewise engaged in missionary activities, the two women were journeying home to Thessaloniki after giving a summer Bible course in the village of Bansko, Macedonia. Suddenly the group was surrounded by a band of guerrillas bizarrely costumed in Turkish military uniforms and Albanian peasant dress. Some had their faces blackened to hide their identity. Brandishing their arms, the bandits separated Stone and

Tsilka from the rest of the group and pushed them quickly up the mountainside, through scrub and over rocks, not pausing until they had brought the women far from the point of capture. A Turkish peasant who had stumbled into their path was summarily shot. The party accompanying the two women included Mrs. Tsilka's husband and several teachers; they were held under guard overnight and released at dawn.

Days went by until the bandits contacted the Protestant missionaries to start a protracted negotiation for the women's release. One night a missionary living in a mission house in Bulgaria was approached by a stranger and handed a ransom note demanding $125,000. Another missionary told the stranger before he fled that such a sum would be impossible to raise. He castigated the bandit for the vileness of the act and tried to dissuade him. But the messenger and the brigands remained determined.

There is no doubt that the ransom money was the principal motive for the kidnapping. Commenting on the revolutionaries' attitude toward ransom, Dr. House wrote: "Their enemies, alas, if known to have wealth, were liable to be carried off for ransom in order to supply the bands with guns and ammunition to furnish them with food and clothing."[1] Dr. House concluded, however, that the revolutionaries had an additional motive in the case of the abduction of the female missionaries: "...to make evident that the Turkish govt. was unable to protect foreign nationals living under that govt. In the case of Miss Stone the reason for her capture was to supply funds for the revolutionaries; arouse American feeling against the Turkish govt., which could not protect American subjects."[2]

Naturally, this incident did focus American attention on the problem of security for Americans in European Turkey. It was the first time in the history of the ABCFM that any of its members had been kidnapped and held for ransom and also the first time in American contemporary history that a female American citizen had been taken as hostage by European brigands. As Stone's brother commented in an article, "The method...which our Foreign Office

---

[1] John Henry House, "The Story of the Capture by a Bulgarian Revolutionary Band for Ransom of Miss Ellen M. Stone, a Missionary of the ABCFM," n.d., AFS Archives.

[2] Ibid.

adopts in settling this international difficulty must stand as a precedent for future action in protecting the lives of American citizens on foreign soil."[3]

How to handle the kidnappers' demand for ransom presented a dilemma for the US government as well as for the ABCFM. At first the ABCFM felt assured that "our government will insist on the Turkish government's duty to provide the required ransom, even though it may be obligated to advance the same in view of the poverty of Turkey."[4] To insure the immediate release of the captives, the ABCFM informed the United States government that it itself "would provide the ransom were it to be provided in no other way."[5]

However the American government refused to recognize the brigands or to deal with them in any manner. When it was ascertained that the bandits were of Bulgarian origin, Teddy Roosevelt thundered at the Bulgarian government that "he would hold Bulgaria responsible to the full measure of proof."[6] This in turn influenced the ABCFM to reconsider paying the ransom, for such a payment would, in the words of Ellen Stone's brother, "put a premium upon brigandage and incite repetitions of this grave offense."[7]

As funds from no quarter were immediately forthcoming, relatives and friends started a ransom fund in America and England that gradually raised $65,000. The international press became tantalized by the sensational aspects of the kidnapping and kept a running calculation of the amount collected. Finally the ABCFM asked that all public statements be discontinued owing to the fact that the brigands, gleefully following the news, were loathe to close negotiations while the amount of ransom was still growing.[8]

Meanwhile the Turks claimed that the captives were in Bulgaria while the Bulgarians claimed that they were in Macedonia. In truth,

[3] Perley Stone, "Precedent for American Honor" (excerpt from magazine article), *Shoe and Leather Facts*, n.d., AFS Archives.
[4] Judson Smith to John Henry House, Boston, 20 September 1901, AFS Archives.
[5] Ibid.
[6] President Roosevelt quoted in letter written by James Beard (missionary in Samakov), n.d., AFS Archives.
[7] Stone, "Precedent."
[8] W. W. Peet (treasurer of American Mission Turkey) to John Henry House, Constantinople, 21 October 1901, AFS Archives.

they were both captured and held in Macedonia, which of course in 1901 was still an Ottoman province. As a result of the international reaction, the Turks set out to hunt down the bandits. The Ottoman government, in its search, arrested innocent individuals, then tortured them to extract confessions of collaboration with the kidnappers. Strangely enough, authorities in the Raslog area concluded that the Protestant communities and the Protestant missionaries were somehow all involved in the ransom plot. Dr. House reported that this theory "seems to have been clung to by some with a curious tenacity."[9] Consequently, several Protestants were arrested and some of them beaten.

Meanwhile the two hostages were forced to live under the most extreme conditions. To keep out of sight, the band changed its hiding places under the cover of night. For five months the captors and captives spent the days hiding in dark, damp, frozen caves, their only comfort an occasional fire built in the mouth of the cave. But without ventilation, near suffocation from the smoke almost erased the sweetness of the heat. Many times the pregnant woman was afraid that they might kill her to rid themselves of her burdensome problems. Yet the women were not treated with indignity. It is significant to note that the bandits forced the women to listen daily to a reading of Karl Marx; but these two women had not spent half a lifetime in the Balkans for nothing. They had learned to strike a good bargain. The women insisted that if they had to suffer Karl Marx then the bandits would have to suffer the word of God. A bargain was struck; the missionaries listened to the Karl Marx readings while the bandits heard the Bible.

The amazing fact is that Mrs. Tsilka, after almost four months of living under such duress, gave birth on 22 December in a freezing mountain cave to a healthy baby girl. Her description of the midwife enlisted by the bandit chief to aid her at the birth gives an added dimension to the bleak and primitive conditions that surrounded her:

> Evidently the bandits had taken her from some hut and, against her will, brought her to us for the emergency. She was an old woman perhaps 50 or 60. Her face was pleasant, her

---

[9] John Henry House, "Report," 20 March 1901, AFS Archives.

features regular though lacking intelligence. It was difficult to ascertain what race she belonged to for she was black with smoke and dirt. The spaces between her wrinkles were embedded with dirt, accumulated there for months and perhaps years. The gray hair hung loosely over her forehead and eyebrows. Her head was covered with a kerchief which had been white once upon a time. She also wore a black garment so patched that there was very little left of the original. Half of a sleeve and the collar were entirely gone.[10]

After a night of agony the baby was born without the aid of drugs or even of water. When the brigand chief came to see the baby, Miss Stone handed it impulsively to him. It was an immediate breakthrough, for apparently the newborn infant touched the guerrilla's heart and at that moment the mother realized that "he was no longer a brigand, to me, but a brother, a father, a protector to my baby... I was relieved. He means to spare my child. He can do it. He is the chief."[11] Naturally at the first cry of the child there had been talk among the bandits of murdering it lest its screams give pursuers a clue to their whereabouts.

The infant was named Elena. The bandits were delighted in the singular honor of having a baby born in their midst, for no other band could claim such distinction. They decided to immortalize this extraordinary event by writing the name Elena on their guns.

Although brave enough in the first few weeks, the women later came close to the limits of despair as shown in a desperate plea written by Ellen Stone after Elena's birth and delivered to Dr. House through the bandits' agent: "Now Mrs. Tsilka and I pray you from the depth of our hearts to do all that is possible to finish the work of our liberation.... For 132 days we have been captives. Is this not enough...?"[12] She went on to tell of the birth of the Tsilka baby and the added suffering: "Eight days ago the little daughter of Mrs. Tsilka

---

[10] Katerina Stefanova Tsilka, "Born among Brigands: Mrs. Tsilka's Story of Her Baby," *McClure's Magazine* 20/4 (May–December 1902): 294. The first recounting of this event in book form was probably Frederick Moore's *Balkan Trail* (New York: Macmillan, 1906) 34–47.

[11] Ibid., 296.

[12] As cited in John Henry House, "Report," 11, AFS Archives.

was found among us. In spite of all the suffering and privations, mother and child progress in health but you must understand that they must be placed in winter quarters…before the winter cold increases."[13]

From the very beginning, the ABCFM had placed the responsibility for negotiating the liberation of Stone and Tsilka upon John Henry House, no doubt convinced that he knew the Balkans better than any other foreigner. It was he who served from the outset as a sort of clearinghouse and coordinator among the Turkish authorities, the missionaries out in the field, the American government, and other cooperative allies such as Sir Alfred Billioti, Great Britain's consul general in Thessaloniki. By December, Dr. House had become the principal agent for the missionaries out in the field, meeting directly with the brigands. During that month the American Legation in Constantinople formed a committee of three, granting it full power to conduct negotiations. The committee consisted of Reverend W. W. Peet, treasurer of the ABCFM in Constantinople, A. A. Gargiulo, chief dragoman of the American legation, and Dr. John Henry House.

From the first, Dr. House's decisions were decisive for the conduct of negotiations. It was he who impressed upon the *vali* (governor-general of Macedonia) and the American consul general in Constantinople, Charles M. Dickinson, that Turkish troops should not be sent out in pursuit of the bandits for fear that the latter might kill their captives.[14]

Just before Christmas 1901, Dr. House was brought a letter addressed to him by Stone with an explanatory cover letter written by Consul General Dickinson's agent, who had conferred with the bandits' representative in Sofia. The guerrillas' representatives had indicated that they would come to terms only with an agent who spoke Bulgarian, someone they might be able to trust. Dr. House was a likely candidate because he spoke the language and because of his known friendliness to all men. Thus Dickinson's agent's cover note appointed him as principal intermediary for the dangerous task of contacting the brigands in the icy reaches of Macedonia's mountains and delivering to them, if they agreed to accept it, the reduced ransom—now fixed at $65,000—raised by friends of the cause.

---

[13] Ibid., 12.
[14] Ibid., 2.

For his first rendezvous with the bandits, Dr. House ventured into the mountains between Serres and Bansko in the beginning of January 1902 to meet the brigands'agent. He described his anxiety as follows:

> It seemed to me as I started along upon the long journey to that mountain region that I know so well that I was undertaking a dangerous and almost helpless task. The road which I had travelled without fear had now become fearful, but duty was duty and with God's help it can always be done. It was a fearful task, as immediately upon the news of Miss Stone's capture I had become aware that originally the plan had been to take me and not Miss Stone.[15]

It took him three days to reach his destination; for the first two days he walked without guards to avoid attracting the attention of the Turkish authorities, who he feared would plunge in to capture the bandits regardless of the hostages' welfare. On the third day, however, in order to protect himself in the most severely bandit-infested stretch, he employed two guards. The missionary work he usually did in this area was a sufficient disguise to conceal from the Turks his real mission. It was not until February during the conclusive stages of the negotiations that the Ottoman authorities learned that he was acting as negotiator for the captives' liberation.

Finally, after eight days, the brigands came to meet Dr. House in the lonely mountains of Bansko. As he told the story, shortly before midnight "three fully armed and uniformed brigands were ushered into my presence, who not only seemed rather formidable, but made it known to me that I was in their power."[16] The leader of the group presented a letters signed by Stone and Tsilka and a letter from Dickinson's agent agreeing to the brigands' terms. Dr. House, with great reluctance and against his better judgment, agreed to deliver the money in coin to the kidnappers ten days before the release of the captives. There followed several other unsuccessful attempts to meet

[15] J. M. Nankivell, *A Life for the Balkans* (London: Fleming H. Revell, 1939) 34. Dr. House's fear that he was the originally intended victim is curious since there is no other mention of it in the dozens of documents on this topic held by the AFS Archives.

[16] John Henry House, "Report," 9, AFS Archives.

with the brigands in the icy mountains to finalize release plans. As time dragged on, Dr. House despaired of a quick end to the negotiations. He soon learned that the brigands had not been able to keep their rendezvous with him because the local Turkish military commander, learning that Dr. House was acting as an agent for the release of the missionaries, and realizing that the guerrillas must be near, had kept hundreds of soldiers on guard throughout the area.

To throw the Turks off his scent and to confuse the scores of newspapermen who hampered the negotiations fully as much as did the authorities, Dr. House devised a diverting scheme. He quietly secured an amount of lead matching the exact weight of the ransom and stuffed it into bags. Then he made a spectacle of loading the lead-filled bags onto horses for the trip from the village to the railroad station. At the station, the bags, soberly and ceremoniously guarded by Turkish police, were loaded onto the train for Constantinople. Observers seeing the bags bound for Constantinople instead of the mountains assumed that the money was being sent back to American authorities in the Ottoman capital because the negotiations had failed. The ruse worked: the press printed cartoons satirizing the bungling of the missionary negotiator.[17]

Waiting for the propitious moment when the press and the Turks were occupied, the bandits finally found the opportunity for an ultimate rendezvous with House. As he describes the event: "One evening while British reporters were having dinner with the Turkish officers, a company of dusky forms suddenly occupied the back yard of the house where we had hidden the bags of gold."[18] Dr. House continues: "The four chiefs present divided it among their men. We shook hands with them and they speedily departed with their pelf and were lost in the darkness of the night."[19]

On the morning of 23 February 1902, a full three weeks after the bandits had received the ransom, the two women victims and the infant were released near the Macedonian town of Stromnitsa after a five-month ordeal.

The founding of the American Farm School is related to this kidnapping in two ways. First, the Illinden massacres that occurred in

---

[17] Nankivell, *A Life for the Balkans*, 138–40.
[18] Ibid., 140.
[19] Ibid.

August 1903 were reprisals by the Turks against Macedonian peasants who had participated in an uprising instigated by the IMRO. Douglas Dakin claims that the IMRO used the Ellen Stone ransom money to buy arms for this insurrection.[20] The first boys brought to the school were orphaned by a brutal Turkish massacre around Monastir. Judging from dates and place names used in school correspondence, it is almost certain that the boys' parents were victims of the Illinden massacres. In other words, their parents' death was caused indirectly by the ransom money. Second, Ellen Stone contributed one of the first donations of $500 toward the purchase of the original 53 acres of land for the founding of the school. The money arrived when the possibility of raising subscriptions from other sources was indeed very slim. There is proof that the amount needed for her ransom was oversubscribed, the surplus being offered to missionary related projects.[21] There is a good probability that at least part of the sum, if not all of it, came from the excess contributed for her ransom.

It is interesting to note that the missionaries almost unanimously assigned the blame for the unfortunate incident to the Turkish and Bulgarian governments and to the Great Powers for not solving the Macedonian problem. Great effort was made by the missionaries to remove the blame from the bandits and place it on the authorities, their rationale being that conditions of the times led the bandits to desperation.[22]

As exemplified by the Stone kidnapping, travel into the outlying mountainous terrain could be dangerous. As a result of the kidnapping

---

[20] Douglas Dakin, *The Greek Struggle in Macedonia: 1897–1913* (Thessaloniki: Institute for Balkan Studies, 1966) 92–106.

[21] "Several large subscribers had informed us that it is their wish that the balance due them should be paid over to the American Board of Commissioners for Foreign Missions, for the prosecution of its work in Macedonia and Bulgaria, the lands in which Miss Stone was a missionary" (Frank Wiggins [treasurer, ABCFM] to Irving W. Metcalf, Boston, 5 April 1904, AFS Archives). After her release, Ellen Stone went to the United States and never returned to Macedonia. Over the years she maintained contacts with the area and continued to support the school for the rest of her life. Mrs. Tsilka, her husband, and the child are mentioned in Moore, *Balkan Trail*, 142. Moore met them in 1906 in Monastir, where the couple was teaching at a missionary school.

[22] W. W. Hall, *Puritans in the Balkans, The American Board of Mission in Bulgaria: A Study in Purpose and Procedure* (Sofia: Cultura Printing House, 1938) 163.

and the disastrous situation created by the intensified Macedonian Struggle, passage became so imperiled that the Turkish government ordered all missionaries to be accompanied by a guard of ten soldiers when touring outside of the city of Thessaloniki. As a military entourage was hardly conducive to missionary activities, the missionaries limited themselves to those villages lying directly on the railroad lines, where they were allowed to travel without a guard.

Certainly a time like this shaped the philosophy and pedagogy of people like John Henry House. To them it seemed a hopeless, tragic moment almost without parallel in history: all law and order had broken down while the strongly armed bands preyed upon the poor. From a sociological point of view it was a stagnant period for Christian peasants in European Turkey, who lived as disenfranchised citizens in a decaying empire. Not much better off than medieval serfs, they endured unrelieved misery. Dr. House commented as follows on the abuses he saw inflicted upon the peasants at the time: "...if a man has raised beans one year and has none the next he may be taxed for the same amount of beans this year as he had last. The fact that fruit trees have not borne this year may not decrease the tax upon them and I have heard of apple trees being cut down because of such taxes."[23] Continuing the subject of taxes, Dr. House wrote: "...most remarkable is the continuance of the war tax upon men who have died. This writer knows of a villager who recently died but it was impossible to get his name off the tax list."[24]

Such extreme circumstances reinforced his philosophy that perhaps the founding of a school for Balkan youths would be at least one step forward. Surely the moment was propitious, since his missionary activity, restricted as it was to villages along the railroad track, seriously shrank the range and effectiveness of his work. He reasoned that the villagers could be served by having a centrally situated institution such as a school where rural youngsters of any ethnic background could gather and live in safety.

In the decade since he had come to Macedonia, Dr. House had found at least two other missionaries who shared his ideals and were willing to participate in the venture. In 1902, with these two other men from the Thessaloniki missionary station, the Reverend Edward

---

[23] John Henry House, untitled notes, n.d., AFS Archives.
[24] Ibid.

B. Haskell and the Reverend Theodore Holway, John Henry House selected a barren stretch of land about 6 miles from the center of the city upon which to found a school for the education of Balkan youths.

Of Holway very little is known. He arrived in Thessaloniki in 1902 and left Macedonia in 1903 to return to Bulgaria. His name appears in some correspondence in 1934 identifying him as the principal of a Bible school in Samakov.[25] Although he was one of the triumvirate who founded the school, in the long view he played only a minor role.

Haskell, however, was a well-known name in missionary circles. His father, Reverend H. C. Haskell, was one of the early missionaries in the Balkans, having served in the Bulgarian section of European Turkey from 1863 through the Balkan Wars. From all accounts, the older Haskell was a conservative thinker of considerable intellectual ability. The younger Haskell inherited his father's mental vigor but was impulsive, at least as a youth, and was considered very advanced in his sociological and theological opinions. His educational pattern was routine. In 1887 he graduated from Marietta College and in 1891 from Oberlin Seminary. Like Dr. House and most of the missionaries of this era, he earned a Phi Beta Kappa key. After seminary, he went to Bulgaria as a member of the mission and in 1894 was reassigned to Thessaloniki with House, where he remained until 1914, when he returned to Bulgaria. He died in 1934.[26]

His advanced theological ideas were formulated early in his missionary career when, three years after his arrival on station in Bulgaria, he went through a period of doubt and spiritual skepticism. His attraction to a freer theology increased through the years. After the turn of the century, he belonged to a group of young imaginative missionaries who tried interesting programs to stimulate communicants. Of his more conservative brethren he wrote: "The whole body of workers up there [in Bulgaria] is so saturated with the theology of seventy years ago that the outlook for progress among people who think in terms of today is not bright."[27]

---

[25] Ann Kellogg House to Dearest Mother, Thessaloniki, 19 June 1934, AFS Archives.

[26] Hall, *Puritans*, 237–38.

[27] Ibid., 235–36.

He found his frustration over political conditions in the Balkans hard to suppress. He wrote quite frankly of his opinion to the ABCFM: "I have no right to advise or encourage them to revolt against the rule whose territory I inhabit. But I have too much Saxon blood to advise or entreat them to endure supinely their intolerable lot."[28] Nevertheless, he tried to restrain himself: "I uniformly refuse to advise them, saying my business is religious and not political."[29]

He indicated that Dr. House used great restraint: "Dr. House urges them not to engage in insurrectionary movements."[30] House, older than his colleague by some years, did not share Haskell's most extreme views, but his philosophy of education, his sociological approach to problems in the Balkans, and his conception of the missionary's role harmonized in many ways with the views of the younger minister.

Their working relationship in Thessaloniki had begun in 1894, eight years before the land purchase for the future school. Together they serviced the territory of Eastern, Central and Western Macedonia, which included about twenty-four outstations in some forty villages. They came and went in turn, changing places every two weeks so that one would be out in the field while the other remained at the Thessaloniki missionary station.

The 53 acres selected by the three missionaries for their school were held by thirteen different Greek peasants. An agent in the city had great difficulty rounding up the landowners and bringing them to agree finally on a price of $1,000. Actually, once the landowners realized that the prospective buyers were serious they were only too happy to sell these parcels of land, which appeared totally unproductive. A title search has revealed that the land belonged at one time to Turks. At some point a Greek named Biliades bought a large stretch of land that included these acres. A group of Greeks in turn bought the 53 acres and other plots from Biliades. However, the buyers, unable to pay Biliades for the entire acreage, sold the 53 acres to the missionaries to pay a part of the amount owed to Biliades.

Still the purchase was a good one in terms of investment. Land speculation was high at this time: everyone knew the city would

---

[28] Quoted in Hall, *Puritans*, 108.
[29] Ibid.
[30] Ibid.

expand but no one knew in which direction. It did in fact expand in the direction of *Kapoudjides* ("Keeper of the Toll Gates" in Turkish), which was a vast area stretching from the city walls as far as the present-day airport, contiguous to what is now Anatolia College and including the missionaries' purchase.[31]

The way to what was then a wasteland lay through old quarries and heaps of rubble, along donkey paths that curved through an area that served as dumping ground for the murdered of Thessaloniki. Moreover, as if to accentuate the desolate quality of the place, an Englishman and later a Turk had been carried off from the area by brigands and held for ransom. The earth itself was totally unpromising.

Thick with dust in the summer, clogged with claylike mud in the spring, and iced over in the winter, there was not even a tree to attest to its potential. The missionaries had been told that although many attempts had been made to discover water on the land, none had ever been found. Dr. House's belief that the school's land should have no more advantages than that of the poorest Balkan farmer was predicated on the conviction that land rich in water and minerals would not relate to average conditions and would be a poor proving ground for teaching young farmers the value of improved methods.

The location was isolated. About three-quarters of a mile beyond, one could see the rough buildings known as the Turkish model farm. About 3 miles in the direction of the city lay the depot where, some two years later, public transportation could be gotten to reach the center of Thessaloniki. The property's redeeming feature was without a doubt the surrounding vista: to the east, acting as a majestic backdrop, towered the green bulk of Mount Hortiatis; directly opposite arched the Thermaic Gulf; some 50 miles beyond loomed, on a clear day, the snowstreaked summit of Mount Olympus.

The Turkish tax records referring to the immediate area used a number of different names, each one descriptive of characteristics of the region. At one time it was included in the general district of

---

[31] Vasilis Dimitriadis (director of Macedonian Archives) interview with author, Thessaloniki, 10 November 1972, written, AFS Archives. Dimitriadis not only found the deeds but also met with some of the descendents of those who sold the land to the missionaries and discovered that they remembered the motive and circumstances of the sale.

*Kapoudjides*, as we noted, an area known generally for its agricultural value; thus we know that rich lands were located at least in the vicinity of the missionaries' purchase, probably down toward the sea. The region was also called *Kömürci yolu*, which suggests that charcoal was made or carried through in large amounts. *Eski Monastir,* another name meaning old monastery, attests to the fact that the fields probably belonged to some monastery during Byzantine times.[32]

Once the location had been selected and the landowners had agreed to the sale, the impecunious missionaries needed to raise the sum in the United States. The mission station in Thessaloniki, although it had no funds to contribute, was enthusiastic about the project. On the other hand, the ABCFM was definitely not interested in helping and was, on the contrary, wary lest duties connected with the new undertaking distract the founders and interested parties from their mission as preachers and pastors. In 1903, the ABCFM voted in favor of founding a school on the land purchased if this could be carried out without essentially diminishing the working force of the missionary station and without using any of its funds.[33] Any new projects—kindergartens, for example, or the farm school—were deplored as displays of extreme individualism. Haskell could perhaps be dismissed as a young whippersnapper, but John Henry House's stability, solidity, and maturity presented the ABCFM with a dilemma.

In an effort to obtain the sum of $1,000 needed for the purchase, Reverend Haskell wrote to a friend, the Reverend Irving Metcalf of Oberlin, Ohio, asking him to lend them the money. He requested that Metcalf cable a simple "yes" or "no" in reply. The latter related afterwards that when the request came he had no means to fulfill it; but the next day a debt of $500 which he had written off as "bad" was repaid to him. Metcalf took this as an act of providential intervention instructing him to cable affirmatively. Then Miss Stone, Dr. House's former associate who had been kidnapped by the brigands and had returned promptly to the United States, offered to donate the other $500. In October 1902 the cable from Metcalf came back "yes"; thus by 27 November, the day before Thanksgiving, the land

---

[32] Vasilis Dimitriadis, interview with author, Thessaloniki, 4 June 1972, written, AFS Archives.

[33] Minutes of annual meeting, ABCFM Archives, April 1903, as cited in Hall, *Puritans*, 190.

transfer was completed. In addition to Stone and Metcalf, a small group of friends in America formed a committee of supporters to raise money. This frail nucleus was essential to the school's birth. Although many problems remained to be solved, the school had become more than just a vision—it had become a possibility.[34]

As a first step, as soon as the purchase was completed, 400 tiny mulberry trees were set out on the uninhabited land. Donkey riders and ox goaders then uprooted the spindly saplings for use as whips. To protect the property and to indicate that it was now inhabited, a caretaker was retained to live on the land. Helped by Dr. House, he built himself a hut of mud and packing cases that he sank into the ground to protect himself from the piercing Vardaris wind that blows mercilessly down the riverbed from the north. This simple, hardy dweller evidenced great faith by writing "The American School" in the mud in front of his primitive establishment. He was also responsible for planting grapevines and two almond trees, which some years later became the glory of the school and the first herald of spring.[35]

If a school was ever to be established, obviously the most pressing need was to find if water existed and in what quantity. In order to determine exactly where to start drilling, the missionaries prayed for guidance. Finally, with the aid of a local diviner, they selected a shallow hollow and instructed the caretaker to direct the workers there. However, in the missionaries' absence the caretaker received his own private inspiration, so that when the workers came he had them drill in a different location. The three founders arrived when the drilling had already reached 175 feet and watched the progress with despair. Finally, to their great relief, water was discovered at exactly that depth.

A short time later a second well was dug. W. A. Essery, secretary of the Bible Lands Mission Aid Society (BLMAS) in London, spoke of the school when lying on his deathbed, saying that he would like to help. It was therefore decided by BLMAS to send money to pay for this well and a windmill. The society requested that it be called Essery Memorial Well in the secretary's memory. Since the well proved so

---

[34] Joice Nankivell Loch, "Thessalonika Agricultural and Industrial Institute: Extract from the Life of John Henry House," May 1956, 15, AFS Archives.

[35] Nankivell, *A Life for the Balkans*, 146.

helpful to neighbors in alleviating thirst in this dry region, in 1906 it was decided that an arbor be built on the roadside with a faucet leading from the well so that travelers and their animals trudging to Thessaloniki from the outlying districts could drink. Carpenters working at the school on other construction volunteered their efforts to make this possible. Eventually, shepherds would often gather around the arbor filling their water barrels. The arbor, octagonal in shape, had carved into it, in Greek, Turkish, and Bulgarian, the words of Christ from the Gospel of St. John: "Whosoever thirsts, let him come unto me and drink." Although the discovery of water in this parched land was of major significance to the future operation of the school and a source of profound encouragement, the well was not adequate for the extensive cultivation that would eventually become part of the school's program.

Reassured by the discovery of water, Dr. House departed for the United States to interest individuals in supporting the school, to form a board of trustees, and to incorporate the institution in New York State. He met discouragement immediately when he approached the foreign secretary of the ABCFM, who flatly stated that it would be impossible to interest people in the enterprise.

For some time it seemed that perhaps the secretary's opinion was correct. After contacting a wide circle of prospective supporters, House found his results so negative that he himself began to doubt that he would find anyone interested in supporting an institution that had as its goal the education of peasant boys in distant, and often unheard of, Macedonia. The founding of the school hung in the balance for Dr. House, the person with the lifelong dream, the person whom Ray Stannard Baker had called "a man on fire," reached his lowest point of discouragement. In New York, viewing the school from a perspective on the other side of the Atlantic, he saw the venture as hopeless.

The encouragement and direction he received at this juncture from his daughter, Grace Bigelow House, may have been the decisive factor in continuing his pursuit. There is no doubt that she exerted a considerable influence on her father.[36] She also served later as a vital link between the American Farm School, on the one hand, and the

---

[36] Ann Kellogg House, tape recorded notes to the author, 1 September 1972, AFS Archives.

Penn School in South Carolina and Hampton Institute in Virginia, on
the other. Grace, a teacher, went to Hampton at the turn of the
century to study the workings of this black college, which, at that
time, was under the directorship of Hollis Frissell. It is worth noting
that the farm school was consciously patterned to a large degree on
the Hampton Institute, Tuskegee Institute, and the Quaker foundation
on St. Helena Island in South Carolina known as the Penn School, all
institutions with strong religious and moral underpinnings that had
been created to help blacks develop not only academic skills but also
practical ones. We know that Dr. House visited Hampton before the
founding of the American Farm School.[37]

Hampton, begun in 1868, was based on the philosophy of
learning by doing. Its founder-principal, Samuel Chapman Armstrong,
son of a missionary, had been selected by the Freedmen's Bureau to
solve problems of former slaves who had gathered behind Union lines
in Virginia. In response to the needs of the freed blacks and with the
aid of the American Missionary Association, he founded Hampton
Institute to train young people "who should go out and teach and lead
their people first by example...and in this way...build up an industrial
system for the sake not only of self-support and intelligent labor but
also for the sake of character."[38]

Grace House had every intention of returning to the Balkans to
employ at her father's school the methods she had learned at
Hampton, but before the school got on a firm footing she was offered
the position of associate principal at the Penn School, which she
would serve for almost fifty years. She was instrumental during this
critical period in persuading Hollis Frissell and Hollingsworth Wood, a
prominent Quaker, to be among the first members of the newly
formed board of trustees of the American Farm School.

In the depth of discouragement, Dr. House contacted a classmate
of his from seminary days, the Reverend Dr. George Payson, pastor

---

[37] "Thessalonika Agricultural and Industrial Institute Annual Report,
1913–1916," AFS Archives. Tuskegee, founded in the 1880s by the famous Booker
T. Washington, stressed the practical aspects of education but also emphasized the
formation of "the whole man." The Penn School was established in 1862 as an
agricultural and industrial normal school for the education of black youngsters.

[38] Samuel Chapman Armstrong as quoted in *The Hampton Bulletin* 100/2
(1972–1973): 17.

of the Inwood Presbyterian Church in New York City. The pastor offered to introduce Dr. House to Grace Dodge, a member of a large philanthropic family, who might lend a sympathetic ear. At this point, totally dispirited, Dr. House felt, "I did not wish to go with him to call upon her, and meet a stranger lest I would meet rebuff."[39] Much to his relief, according to House, "she expressed great interest in it and awakened new hope and courage by saying that she would help us with an annual contribution as soon as I could get a board of trustees organized."[40]

Frissell of Hampton, who had also been persuaded by Grace House to commit himself to the board, suggested that Dr. Leander Chamberlain might be interested in serving as chairman. Chamberlain was a distinguished Princeton graduate who had been president of the Evangelical Alliance for the United States and was a prominent member of the American Foreign Christian Union and the New York State Colonization Society. His acceptance of this position was clearly an asset for the farm school since his prestigious name, solid reputation, and exceptional ability established the new undertaking's credibility. The new chairman of the board calculated that the institution would need a minimum of $4,000 each year and promised the founder that the school could count on that amount to be forthcoming annually, with the board of trustees assuming the responsibility for the raising of funds.

By autumn 1903, a full board of trustees had been formed. Its members were listed as: John Henry House, Salonika, Macedonia; Charles Cuthbert Hall, Borough of Manhattan, City of New York; Josiah Strong, Greenwich, Connecticut; Leander Chamberlain, Borough of Manhattan, City of New York; Hollis Frissell, Hampton, Virginia; Irving Metcalf, Oberlin, Ohio; William Isham, Jr., Borough of Manhattan, City of New York; Edward Haskell, Salonika, Macedonia; George Payson, Borough of Manhattan, City of New York; Lucius Beers, Town of South Hampton, Suffolk County, New York; Eli Partridge, Borough of the Bronx, City of New York. The first meeting was held around Lucius Beers's dining room table. Mr. Beers, the brother of Susan Adeline Beers, was a prime mover in the formation of the board and remained an active member until his

---

[39] Loch, "Extract," 20–21.
[40] Ibid.

death. A man of means with a deep philanthropic impulse, he contributed to the school throughout the years. Indeed, "when things got really difficult financially at the Farm School and no one knew where the next penny would come from, Mother House would sit philosophically in her chair and say, 'if God wants this work to go on, He will provide.' He did provide, but it was not God, it was Lu Beers."[41]

A few days before the Houses were to set sail for their return to Thessaloniki, the board sent its petition for the incorporation of the school to the Supreme Court of the State of New York. On 10 October 1903, the eleven board members obtained the act of incorporation for the non-denominational Thessalonica Agricultural and Industrial Institute "...for the purpose of providing agricultural and industrial training, under Christian supervision, for the youth in the province of Macedonia of European Turkey; in order that they may be trained to appreciate the dignity of manual labor and be helped to lives of self-respect, thrift and industry."[42]

With this act the school became a legal body according to the laws of the state of New York—a tax-exempt, fund-raising corporation. In 1907 it was to become recognized by the Ottoman government and later granted tax-free privileges.

While Dr. House had been involved with these activities in the United States, Haskell, remaining in Thessaloniki, was concerned with developing the land and finding prospective students who could benefit from the atmosphere and course of instruction that the school was ready to create. The Illinden massacres, which had left so many children orphaned, were a spur to Haskell to complete the

---

[41] Polly Beers Dillard, interview with author, Ekali, 18 April 1972, written, AFS Archives.

[42] Taken from Actual Certificate of Incorporation, as cited in "By which Law the Operation of the School Has Been Authorized," secretary of state of the state of New York, 10 October 1903.

construction of a building that could serve as a shelter for homeless boys who had been dislocated by this violent incident.

# Chapter 4

# The Early Years

While Dr. House was in the United States, Reverend Edward Haskell had been planning and actually setting up some projects on the land in the summer and fall 1903. He wrote to his colleague about current conditions and projected activities:

> We had better build a two-room house and stable and get a couple of yokes of oxen. We might as well get an income from the place as to let it lie. Pano of Mourtino went to it with me the other day. He thinks he can get a couple of capable protestant [sic] farmers of Mourtino to work on its shares. Almost all of the mulberry trees have lived. The potatoes are to be dug this week. Demeter says he had splendid watermelons but as no one went out to get them and as he has no donkey to get them into town, some rotted and he gave others to shepherds.[1]

At the time Haskell was writing, Macedonia was convulsed with Christian insurrections and Ottoman reprisals. In response to the Illinden uprisings, the Ottomans laid waste to 150 villages around the Monastir area, slaughtering a great number of peasants whether or not they had taken part in the uprising. Working in his capacity as a missionary and also as an agent for the Macedonian Relief Fund of England, Reverend Haskell traveled to the Monastir vilayet to take stock of the tragedy. He noted: "In the twenty or thirty burned

---

[1] Edward Haskell to John Henry House, Thessaloniki, 14 September 1903, AFS Archives.

villages with which I dealt, I naturally came across a number of orphans. One or two villages asked whether we had no orphanage where we could gather them."[2]

Before he started on his trip, Haskell had asked an indigenous preacher, Gratchenoff, who was also an experienced builder, to supervise the beginnings of construction for the first building on the newly acquired property. Upon his return from Monastir, he found that the building had made great progress and appeared to be "a good thorough piece of work."[3] For their materials the men had utilized mud brick, a local substance, one that was used throughout Macedonia. The dwelling was a house, 11 by 6 meters large, with a stable of 9 by 6 meters attached to it. At work on the construction was a farmer named Vassil Varvaritza, destined to become the first farmer-teacher in residence at the American Farm School. Of him Haskell wrote, "He is an excellent practical farmer and at once joyfully accepted my proposition to work on the land.... When I asked Vassil whether he and his good wife could not gather a few orphans into the new house with him, he cordially assented."[4]

Reverend Haskell admitted that the shelter would be primitive, yet the children would have food and clothing for the winter. Vassil promised to hold family worship with the boys and teach them the alphabet. His hope was to have the orphans gathered and settled in before Christmas. The exact date when the first boys were installed does not appear in any document, although it is certain that ten orphans were in residence by the beginning of 1904.

In order to avoid dealing with the slow and arbitrary bureaucracy of the Ottoman state, Haskell decided not to apply for a building permit, relying on the turbulent situation to keep the Ottoman authorities' attention turned to more crucial matters than the construction of a small building on a barren farm. His judgment was correct and officials ignored the missionaries' activity.

[2] Edward Haskell to John Henry House, Thessaloniki, 12 November 1903, AFS Archives. For a synopsis of the Monastir massacre see L. S. Stavrianos, *The Balkans Since 1453* (New York: Dryden, 1958) 523–24.

[3] Ibid. Haskell was correct in his comment on the quality of the construction. The building called Haskell House still stands. In 1974 funds were raised for repairs and remodeling.

[4] Ibid.

The real problem, as both Haskell and House realized, was financial. While House was in the United States, Haskell outlined to him and the newly formed board the immediate financial needs if the boys were to be settled on the property. True to the original philosophy of the founders, his plans included employment for the children: "Simply to clothe and feed the children in idleness would cost comparatively little. But if they are to be given employment on the farm (and they are all farm boys, remember) and thus make useful citizens while incidentally supporting themselves—in part at least—there must be equipment for them to work with. This will cost more at the outset, but will be far more satisfactory (if not cheaper) in the end."[5] He itemized the immediate, basic needs for the boys' welfare and education as follows:

1. The house and stable (which later can be altered into a dwelling if the number of orphans demands it) will cost about $500. Of this sum I have but $150 in hand. I need the balance of $350 as soon as I can get it.

2. In order to bake for so large a family we must have a brick oven in the yard which will cost $20.

3. A horse and two-wheeled cart will be necessary for hauling provisions, etc., from the city, as the farm is 5 miles from the market part of the town. The outfit will cost about $50 and the keep of the horse for six months or till the harvest will cost $50 more.

4. One yoke of oxen should be bought at once for the winter plowing; (as you know) in this climate plowing is done most of the winter. The yoke will cost $40 and its keep for six months about $65 more. In Spring another yoke should be bought at the cost of $40 and its keep for three months being $33.

5. Two cows for milk would cost about $22 and their keep for six months about $65.

6. Plows and other farm implements to the amount of at least $50 needed.

7. Kitchen utensils, furnishings, bedding, etc., say $150.

---

[5] Ibid.

8. Wages of farmer in charge, $9 a month (bargain made until May for experiment).

9. Keep of orphans at 12 cents each a day for food equals $36 a month for 10, or twice that for 20. This is a total of $1209 to establish the orphanage and run it six months. The produce of the farm ought to help out materially in the following six months' expenses.[6]

The first ten boys brought to the school in 1904 were all Protestant, Slavic-speaking, and with Slavic surnames.[7] From the first, the boys were taught English so that they could use British and American textbooks, local textbooks being inferior or nonexistent. Since the school was founded on the principle that all nationalities would be gathered there, the founders thought that English would serve as a practical common language for the polyglot student body as it enlarged. In addition, boys were instructed in their native tongue; both a Bulgarian and a Turkish teacher were on the staff in 1907.

During the early months, a young Protestant man was hired as a teacher to supplement the strained academic talents of the farmer and his wife. Students and staff were all crowded into the small mud-brick farmhouse that served as dormitory and schoolroom. A large porch was built around the front of the building to be used as an eating place. The little barn that had been connected to the house sheltered a pair of oxen purchased to pull the Hesodian plough, an instrument used since Biblical days. A short time later, money was donated to build a

---

[6] Ibid. Funds to aid the refugees were more difficult to raise in America because many donors had been prejudiced against the Macedonians as a result of the Stone kidnapping. W. W. Hall, *Puritans in the Balkans, The American Board of Mission in Bulgaria: A Study in Purpose and Procedure* (Sofia: Cultura Printing House, 1938) 169.

[7] "Thessalonika Agricultural and Industrial Institute Annual Report, 1913–1916," 7, AFS Archives. In this regard, it is relevant to note that on the first tax sheet recorded by the Ottoman officials in 1907, the school was listed first as the Bulgarian Agricultural School. Obviously, since the students were Slavic speaking, the Ottomans misunderstood the American base of the school. In another entry on the same sheet, the name was corrected by pencil to read "American School" (Tax Entry 1907, Macedonian Archives, Thessaloniki). For reasons explained in chapter 1, Greek boys did not enter until later. Sizeable numbers of them entered after 1912, when Macedonia became Greek. By that time, Father House was too old to learn Greek and always communicated with the Greek boys in English.

second story over the barn for use as a dormitory for the boys. When they moved into their new quarters they were provided with the luxury of proper beds. Soon after, a room at one end of the porch was constructed for the teacher, and a kitchen at the other end for the farmer's family.

A small brick oven was built in native fashion so that the boys could bake their own bread from the wheat they would raise with their own hands. The older boys were taught to make the dough and bake the bread. Later, broom corn was grown from which the youngsters learned to make brooms for sweeping out their quarters. As Mother House stated in her own practical manner, "the smallest expense had to be looked after in those days."[8]

One of the most basic of the early acquisitions was a donkey and cart. The school's location, isolated from the city, created a serious transportation problem. With the donkey and cart the older boys could at least ride into Thessaloniki to pick up supplies. Dr. House felt that even this assignment provided a lesson from which a student could learn responsibility.

During the first two years, Dr. House was able to spend only two days a week at the school since he was fully salaried by the ABCFM, the school's status being still too tenuous to provide the director with a salary. Dr. House bought a bicycle; sometimes he and his daughter, Gladys, would pedal out to the school from Thessaloniki. Often they would walk. In the absence of the missionaries, the boys were difficult to handle. The teacher made no progress with the undisciplined youths and was replaced by another Protestant teacher, Mrs. Busheva, who seemed to manage them a little better. Worse still, the farmer got on very poorly with the boys, whom he considered lazy and poorly equipped to be farmers. A farmer he was; a teacher he was not.

Initially, the youngsters were taught the simplest farming procedures such as how to plough, how to sow seeds for wheat and other grain crops, and how to reap with a sickle, all methods indigenous to Macedonia. But the farmer, misunderstanding the purpose of the training, wanted no lazy, untrained boys to spoil the pitiful harvest by being allowed to reap. Dr. House, on the other hand, insisted that they learn by doing. The farmer grudgingly acquiesced.

---

[8] Susan Adeline House to her children, Thessaloniki, 10 December 1934, AFS Archives.

The students were also taught to thresh, employing the ancient method of driving the oxen round and round over the threshing floor. Regarding the agricultural success, Dr. House wrote, "At first the ground was so hard and barren that our grain harvests were pitifully small. I remember how I was almost ashamed to report only six to seven bushels of wheat an acre."[9]

The small farmhouse that provided such cramped and primitive living conditions for the school occupants; the youngsters' undisciplined state; the inadequate, untrained personnel; the isolation; and the violent struggles that raged around Thessaloniki, were all unpromising, depressing internal and external features that must have made the infant institution's chances for survival seem dim to even the most optimistic observer.

Finances continued to be a crucial problem, but money did trickle in from individuals who had been touched by the shaky beginning. Commenting on finances, Mother House wrote, "I cannot remember where the money came from that paid for the first improvements of the place. But I am quite sure that there were few debts. I was looking over some of the reports that Father sent to the trustees...and was much amused at the regret expressed in one of them because there was a debt of $13."[10] Since the boys were all orphaned, the full responsibility for their support rested with the school.

Yet, if observers looked closely, they would see that cultivation had indeed begun and that the farm was taking on the appearance of an "oasis," as it was called so often in later years. More land was planted with vines to increase the 2 acres of vineyard planted by the farmer when he had first arrived. Together with the almond trees that had been planted by the caretaker in 1902 and the mulberry trees, they gave the school a somewhat cultivated appearance. Some pine trees were donated, while the missionaries planted locust and other trees. Dr. House, an amateur botanist with a real ability to grow things, set his hand to raising peach and cherry saplings from pits.[11]

The first two years of the school's actual functioning are rather well documented; thus we can obtain a clearer picture of the physical plant and the areas under cultivation. What is lacking are facts that

---

[9] John Henry House, as quoted in ibid.
[10] Ibid.
[11] Ibid.

would give us insight into the boys' character, their problems of adaptation, their hopes and disappointments. As the House family did not move to the school until 1906, the family-like intimacy that developed after their arrival was naturally absent during the initial two years. After the Houses were installed in James Hall, their first home (James Hall still stands after successive rebuilding and remodeling), that we begin to get sketches of the youngsters and their warm relationship to the House family.

In 1905, the founders took the first steps toward gaining recognition of the school's operation from the Ottoman authorities. Until this time, officials focused their attention on the widespread political unrest throughout most of the empire; hence the operation of a penurious school run by a Protestant foreigner was of little interest to them.

A political crisis involving the French and Ottomans erupted in 1905, resulting in the French occupation of the island of Lesvos (Mytilini). France held this island until some of her demands were met. Among these was the call for an *irade* (decree) from the sultan recognizing French schools throughout the empire. When the Ottoman government acceded to this demand, the British and American governments, seizing the opportunity of the moment, demanded the same concession for their educational institutions under the "most favored nation" article of a treaty that had been previously ratified by these nations. Accordingly, in July 1905 the American Farm School asked that it be added to the list of American schools requesting recognition. When the *irade* was granted in 1907, the school became a legal institution in Macedonia with the privilege of operating tax-free. This *irade* was an extremely important factor in the institution's history and survival, for when the province became Greek, the Greeks recognized the school under these long-established terms.[12]

The year 1906 was another landmark in the development of the American Farm School. Just before Thanksgiving of that year, Dr. and Mrs. House and their daughter Ruth took up residence on the campus. Ruth Eleanor House, their fifth child, was born in Samakov in 1883. She spent her adult life in service to the school from the time

---

[12] Leander Chamberlain, "A Report on the Thessalonika Agricultural and Industrial Institute in Macedonian Turkey," 12 November 1906, AFS Archives.

of her graduation from Lake Erie College until she returned to the United States with her mother in 1937. She gave vital support first to her parents and then to her brother, Charlie, and his wife, Ann, acting as matron, nurse, teacher, hostess, curriculum coordinator, as well as in a number of other capacities. In addition, she shouldered so many of the domestic and routine burdens that one wonders how she generated the energy to maintain the level of her devotion. It was not uncommon among missionary families, which were often large, for children to follow their parents' path. Even within this pattern, her self-effacement and conception of the "Christian mission" were extreme. Her personality was such that she worried excessively and possessed a vivid imagination. Given conditions that would most likely drive the most stable person to uncertainty—oil lamps, no telephone, no transportation, no doctors, little diversion, violence on every path—her constancy was all the more remarkable. For the first seven years, she accepted no salary for her work at the school.

Dr. House received one-third of his salary from the board of trustees after his move to the campus, collecting the other two-thirds from the ABCFM. He distributed his time proportionately between the school and the mission.

Early in the year, Arthur Curtis James had donated a substantial sum in memory of his mother, Mrs. D. Willis James, to cover the construction of a large building to serve as a dormitory and residence, as well as to provide classroom space. The construction of this building, the constant presence of the House family, and more substantial financial support from the board of trustees contributed to the phenomenal development that took place in 1906. That year an instructor was hired to teach carpentry and a cook and housekeeper were installed for smoother operation of the daily schedule. By November, a carriage house, a granary, a new bakery, and a substantial building of 90 by 50 feet to house the industrial arts program were underway. Dr. Chamberlain, chairman of the board of trustees, proudly reported that the farm livestock now consisted of three horses, two yokes of oxen, two donkeys, and abundant poultry. The cultivation statistics for this year are astounding if one considers the condition of the land in the year of purchase. Chairman Chamberlain reported that the school had 6,000 young grape vines, 2,000 young mulberry trees, 150 shade trees, 35 English walnuts, plus a quantity of apple, cherry, quince, almond, and fig trees. By this year, some cocoons were sold from the mulberry trees for the cultivation of silk, since Macedonian silk was developing into a sizeable industry.

To enlarge the watering capacity of the original well, a 50-foot high steel windmill was constructed over it. This structure not only pumped water; it had sufficient power to grind wheat and animal feed.

This same year, 1906, the real estate holdings of the school were increased when 3 more acres of land along the Thermaic Gulf were purchased for $800. The chairman reported, "The city of Thessaloniki...which has not more than one hundred and fifty thousand inhabitants, is steadily growing in the direction of the Farm, and the shore prices would soon have become prohibitive. The value of the Farm itself has materially increased since its purchase."[13]

As the school expanded its facilities, its reputation widened apace. The province of Macedonia was beginning to respond to the institution in a small way with gifts of money donated by the local population. Also, local tailors gave materials for the boys' uniforms, and village women wove cloth to sew into underwear for the youngsters. The school was becoming a recognized feature on the Macedonian landscape.

Clearly the Houses' presence was bringing about a dramatic change in the boys' lives. The students no longer followed a haphazard program. Subject now to a demanding schedule, they studied half the day and performed farm chores the other half. As Dr. House often had to be absent from the school owing to his missionary duties, the regulating of the program was left to Mother House and Ruth. They demanded a strict code of ethics from the boys. Drinking and smoking were grounds for suspension; Protestant church services were mandatory. The boys were expected to treat the Houses and school staff with respect and to regard dishonesty as sinful. The animals were to be treated humanely. A boy who kicked a cat was forced to kick a pile of dirt repeatedly as punishment. But the strict ethical code was tempered with love. The boys were allowed free entrance into the Houses' quarters; they came and went as they desired. From the beginning, a familial relationship was established between the boys and the director's household. Once a week the boys brought their clothes

---

[13] Ibid.

for mending. Ruth and Mother House would do the complicated sewing while the boys gathered round to sew the simple things themselves. This gathering, plus Sunday night get-togethers for songs and reading, established a relationship with the director and his family that had a formative influence on the boys' development. The relationship was multileveled. For instance, to teach the boys proper manners and to strengthen the bonds, the Houses had a few boys at table with them every evening, rotating places so that at each evening meal at least two students were present.

The prospect of moving to the school from the city of Thessaloniki was not an attractive one for Mother House. The isolation and lack of companionship, not to mention the dangers, were considerable. Years later she described the foreboding she entertained at the time of the move: "It did not seem at all easy to go to the lonely farm where in case of sudden danger no earthly help was near and it was especially hard to take our young daughter, Ruth, there.... When at last we moved to James Hall there were no outside doors. We slept above in the empty house except for three or four strange workmen."[14]

Although the facilities were primitive, thirty years in the Balkans had conditioned the House family to privation. The family had running water at least in the kitchen (cold, of course); their quarters also had an outdoor toilet, something hardly common in an ordinary Macedonian dwelling, since the fields served well enough for this purpose. As for the boys, they used "the fields" in the very beginning. The students washed themselves in containers made from kerosene tins and bathed in an old iron tub to which water had to be carried from the kitchen tap. The number of baths they took in a year is anyone's conjecture. There was no electricity, of course, so kerosene lanterns and candles were used for light. During the cold winters, the family had heat from kerosene stoves, but the boys slept in unheated rooms.

Times were lean and the food at the school was not plentiful; yet it was sufficient to keep the growing boys well nourished. The youngsters' fare was standardized. Breakfast consisted of tea brewed from local herbs, feta (white cheese made from sheep and goat milk),

---

[14] Susan Adeline House to her children, Thessaloniki, 10 December 1934, AFS Archives.

and bread. For their noon meal they had potatoes, onions, tomatoes or other vegetables, and fruit, depending on the season, plus bread and a small portion of meat once or twice a week.[15] Milk was basic to the boys' diet.

A Jersey bull brought from England by the Reverend Gentle-Cacket was bred with two local cows, and other animals were added to the herd. Gentle-Cacket was a member of the Bible Lands Missions' Aid Society (BLMAS), an organization that supported philanthropic organizations mentioned in the Bible. From his first gift, milk production developed over the decades into one of the School's strongest features, providing not only enough for the students but also a surplus that could be sold for profit.

The school herd became one of the finest in Greece, the dairy having the distinction for many years of producing the only pasteurized milk in the country. The model dairy became a showplace for visitors to watch as white-clothed students managed the milking and pasteurization process.

In the beginning visitors came infrequently. Although years in the Balkans had strengthened Mother House's ability to endure hardships, she found the school's isolation almost unbearable. Ann House, her daughter-in-law, would echo this same sentiment nearly two decades later when she took up residence on campus with her new husband.

An Albanian guard, Hassan, was hired to protect the school and to escort Mother House and Ruth on their journeys to and from Thessaloniki and through the winding streets of the city's marketplace. An old photo shows him mounted proudly and solidly on a dark horse, clothed splendidly in white Albanian dress.[16]

Happily, no brigands raided the farm school during these dangerous times. It was not until the German occupation in 1941, and then again during the civil war in 1949, when the communist *andártes* kidnapped the senior class that the school was violated.

The next two years, 1907–1908, the school developed so substantially that many of Dr. House's educational precepts could be

[15] Catherine Owen Pearce, "Lighthouse on a Grecian Shore," 47, AFS Archives. The evening meal is not recorded.

[16] "Thessalonika Agricultural and Industrial Institute Annual Report, 1907," 7, AFS Archives.

put into practice. The teaching staff was expanded and its duties were clearly defined. Dr. House gave scheduled lessons in agriculture. Mother House took over Bible studies, and Ruth served as registrar and taught drawing and English. Miss Busheva remained to teach general studies together with Mr. Izeff, who also taught carpentry. Teachers were obtained for masonry, shoemaking, and horticulture. As the school could not afford salaries for this expanded staff, Dr. House worked out a plan of barter whereby the men contributed their services free and he gave them theology lessons in return. This was a satisfactory exchange since the teachers were from Protestant communities and hoped to become village preachers.

Given this enlarged staff and the House family living on the premises, systematic order became much easier to maintain. In agricultural terms, the school was emerging as a true farm. Not only were new crops planted, more methods of cultivation were introduced to replace ancient farming practices. In this regard, the straight tooth harrow brought to the school in 1907 was probably the first one ever seen in Macedonia. The age-old method of sowing the grain and then ploughing it under with a wooden plough was now supplanted by the use of the harrow; upon the arrival of the harrow, the Hesodian plough became a museum piece. A cultivator was also introduced to the farm school, it too presumably the first ever seen in Macedonia. Eventually these tools and the method of dry farming were to change the whole course of agriculture at the farm school and certainly to have a large impact upon the methods of agriculture in what was to become Northern Greece.[17]

One of the most useful, innovative additions to this area was a maximum-minimum dry and wet thermometer and rain gauge, the gift of Dr. Chamberlain in 1907. This instrument made possible the start of a farmers' meteorological station that became the most important weather recording station in Northern Greece. Rain and temperature records have been kept on campus until this day.[18]

---

[17] "Annual Report, 1907," 12, AFS Archives.

[18] When the Greek government introduced meteorological stations in Macedonia, the farm school station became a link in that system. In 1924 the Greek government integrated the American Farm School station totally into its system by installing a telephone to transmit its weather observations to the observatory in Athens three times a day. Even during the war and German occupation, a teacher,

Throughout its existence, the farm school has had a close working relationship with the agricultural sectors of both America and Greece. Its first project in cooperation with the United States Department of Agriculture dates from 1907, when Washington sent several types of California grapes so that the school could experiment to determine their value in the Balkan climate. The indigenous grapes in Macedonia were highly susceptible to phylloxera, a destructive aphid-like insect. By grafting American grapes onto the indigenous stock, Dr. House was eventually able to develop a grape entirely resistant to phylloxera.

Although the implementation of modern agricultural methods made Dr. House very proud, his main emphasis remained training and shaping the students' character. In his 1907 report he stressed their industriousness, noting that "in the farm work all has been done by the farmer and the boys with the exception of grafting our mulberry trees. The fields of grain were all reaped by them... As we have no reaper, it had all to be done with the sickle."[19] The boys were also set to work in the vegetable garden: "our only success this year in gardening was the raising of our own onions, a few tomatoes, peas, beets, and sweet potatoes; the latter were probably the first ever raised in this region."[20] Learning by doing, the boys were becoming true farmers—manual laborers with a sense of accomplishment. They were indeed pioneers.

By the end of 1907, sixteen orphans had been gathered under the farm school roof. It is noteworthy that this tiny, remote operation had already attracted international attention. Dr. House ended his 1907 report with this positive statement: "We count among our helpers friends from California on the Pacific to Madras on the Indian Ocean: Americans from one end of the United States to another; Britons from various parts of the United Kingdom; Swiss from

---

John Boudourouglou, maintained the records, a good part of which were mistakenly destroyed by British troops right after the Germans' departure. John Boudourouglou, interview with author, Athens, 4 September 1974, written, AFS Archives.

[19] "Annual Report, 1907," 8, AFS Archives.
[20] Ibid.

Helvetia; Bulgaria and Turkey; German, American and Swiss friends in Egypt and India."[21]

Outside the farm school gates, political crises were mounting. In 1908 it was the Turks themselves who were revolting against their own authorities, their center being Thessaloniki. The chief actors in this drama were a group of liberal army officers called the Young Turks. Their branch in Constantinople forced Sultan Abdul Hamid to reactivate the defunct constitution of 1876. The Young Turks hoped that this action would shore up the Ottoman Empire by uniting warring factions and providing forceful leadership. In the first few days after the revolt and the declaration of the constitution, most inhabitants throughout the empire were in a state of euphoria. Dr. House described the background to this jubilation in his 1908 report:

> ...a word of the wonderful events which have recently occurred in this historic land.... A year ago none of us had even dreamed of the far reaching changes which were so soon to come over the government of the country in which we live. Then Turkey was an absolute monarchy under a system of espionage, corruption and terrorism which for years had made it the nursery of revolutionary movements amongst its various nationalities: Turks, Bulgarians, Greeks, Albanians, Roumanians and Serbians.[22]

Dr. House went on to explain that "the whole province of Macedonia was in utter gloom and despair owing to the reign of anarchy which had terrorized the land with bloodshed, fire and rapine."[23] "Now," he continued, "all things are changed and we are living under a constitutional government with a national parliament now in session in Constantinople. Our city of Salonika from the first has been the center of the great revolutionary movement."[24]

---

[21] "Annual Report, 1907," 7, AFS Archives.

[22] "Thessalonika Agricultural and Industrial Institute Annual Report, 1908," 7, AFS Archives. For an account of the Young Turk revolt see Bernard Lewis, *The Emergence of Modern Turkey* (London: Oxford University Press, 1968) 211–18.

[23] Ibid.

[24] Ibid.

The city rocked with celebrations that echoed all the way out to the farm school:

> For three consecutive evenings immense processions with many bands and thousands of Japanese lanterns paraded the city. Joy was exhibited on every hand and even the Turkish women were talking to the officers and others in the street-cars—an unheard of thing. As I rode along, I heard the cries of newsboys—also unheard of before.... It was, indeed, too good to last long, but it did last for some time, and all revolutionary bands were called in bringing their arms, because the Turks informed them that they too were revolutionists and "we are all brothers."[25]

In 1909, Sultan Abdul Hamid was brought from the capital to Thessaloniki, and placed under house arrest in the Villa Allatini, not far from the school, where he remained until his death.

Most individuals throughout the empire hoped for radical changes to occur in every segment of Ottoman life. The director of the American Farm School saw widening prospects for the institution's growth under the new conditions: "The remarkable change in our surroundings gives a new emphasis to the work our institution is striving to do for the peoples of Macedonia. Never, perhaps, was there a larger opportunity for an institution of this kind, which looks towards the education of the mind, heart, and hand."[26] Father House's optimism sprang from two programs that the Young Turks hoped to effect. The new leadership's first goal was to improve the educational system throughout the empire. Secondly, the Young Turks expressed high hopes of defusing the tensions that existed between embattled nationalities, particularly among the Christians of the European portion of their empire. Such a relaxation of hostilities would have been a boon in itself, enabling missionaries to circulate throughout the Balkans without risking their lives.

But all those who looked forward to improvements were quickly discouraged. Few events in European history have generated such high

---

[25] John Henry House, Thessaloniki, letter to his wife, 1908, as cited in J. M. Nankivell, *A Life for the Balkans* (London: Fleming H. Revell, 1939) 165.

[26] "Annual Report, 1908," 8, AFS Archives.

hopes that were so rapidly dispelled. Macedonia was destined to become the battleground of the two Balkan Wars before its ethnic problems could be reduced to any degree. Actually, the reforms of the Young Turks soon embittered their Christian subjects and served as a unifying force among Bulgarians, Greeks, and Serbs. Significant causes of irritation were the imposition of military service upon Christians (as well as Jews) and the closing of many Christian schools in an effort to Ottomanize the empire's non-Muslim population. Such irritants led to a momentary reconciliation in Macedonia between Bulgarians and Greeks. This rapprochement reached its logical conclusion when the two ethnic groups united in the first Balkan War to eject the Turks once and for all from a large portion of the lands to which they laid claim.[27]

In spite of the political spasms occurring in its immediate neighborhood, the school was still growing. In 1908, thirty-nine students were enrolled. As if the political problems were not enough for Macedonian peasants in that year, they also had to cope with a serious drought. The farm school wells did not dry up, but many in the countryside did run dry. Some fields under cultivation and also some agricultural experiments at the school were ruined owing to lack of rain, while trees and extensive areas of cultivation in the countryside fell victim to the drought. It was this condition that confirmed Father House's earlier hypothesis that "the region about us here really belongs to the semi-arid countries and needs to be cultivated according to the scientific methods now in vogue in the Great Western Plains of the U.S.A. and similar regions of Australia."[28] He was referring, of course, to the method of dry farming that he would refine with ever-increasing success through the years. The drought notwithstanding, the year 1908 was anything but a disaster for the school. The wheat improved in quantity as well as quality, as did the barley.

New equipment received that year did much to modernize farming procedures. Horses replaced oxen; benefactors in the United

---

[27] Edward S. Forster, *A Short History of Modern Greece* (London: Methuen & Co., 1960) 42.

[28] "Annual Report, 1908," 10, AFS Archives. Dry farming is a method of conserving water in semi-arid regions to facilitate the cultivation of crops without irrigation or with limited irrigation. Certain tillage methods render the soil more receptive to moisture and reduce evaporation.

States donated machinery. The carpentry department became second only to the agricultural. Since the school was trying to furnish its buildings, the boys learned to make their own furniture. Storage sheds were erected to house the farm equipment. There was hardly any aspect of rural or agricultural life to which the young farmers were not exposed.

Shoes for the students were such a problem that a shoe-making department was set up soon after the school's inception. While a kind of simple moccasin was designed for fieldwork, proper shoes were fashioned for Sundays; the shoes were so successful that a limited number were sold outside of the school.[29]

Along with the demanding school schedule, the boys were exposed to Dr. House's efforts to enhance their spiritual lives. Religion became an integral part of the academic and social program; Bible was taught as part of the curriculum. Services were held every Sunday and the whole day was spent quietly in accordance with the old-time Protestant observation of the Sabbath. In the morning the boys attended Sunday school and in the evening someone talked to them on a subject of moral content followed by hymn singing. Every Thursday evening there was a half-hour prayer meeting.

Considering the primitive state of the plant and the manifold problems involved in running such an institution, it is remarkable that Dr. House's precept of educating the "whole individual" was translated into reality in these first years. As early as 1908 a library was started and considerable effort made to enlarge the holdings. The fire that destroyed much of James Hall in 1916 ruined most of the original collection; nevertheless the library has always remained an intrinsic part of a farm school education. Alexander Andreou, a 1920 graduate and later a member of the staff, claimed that his first real contact with literature came when a teacher handed him a book.[30] He also heard good music for the first time in the House family's living room, where the boys gathered for informal get-togethers. As early as 1908 a guitar and mandolin were purchased so that the youngsters could take music lessons. An organ, so essential for hymn sings, was borrowed. And Dr. House expectantly reported, "Ever mindful of the

---

[29] Ibid., 14.

[30] Alexander Andreou, interview with author, Thessaloniki, 6 September 1972, written, AFS Archives.

refining and uplifting influence of music, we are hoping soon to have the beginning of a school band."[31]

The three years preceding the Balkan Wars, 1909–1911, constituted a period of growth and stabilization. It is fortunate that the institution had that time to translate its goals into reality, to attract indigenous and foreign support, to expand its acreage and student body, and to settle some of its major problems. If this had not been the case, surely the school would have been unable to withstand the shock of the Balkan Wars.

Perhaps the most significant factor in the School's stabilization was Dr. House's decision in 1910 to end his official duties with the ABCFM and to assume a full-time position at the school as principal and teacher. The financial support filtering in from an interested public stimulated by the board of trustees in New York had reached an adequate level to provide him with a full salary. His constant presence assured that the institution would enjoy continuity and be steadily strengthened.

The scholarship program, something that was to become a permanent feature of the farm school operation, was instituted in 1909. The school raised the money to cover complete tuition costs for all students as well as a major portion of their room and board. That year, all boys were asked to pay a fee to cover the remaining portion of their boarding costs, a sum of $17.60. Those who could not manage even this fee were given additional financial aid. Enough money was donated in 1909 so that twenty students could receive this added financial support. Dr. House felt that the half day these students spent doing manual labor would counterbalance the negative effects of a completely free education, for the boys were actually paying off their debt by working. The scholarship program, and the added financial aid that was made available, constituted one of the stabilizing elements of the pre-Balkan War period and did much to help poor and orphaned farm boys.

Another element enhancing the school's stability in the prewar period was the fact that it had become known throughout the area. In 1909, for the first time, a sizeable number of applicants were turned away because the school could not admit all the youngsters who

---

[31] "Annual Report, 1908," 15, AFS Archives.

wished to attend. Dr. House registered his own surprise at this success: "...the unique character of the institution in making manual labor obligatory upon all, which is not especially popular with young people in this country, made it doubtful in our minds whether the school could grow rapidly in numbers until several classes had been graduated and the character of education given had been practically tested."[32] But fifty-one students enrolled that year, filling the facilities to overflowing. Classes had to be held in corridors and the director's office.

To accommodate a larger enrollment, 8 more acres of land were purchased, increasing agricultural production. By 1910 the farming operation had become relatively more profitable and efficient, even though the farm's central function was educational and experimental.

The fields' productivity served as an advertisement to attract the attention of local farmers. The harvest was now yielding large increases over earlier years and the cultivation was so intense in contrast to other farms that it constituted a convincing example of successful agriculture.

The year before the outbreak of the first Balkan War, the American Farm School graduated its first class. On 10 June 1911, four young men—Anastas Vasiloff, Joseph Anastoff, Angel Kremenlief, and Todor Krusteff—were photographed in their graduation finery, stiffly posed and dressed in proper suits and collars, each one sporting a flower in his lapel.[33] The fact that the institution was able to graduate its first class before the war lent it an air of progress and stability.

Yet crippling problems remained. Serious health conditions throughout the Ottoman Empire were particularly grave in Macedonia. Plague and other epidemic disease were commonplace in the history of Thessaloniki. Malaria was especially virulent in most of Macedonia for topographical reasons. Directly outside of the city's Vardar Gate, for instance, lay pits of standing water, breeding grounds

---

[32] "Thessalonika Agricultural and Industrial Institute, Annual Report, 1909," 9–10, AFS Archives.

[33] "Thessalonika Agricultural and Industrial Institute, Annual Report, 1910–1911," 10, AFS Archives.

for the swarms of mosquitoes that infested the city.[34] A number of boys came to the school with malaria already in their systems. Dr. House was also a victim of this debilitating disease, as were his son, Charlie, and his daughter-in-law, Ann.

Dr. House established contact with local doctors who rendered their services free of charge or for token payment. These physicians treated not only the students who had malaria; they improved the general health of the entire student body. Most of the youngsters left the school in much better health than when they had entered.

An equally serious concern was the water supply. Early in 1911 the water problem reached catastrophic proportions when the school windmill failed to draw water from the most productive well. Water had to be transported from the farm of a neighbor, Ali Bey, a mile and a half distant. The struggle to supply the large acreage under tillage dissipated the energy of the staff and student body to such an extent that the school's farming operation almost collapsed under the strain. Donations from the United States had to be found to purchase tools and manpower to determine the cause of the failure. As a stopgap measure, three experienced diggers were put to work in quest of another well. They dug without success until they reached 100 feet, finally discovering what proved to be a rich source of water.[35] This discovery and the rehabilitation of the old well settled the water problem for many years. Moreover, the abundant water supply increased the value of the property in this waterless, rain-starved area where farms with ample water supply were in great demand.

The school's steady progress and the financial support from a small but continually increasing public were clearly factors leading to stabilization. The board of trustees in New York remained steadfast throughout the institution's first decade. The board's minutes during this decade reveal a remarkable lack of tension between the trustees and the director. Furthermore, harmonious relationships existed among the board members themselves, for they shared similar goals and agreed upon common methods. It would be correct to assign to

---

[34] Apostolos Vacalopoulos, *History of Thessaloniki* (Thessaloniki: Institute For Balkan Studies, 1963) 108.

[35] "Thessalonika Agricultural and Industrial Institute, Annual Report, 1911–1912," 19–20, AFS Archives.

the board a major portion of responsibility for the school's continuing success and even its survival.[36]

By the outbreak of the First Balkan War on 18 October 1912, the school had sunk its roots deeply into the Macedonian soil. Obviously, the war created new problems; yet the institution emerged healthy and strong, ready to take its place as a contributing force in Greek life.

---

[36] Minutes of the Board of Trustees of the Thessalonika Agricultural and Industrial Institute, 1903–1913, AFS Archives.

# CHAPTER 5

# THE BALKAN WARS:
# MACEDONIA BECOMES PART
# OF MODERN GREECE

The modern state of Greece was born with the War of Independence against the Turks that began in 1821. The boundaries of the infant state drawn as a result of that war left nearly three-fourths of the Greeks in the Aegean area outside of the newly liberated territory. Inhabitants of what is known as "old Greece"—that is, the Peloponnesos, Central Greece, and the Cyclades Islands—were freed of the Turkish yoke. Hellenes residing above the northern border of "old Greece" in Epiros, Thrace, Thessaly and Macedonia looked forward to the time when they, too, would become part of modern Greece. The *Megáli Idéa* (Great Idea), the cornerstone of Greek internal and external political activity until 1923, expressed the Greek passion to annex extensive territories that for historical and demographic reasons were considered by Greeks to be their unredeemed lands. The *Megáli Idéa* included foremost within its scope the holy city of Constantinople and the bulge of western Turkey in Asia Minor known to the Greeks as Anatolia. In reality, the truncated Greek state as first constituted, particularly before the redemption of Thessaly in 1881, was hardly an economically viable entity unless it could annex fertile lands and a larger population.

In 1864, the English ceded the Ionian Islands to Greece. In 1881, the Turks, responding to French and British pressure on the Sublime Porte, relinquished Thessaly, the richest agricultural plain in modern Greece. The northern borders were thus redrawn from the Valley of

Tempi westward to Arta, and north to the border of Macedonia. Nevertheless, Greek ethnic and national desires remained unsatisfied. Eleftherios Venizelos, who became prime minister in 1910, an ardent exponent of the *Megáli Idéa*, realized that the moment was propitious for welding together other Christian blocs in the Balkans into a concerted military effort to expel the Ottomans from Europe. The Young Turks' plan of strengthening their empire had misfired. On the contrary, disintegration of the empire had accelerated as the Young Turks, in their zeal, pressed upon the Christian subjects a program to "Ottomanize" the Balkans. Under this intolerable pressure, the Christians of Bulgaria, Greece, and Serbia overlooked their acrimony for a few moments in order to join forces.[1] The Macedonian Struggle was over; outright war was about to begin.

A skilled statesman, Venizelos, entered into cautious dialogue with the Bulgarians. Perceiving that no solution for Macedonia was possible with them on a diplomatic level, he decided that Greece should join Bulgaria and Serbia in a military alliance against the Turks. The delicate question of the division of Macedonia was left in abeyance, to be decided after victory. Between Serbia and Greece there were no irreconcilable issues, for although the Serbs had claimed certain sections of Macedonia, they were on the whole less militant. Therefore, no formal agreement was signed between Serbia and Greece. But Serbia, Montenegro, Greece, and Bulgaria agreed verbally to march together as members of the Balkan League. War erupted in mid-October 1912. Within weeks the Greek army marched triumphantly into Thessaloniki, led by Crown Prince Constantine, commander-in-chief of the Greek army.[2]

If ever a city was coveted by two claimants, that city was Thessaloniki. Indeed, the primary phase of the First Balkan War was characterized by a race between the Greek and Bulgarian armies to enter the city first. Although two Bulgarian battalions had entered, numerically and emotionally there was never any doubt that Greek strength held sway.

---

[1] Douglas Dakin, *The Greek Struggle in Macedonia: 1897–1913* (Thessaloniki: Institute for Balkan Studies, 1966) 403–406.

[2] Apostolos Vacalopoulos, *History of Thessaloniki* (Thessaloniki: Institute For Balkan Studies, 1963) 9.

Despite the smoldering hatreds that threatened the unity of the Balkan allies, they did manage together to defeat the Turks by 16 April 1913. The Treaty of London, signed on 30 May 1913, terminated the First Balkan War.

The Second Balkan War, begun on 30 June 1913, was fought this time with the Serbs and Greeks aligned against the Bulgarians to settle the division of Macedonia. Bulgaria's defeat came quickly, on 31 July 1913. As a consequence of this second conflict, Greece received the largest portion of Macedonia, and also Epiros, Crete, and many Aegean islands for a gain of 1,979,000 in population. According to the Treaty of Bucharest, ratified on 10 August 1913, Macedonia was divided among Serbia, Bulgaria, and Greece, with Greece receiving the lion's share.[3] As a result of the Balkan Wars, Greece's total population increased by 67 percent.

The impact of the wars on the farm school boys was enormous in psychological terms. Father and Mother House had departed for the United States, leaving Reverend W. C. Cooper in charge. Considering the institution's stability and viability during the crisis, we must assume that Reverend Cooper had the stamina and clear judgment demanded for the task of maintaining order and tranquility. The survival of the school throughout its development has hinged upon capable, dedicated people who rose to the task during various crises. Reverend Cooper can certainly be counted among that number.

His report of 1913 reflects the view from the campus: "It was a peculiarly exciting crisis in Macedonia, because our people saw, or thought they saw, the dawn of freedom for which they had waited and longed for through centuries.... The air was full of rumors; the roads were crowded with soldiers and munitions of war; the whole land was aflame with martial excitement." He added that the boys could not concentrate on arithmetic or grammar. The battleground was not far from the campus: "On 1 November only 25 miles away in Yenyr Varde, the cannons could be heard booming all day. News of that terrible battle trickled in to the School. Eyewitnesses reported that unburied corpses lay about for days, then thrown into trenches so

---

[3] Douglas Dakin, *The Unification of Greece, 1770–1923* (London: Ernest Benn, Ltd, 1972) 314; L. S. Stavrianos, *The Balkans Since 1453* (New York: Dryden, 1958) 537–43.

shallow that arms and legs protruded."[4] Then the war moved to the school's very perimeter. Entrenchments were thrown up around the campus, but no actual battles took place there. Since the Turks decided to surrender the city quickly, violence, at least on school grounds, was avoided.

On 16 November 1912, when it was clear that Thessaloniki would remain in Greek hands and the battle to eject the Turks was going forward, a grand celebration took place in the city: the "clang of bells and shriek of whistles told us that the Turkish rule had forever ceased in ancient Thessalonika. We all turned out to see the troops enter the city."[5] Thus the youngsters, although they had seen extensive cruelty and violence in their short lifetime, were spared the immediate horrors of the Balkan Wars. The institution had become an oasis—a place of peace, continuity, and relative normalcy—while all around it the countryside seethed with armed conflict and thousands of refugees displaced by the violence roamed throughout Macedonia seeking food and shelter.

However, some older boys who did leave the school suffered greatly. On 3 February 1913, when the Bulgarian government called for volunteers, boys and teachers of Bulgarian heritage were quick to respond. Reverend Cooper refused permission for the younger boys to leave, all the while trying to dissuade the older ones. Finally, with great reluctance, he granted permission for the most mature to leave. Eleven students and three teachers reported to call-up stations, but were not given arms or uniforms. Disillusioned, they struggled back through the deep snow and extreme cold only to return after a four weeks' absence in poor health and overwrought.[6]

In the spring, after the end of the First Balkan War and before the outbreak of the second in late June 1913, Bulgarian schools throughout Macedonia, in an effort to return their students to their homes before transportation became totally disrupted, decided to close quickly, forgoing the usual commencement exercises. When it was apparent that no agreement could be concluded between Greece, Serbia, and Bulgaria over the question of Macedonia, further armed

---

[4] "Thessalonika Agricultural and Industrial Institute Annual Report, 1913," 19, AFS Archives.

[5] Ibid.

[6] Ibid., 20.

conflict seemed inevitable. All traffic was suspended between Thessaloniki and surrounding areas. Special trains, however, had been scheduled to insure that students would be moved without mishap to their villages. The farm school decided to send its Bulgarian students on these trains on 26 June without conducting commencement exercises. Four days later the Second Balkan War broke out. Some boys, whose villages had been decimated by the first conflict, had no home to which they could return and remained at the school.[7]

Permanent changes in the structure and range of the school came about as a result of the Balkan Wars. Since it had become a recognized fixture by 1913, agencies engaged in philanthropic programs for the care of refugees recognized immediately the school's potential ability to contribute broadly. The Balkan Wars mark the beginning of a new period in farm school history—that of a relief organization for refugee and reconstruction work. Such activities naturally brought the school into wider contact with the population of the newly enlarged Greece.

The following schedule of receipts and disbursements indicates the institution's involvement in refugee activity.

*Receipts*
From the American Red Cross Society
through State Department $500.00
From sundry donors as per printed report
of the institute $110.75
Items sent in care of W. C. Cooper
through the ABCFM $368.11

———————

Total receipts $978.86

*Disbursements*

Given to the international Relief Committee
through the U.S. Consul, Salonika $250.00
Given to the Relief Committee of which the
Rev. E. B. Haskell was member $250.00

———————

[7] Ibid., 22.

Wages paid for 136 days work to destitute Turks $394.00
    Food given to above workers $67.36
Gift to sick Bulgarian $2.11
Gift to two destitute families in Albania $17.60
Balance on hand $ 41.86

_____

Total disbursements $978.86[8]

In explanation of the above we learn: "The International Relief Committee took care of 7,000 Turks just beyond our Institute and the other committee fed some 20,000 souls in Salonika. The $368.11 which passed through Mr. Cooper's hands was used to supply refugees in making a stone wall."[9]

By giving refugees work through relief funds the farm school profited since it obtained free labor to accomplish tasks for which the school had no funds: "We hired a lot of Turks from the camp to quarry stones and dig *trosusk* or wild grass. They brought bread from the camp and we gave them a warm stew at noon and 22 cents each when they went away in the evening."[10] Through this project Turkish labor built a wall for a barn lot, a large stone water trough and a stone stall for the Jersey bull.

Lest we forget that Dr. House's impulse was essentially religious, the report assures that the opportunity was not lost to instruct the Turks: "Once a week we held a meeting with the men at the noon hour and explained some of the fundamentals of Christianity."[11]

The central and most significant shift in the school's total structure and outlook was its new orientation toward a Greek Orthodox state whose character, although fraught with catastrophe and perplexities, would be dynamic and outreaching. The school's redirection was in some phases immediate, in others delayed, but in general the institution adapted successfully to the new reality.

The population of Greek Macedonia—that part of Macedonia ceded to Greece by the Treaty of Bucharest in August 1913—consisted of a clear Greek plurality. The rest of the population

_____

[8] Ibid., 29.
[9] Ibid., 30.
[10] Ibid.
[11] Ibid.

consisted of Muslims and Bulgarians, with a minority of Jews and Vlachs. The League of Nations's Refugee Settlement Commission later analyzed the 1912 population of Greek Macedonia as follows:

| | | |
|---|---|---|
| Greeks | 513,000 | 42.6 percent |
| Muslims | 475,000 | 39.4 percent |
| Bulgarians | 111,900 | 9.9 percent |
| others | 98,000 | 8.1 percent |
| | | |
| Total | 1,197,900 | 100.0 percent[12] |

As part of Greek Macedonia now, the school, too, changed its ethnic make-up. With the erection of new national boundaries and the virtual disappearance of the Ottoman Empire in Europe, populations could no longer freely flow from one place to another. Under Turkish rule, Thessaloniki had been an open city into which Serbs, Turks, Greeks, Albanians, Bulgarians, and Jews entered for education and commerce. Under these old conditions, boys of various ethnic orientations could float into Macedonia without crossing national boundaries. But now, as a part of the Kingdom of Greece, the school would be restricted by the new situation to enrolling mostly Greek boys. Since many Bulgarians had fled Macedonia upon the Greek victory, the school gradually oriented itself to a Greek environment.

Missionaries, whatever their projects and interests, all faced the same general problem in Greek Macedonia. Historically they had moved among the Bulgarian and other Slavic populations with some success. Now a new code of behavior and a strong set of rules were demanded by the Greek state. In Thessaloniki, for example, the Greek authorities immediately forbade Bulgarians to preach or conduct meetings in their native tongue.[13] Fortunately, the school had always stressed English, although the boys used a Slavic tongue when speaking among themselves.

---

[12] George Zotiades, *The Macedonian Controversy* (Thessaloniki: Institute for Balkan Studies, 1961) 39.

[13] W. W. Hall, *Puritans in the Balkans, The American Board of Mission in Bulgaria: A Study in Purpose and Procedure* (Sofia: Cultura Printing House, 1938) 249.

The atmosphere outside the gates of the school was now distinctly Greek. From the first day Greek troops entered the city, non-Muslim groups threw off the fez, and wealthy Muslims and their families fled along with Turkish officials for lands that were still in Ottoman hands. The Turkish language was barely heard in the streets and all the signs in old Turkish script were replaced by Greek. Ancient Byzantine churches, which had been converted by the Muslims to mosques, were reconverted to Christian churches. Greek soldiers in Greek uniform crammed the streets and sat in the coffee houses with a proprietary air.[14] No one mistook the city as being anything other than part of the modern Greek state.

In spite of the fact that the Macedonian political problem was settled and the borders fixed, a population problem continued to exist; in fact, it had been exacerbated by the Balkan Wars. At the onset of hostilities, a great migration of peoples had begun: Bulgarians and Turks were leaving Serbia and Greek Macedonia, while Greeks moved out of Thrace and Asia Minor. Dr. House's account of this dislocation reported that "it would be a small estimate to say that this migration has swept from their homes 600,000 with an immense loss of life and property, the former from lack of food and clothing."[15] He observed quite rightly that "these migrations have changed undoubtedly the nationality of the inhabitants of large tracts of country, while they have thrown upon the governments which have received the refugees immense financial burdens which will tax their subjects perhaps for generations."[16]

Although the migrations connected to the Balkan Wars were not fixed by treaty and in this sense not forced, the settlement and relief problems of these people presented a prototype for the later swell after 1922. Such turmoil was bound to place tremendous strain on the school. Donors were fearful that perhaps their money was invested in an institution that would never survive the new conditions. The departure of Slavic communities meant the diminution of evangelical converts, for it was well known that Greeks had resisted conversion and might be indifferent to, or at worst antagonistic toward, having

---

[14] Vacalopoulos, *Thessaloniki*, 130.

[15] John Henry House, "A New Era for 'Hampton' in Macedonia," 4–9, reprinted from the *Southern Workman*, December 1915, AFS Archives.

[16] Ibid.

their children attend a school run by Protestants. Yet from the beginning of the migration of the Greeks from Asia Minor to Macedonia, refugees appeared at the gates to enroll. Dr. House noted that Greeks who had heard of or attended Protestant schools throughout Anatolia sought them out when they came to Greece.[17]

Not only did the American Farm School survive the Balkan Wars, it even showed that it could be agriculturally bountiful. In addition, it boasted that it was the only educational institution in Thessaloniki that managed to stay in operation throughout 1912, when all the others closed down because of the problems created by the war. And in 1913, despite the violent collision of the Second Balkan War, it turned out thirteen graduates. The American flag flew over the campus during the war years; it has been suggested that the flag was respected by warring groups, and that therefore the school was not violated. Whatever the reason for its continued health, it now stood firmly implanted on Greek soil. The more radical changes that were to occur as the result of the school's absorption into the new host nation were retarded by the outbreak of World War I in the Balkans. This new calamity, coming directly on the heels of the Balkan Wars, absorbed the full attention of the Greek government.

What did the acquisition of territories mean in terms of agricultural development? With the Treaty of Bucharest, Greece received a huge increase in cultivable land: it now owned 3,325,000 acres as opposed to 2,150,000 before the Balkan Wars. With this addition the value of agricultural products increased some 60 percent.[18]

Unsurprisingly, the Balkan Wars brought about a drastic change in land distribution. Under Turkish rule, land was held to a large degree in estates, called *tsifliks*, owned by Muslims, with Christian peasants working as sharecroppers. However, as Muslims moved away, Greek peasants gradually bought up parcels of these estates. By 1918, for instance, about 150 of the original 466 *tsifliks* had disappeared as peasants purchased plots with the aid of legislation and loans.[19] The Greeks became small proprietors as the huge estates were split into

---

[17] "Thessalonika Agricultural and Industrial Institute Annual Report, 1913–1916," AFS Archives.

[18] Dakin, *Unification*, 201–202.

[19] Ibid.

fragments. Production was hampered by reducing the unit of cultivation, especially when modern mechanized means were introduced. Yet, the new landowners were more strongly motivated than the indolent former landowners, who were often absent and ran their production on a sharecropping basis. That the peasant landowners could not amass capital to buy machinery or fertilizer and that they were deeply conservative was bound to militate against progressive farming methods; nonetheless, Greeks coming into Northern Greece were attracted to farming through government stimulation. Moreover, Greeks who had served as sharecroppers and now owned their own lands felt a new pride at being farmers. The Greek movement to the land confirmed beyond a doubt that the school could be useful to the swelling population of Greek farmers, for the Greek, a hard worker, possessed a work ethic that was not at odds with the Protestant work ethic.

So far we have viewed the school only in its chronological development, with sociological and historical implications. Now it will be helpful to visualize the students in their home environments operating as farmers on their land. Macedonia, although having its own particular set of characteristics, possesses enough problems agriculturally in common with the rest of Greece to enable us to examine it within the framework of the country as a whole. In addition, as the school developed after 1913, youngsters started coming from all parts of Greece and had to be trained to handle problems in their particular areas.

Greece is no farmer's paradise: the geography of the country explains certain basic disadvantages. The center of the Greek peninsula is dominated by a mountainous mass extending the entire length of the country. The mountains are braced on either side by high ridges that rise abruptly from flatlands. Locked in among the mountains are areas of lowlands, plains, and valleys that hold populations separated, and in many cases actually isolated, by intervening walls of mountains. The mountainous nature of the landscape presents a serious difficulty for the economic life of the farmer. Roads built through mountainous terrain offer almost insurmountable expenses for a poor country and deteriorate rapidly under the extreme weather conditions prevailing in such topography. In the days prior to World War II the scanty and poor road network meant that the farmer usually had difficulty reaching a market to sell his produce. Small villages, cut off from business centers, were obliged to provide for most of their own needs. In the days when little cash could be earned, the peasant lived on an economy that was not far removed from barter.

Some perspective may be gained if one notes that Greece is no larger than Illinois yet has only one-fifth of that state's arable land.[20] The mountainous areas, aside from creating the disadvantages already cited, offer only meager grazing ground for goats and sheep, except for a few slopes where cultivation is marginal. The goats and sheep, grazing in the open countryside, do inestimable damage to tree and shrub growth, which in turn produces vast areas of erosion.

Aside from the topographical features, a grave factor inhibiting agricultural production is the fragmented division of landholdings caused by inheritance, government redistribution, and, most importantly, in those days, the dowry tradition. Although the dowry in most places has disappeared, fragmentation persists as a major disadvantage. Also, the farmer's plots, sometimes several kilometers apart from each other, present insurmountable problems. Men and animals dissipate their strength in moving from one field to another. Severe land fragmentation precludes use not only of mechanized aids but also the installation of sensible irrigation systems and rational crop rotation. Moreover, land fragmentation also has the disadvantage of using up scarce land for pathways, boundaries, and roads, besides making inefficient the tasks of sowing, planting, and supervising the crops.[21]

As a result of land fragmentation, the size of a farmer's holdings in Greece before World War II made for large underemployment where farmers were busy only part of the year and sat in idleness the other months. It also meant that children who could not find adequate employment on their parents' land had to find another means of livelihood in the countryside or undertake a difficult move into a city.

---

[20] Script for the American Farm School film, 1951,, AFS Archives. For background on the Greek countryside see John Campbell and Philip Sherrard, *Modern Greece* (London: Ernest Benn, 1969) 322–63.

[21] A. N. Damaskenides, "Problems in Greek Rural Economy," *Balkan Studies* 6/1 (1965): 25. For the standard work on the Greek peasant and his environment see Irwin Sanders, *Rainbow in the Rock* (Cambridge: Harvard University Press, 1962).

An evaluation of Greek agricultural problems written in the mid-1960s still stressed that one of the basic problems in the country's approach was the lack of proper education for farm youths:

> Elementary schools in rural areas function on the basis of the general program of elementary education which is valid for all the schools of the country. Not many lectures of more or less technical character are given to the farmer's children, little effort is made to instill in the boys' and girls' souls the love of countryside and their determination to avail themselves of opportunities existing in rural areas and to improve their standard of living.[22]

Still another grave disadvantage to cultivation in Greece is the weather. In Macedonia, for instance, the winters are severe. Cold rains and snow soak the earth, while in summer little if any rain occurs. The intense summer heat bakes the earth until it cracks. The violent Vardaris wind causes havoc, drying out the soil all year round.

In spite of these problems, the country does yield wheat and other cereals, tobacco, cotton, fodder plants, vegetables, grapes, raisins, olives, and citrus fruits that are some of the best in the world. Chickens, ducks, turkeys, cattle, sheep and goats, plus horses, mules, and donkeys are found in all regions of the country.

Certain traditional attitudes that impede efficient organization of agriculture prevail in many areas of Greece. Anthropologist Ernestine Friedl notes in her valuable study of the village of Vasilika in Central Greece that the inhabitants

> rarely make repairs on their houses or on their equipment; they rarely make shelves or simple pieces of furniture. In short, they do not have the habit of pottering and tinkering about the farms whether from necessity or because they enjoy it. Vasilika farmers and shepherds consider themselves specialists; they view themselves with pride as

---

[22] Damaskenides, "Rural Economy," 29–30.

experts in farming and sheepherding and assign the same dignity to the expertness of others.[23]

The crippling effects on the farmer of such an attitude are obvious:

> If his saddle strap breaks, [he] does without a saddle or borrows someone else's until the saddlemaker comes by, even if it means several weeks of inconvenience. In one of Vasilika's households, the stone wall of one side of an oven was accidentally broken by a tractor plow one day in April of 1956. At the end of July the wall had still not been repaired—even temporarily—because the stonemason with whom the farmer was accustomed to deal had not yet had the time to come and fix it.[24]

But the Greek farmer is, if nothing else, resilient and independent in spirit. His long struggles with poor land and years of war and devastation have made him one of the most buoyant and philosophic creatures on earth. Charles E. Greene, a young American volunteer who worked for some months at the school right after the Greek civil war, noted that the Greek farm boys exhibited, among other characteristics, "an unexampled courtesy" and "an unflagging talent at conviviality." Among other traits he noted "the continuance in all vigor of the family and village forms," "startling composure" and "quickness of complicated thought."[25] To illustrate his analysis, Greene told of the following incident:

> I sometimes felt, when matching wits with the student body, that I was gaining a point less by skill than by reliance on their qualities of good will and good cheer. I once called a meeting in which I explained to the students that the spirit of independence which I had noticed rising up against the athletic

---

[23] Ernestine Friedl, *Vasilika: A Village in Modern Greece* (New York: Holt, Rinehart and Winston, 1964) 35.

[24] Ibid.

[25] Charles E. Greene, "The American Farm School Report to the Department of State," July 1951, AFS Archives.

program was no longer to be tolerated, and outlined the drastic discipline I should now take up against those not cooperating. I ended my speech with a sharp stroke of harsh, authoritarian sarcasm, crying out, "Long live the athletic program!" The students then rose as a man and carried me from the hall on their shoulders. The program continued in much the same force as it had the day I called the meeting.[26]

This, then, briefly characterizes the Greek farmer and his land to which the American Farm School had to attune itself in the first decade of its existence. However, prohibiting any systematic program of adjustment, the First World War burst upon Greece with great intensity. The upheaval forced the director to handle problems on an ad hoc basis, which ironically in the long run made the transition unstudied and successful.

---

[26] Ibid.

# CHAPTER 6

# WORLD WAR I

The explosion of World War I that burst upon Europe in 1914 was ignited in the Balkans. For four years its shock waves reverberated throughout Northern Greece. Villages that had been leveled and depopulated during the Balkan Wars were again exposed to devastating warfare, the refugee upheaval augmented and exacerbated by yet another catastrophe. Young people maturing in the former Turkish province during the second decade of the twentieth century knew no more of normalcy than their fathers or forefathers had known.

The school was affected profoundly by the war, for the Macedonian front pivoted around the center point of Thessaloniki, which was used as homeport, garrison, and base for the Entente troops. Indeed so much Entente effort was sunk into making Thessaloniki an impregnable city that Clemenceau, ridiculing the defensive digging-in mentality of French General Sarrail, dubbed these forces "the gardeners of Salonika."[1]

Although a review of the campaign is well beyond the compass of this book, some points should be highlighted so that the farm school experience may be viewed against the background of the times.

In October 1915 the armies of the Central Powers—Austria, Germany, and Bulgaria—attacked Serbia and succeeded in routing the Serbs. That same month France, Serbia's ally, at the invitation of

---

[1] Alan Palmer, *The Gardeners of Salonika: The Macedonian Campaign 1915–1918* (London: Andre Deutsch, 1965). See Stratis Myrivilis, *Life in the Tomb*, trans. Peter Bien (Hanover NH: University Press of New England, 1977) for a novel depicting the Macedonian front from a Greek point of view.

Greece's prime minister, Eleftherios Venizelos, sent three divisions that debarked at Thessaloniki to be joined by a British division in a belated attempt to halt the progress of the Central Powers. After several weeks of heavy fighting, the Entente forces were compelled to retreat back to Thessaloniki. The original Allied army was later augmented by thousands of additional troops, among which were the British and some French components of the Gallipoli disaster. By the end of January 1916 the total number exceeded 160,000 men. Four-fifths of this force was concentrated within a 20-mile radius of the port of Thessaloniki.[2] Before the conclusion of the campaign in 1918 there were over 500,000 soldiers, a colorful, disparate group including—besides English, French, Canadians, and Serbs—Australians, New Zealanders, Indian Mohammedans, French Cochin Chinese, Tunisians, Senegalese, blacks from Madagascar, Italians, Russians, and Albanians.[3]

A cursory glance at these figures suggests immediately the staggering economic, sociological, and logistical problems that arose from the infusion of such a population mass within a compressed time frame. The school was affected in all aspects of its operation as it attempted to cope in the wartime environment.

Greece's equivocal position during the first years of the war added another element of strain that was to impact strongly on Thessaloniki and environs. Ostensibly Greece's stand was neutral but, behind this stance, the country was being pulled apart by two opposing factions. King Constantine, married to the kaiser's sister, represented that group in the country that wanted to remain neutral and was accused by its critics of being pro-German. On the other side, the Liberal Party leader Venizelos spoke for those who were sympathetic to France, Great Britain, and Serbia, and were willing to commit Greece fully and immediately to the Entente cause. In fact Venizelos had offered the Allies every facility in Thessaloniki from the beginning. In 1915 he split from the king on the issue of choosing sides and resigned. Early in autumn 1916, he established a provisional government in Thessaloniki, leaving the king to rule through another

---

[2] Palmer, *Gardners*, 51.

[3] "Thessalonika Agricultural and Industrial Institute Annual Report, 1913–1916," 9, AFS Archives.

prime minister in Athens. Greece was thus divided into what amounted to two separate states.[4] As historian Vakalopoulos explains it:

> Thessaloniki became a center of inspiration that roused the spirits of the Greeks. Local men from Thessaloniki and Greeks from elsewhere came together and signed an act which determined the composition of a committee and fixed the committee's sphere of activity in facing the dangers which could materialize to the detriment of the independence of Greek Macedonia. As a result of this initiative on the part of the above men the Movement for National Defense sprang into being around the 30th of August 1916. Afterwards the Army of Defense was formed and thus ranged itself on the side of the English and French.[5]

Meanwhile, before the Provisional Government was set up, the rupture between Greek antagonists on the foreign policy issue had left the country weak and a prey to both the Allies and the Central Powers. In May 1916, when a combined Bulgarian and German force attacked Fort Rupel, an impressive Greek stronghold on the Greek-Bulgarian border, the Greeks under orders from the king's government in Athens surrendered it with little resistance. The Allies were horrified. Rupel, only 50 miles from Thessaloniki, commanded the central access from Bulgaria into Greek Macedonia. Fearing that the enemy would advance into the city, the Allies took drastic measures. On 31 June 1916, French General Sarrail gave orders that put Thessaloniki and surrounding areas virtually under Allied occupation. He proclaimed martial law in the city and in all other areas where Allied troops were positioned. They took control of the police and of postal and telegraphic operations. For all intents and purposes, Greek sovereignty in this area was revoked; inhabitants who had just been freed from foreign occupation during the Balkan Wars found themselves once more under the rule of an alien administration. Additionally, the Allies imposed a blockade on all ports. They

---

[4] Douglas Dakin, *The Unification of Greece, 1770–1923* (London: Ernest Benn, Ltd, 1972) 206–14.

[5] Apostolos Vacalopoulos, *History of Thessaloniki* (Thessaloniki: Institute For Balkan Studies, 1963) 134.

demanded demobilization of the Greek army, the replacement of the
allegedly pro-German government by one that the Allies could
approve, fresh elections throughout Greece, and the firing of all
police officials who had allowed Allied diplomats to be insulted.[6]

Within a year, the matter of Greece's disastrous schism was
settled in the following manner: King Constantine, responding to
irresistible pressures exerted by the Allies, although he did not
abdicate, went into exile leaving his son Alexander in his place. On 27
June 1917, Venizelos left Thessaloniki to be installed once again in
Athens as prime minister. The Allies, pleased with the sympathetic
leadership in Greece, raised the blockade; foodstuffs and matériel
began to flow into Greece. With a loan granted to him by the Allies,
Venizelos was able to equip ten divisions that were to fight along with
the British and French in the final campaign in Macedonia in
September 1918.[7]

On 30 December 1917, the Germans bombed Thessaloniki for
the first time; after that they bombed frequently.[8] From the beginning
of hostilities, officials advised that it would be wise to relocate the
farm school as soon as possible since it was in a threatened area. For
Mother House, however, the likelihood of a world war exploding
outside the school's gates was not a reason to move. She wrote, "Of
course that was out of the question without money, and we should not
have moved even if we had the money," reminding her readers that
she and Father House had been "in tight places before." Apparently
the years living under Turkish rule had left such an impression on her
that she felt relieved that "the armies around us are not Turkish!
After passing through so many years where the Turks were either our
enemy or our defender, this one does not seem so hard to endure." She
felt secure in the premonition that "the fighting will not reach
Salonika."[9]

[6] Palmer, *Gardeners*, 68.
[7] Dakin, *Unification*, 216.
[8] Vacalopoulos, *Thessaloniki*, 134.
[9] "Thessaloniki in War Time, 1917," 14; 17; 23, AFS Archives (an American Farm
School brochure used as house organ to keep friends and contributors informed.
Much of the contents was extracted from letters of Susan Adeline House.).
Mother House was somewhat premature about the Turks. When Greece entered the
war they did become the "enemy." They were credited by populations in eastern
Macedonia of atrocities in the region of Pangaeon.' see First Lieutenant G. C. Berry,

Since General Sarrail, according to historian Alan Palmer, had transformed Thessaloniki into "one of the best fortified cities in the world," it was only a remote possibility that the Central Powers would attempt to capture the city. Yet the Allied occupation of Northern Greece made all inhabitants, both Greek and foreign, feel insecure.[10] Mother House registered the general uneasiness: "Everything is very uncertain for the future politically. We have changed the political language of the school from Turkish to Greek and it is possible that we may be obliged to change it once or twice more before things get settled. No one can guess what will happen in the future but we trust that the fighting will not reach Salonika."[11]

The very atmosphere and landscape seemed at times like correlatives to the ominous events of the moment that were destined to become chapters in history books. If heavy fog hung over the Western front, so it did also on the Macedonian front. In 1916, during the Allied occupation of Thessaloniki, the fog became so thick that visibility was reduced to nearly zero. Mother House described what it was like: "We feel almost like prisoners. We see the boys dimly, as they move about the playground; beyond that we are quite shut in."[12] It was during that deep fog, which lasted for weeks, that Mother House heard the Greeks march invisibly like ghosts past the school as they fulfilled General Sarrail's order that the Greek army demobilize.

The school's main task during the war years was essentially twofold. First, the most crucial challenge for the director was to keep the school in operation despite the almost insurmountable difficulty of supplying the boys with the necessities of life. It was feared that failure to do this might spell the institution's permanent demise. Secondly, the school's other role, one that had begun with the Balkan Wars, was to administer aid and comfort to communities beyond the campus. This broader mission and consequent wider usefulness was noted by the largest philanthropic agencies; requests from them for cooperative efforts to aid war victims filtered in from many corners.

---

"Relief Work Among the Villages of Mount Pangaeon; Report of the American Red Cross Commission to Greece," 11, AFS Archives.

[10] Palmer, *Gardeners*, 52.

[11] "War Time," 14.

[12] Ibid.

As for the first concern, that of keeping the school in operation, the main problem was the severe food shortage caused by the immense population influx. The newcomers exhausted all local food stocks and placed such a burden on transportation that conditions of real famine existed for all who did not enjoy self-sufficiency owing to land of their own. The troops were given rations, of course, but at times even they lacked proper nourishment. Prices for all articles, especially foodstuffs, reached inflationary levels, sometimes five to ten times the normal price. Worst of all, food could sometimes not be found at any price.

At the time, the farm school family consisted of seventy persons who lived on campus. The farm and dairy operations were doubtlessly the main contributors to the institution's survival. By the time of the First World War, the agricultural and dairy departments were well-functioning modern enterprises. In addition, the work ethic and the ideal of cooperation laid down by the founder met the test during this time of great stress. The boys learned that they could perform successfully even under the most dire conditions. If anything, the necessity of coping with wartime exigencies actually deepened the conviction of students and faculty that the American Farm School's practical methods and moral attitudes were unquestionable strengths.

As a result of the difficulties, the enrollment of pupils was carefully regulated during these unpredictable years. For the most part, only former students were taken back at the beginning of each school year, new ones being accepted only under extenuating circumstances. As early as the fall term of 1915, the school had to turn boys away because of the scarcity and high price of food. In November of that year, Mother House wrote that Father House was "afraid that money might not hold out, prices are so high. We have to pay three, four, five times what we used to. Father bought 422 pounds of cheese for about double what we used to pay."[13]

There were many other items that the school had to buy at exorbitant prices with limited funds. But the basic foodstuffs were procured from the school's own farm. Although meat was extremely rare, the boys had fresh fruit and vegetables from the fields for at least eight months of the year; milk and butter came from the Jersey herd;

---

[13] Ibid.

the poultry operation provided fresh eggs and an occasional chicken dinner; harvested wheat provided home-baked bread; barley and oats kept the horses well fed and straw provided them with fodder. Meals were not always well balanced, yet it was certain that no one would starve provided that the enrollment remain reduced. And the sale of an occasional surplus enabled the school to raise much needed cash.

But the history of the American Farm School during the First World War was not just a struggle for survival. War proved to be an exciting event in the students' lives, for it brought the school into the mainstream. Thessaloniki had always been outside the "grand tour" for those visiting the "continent" and the Middle East. Furthermore, travelers who did come to Thessaloniki found the farm school miles beyond the reach of public transportation. Now, however, the opening of the Macedonian front brought thousands of foreign and Balkan people to the city, changing it from a quiet backwater into a pulsating metropolis several times its peacetime size. Alan Palmer gives one of the best descriptions of the city:

> Salonika in 1915 was an ancient city almost mocking her own history. Sixteen hundred years ago the Emperor Galerius had erected a triumphal arch commemorating his victories over the Persians and spanning the Via Egnatia, Rome's highway to the East. Now the bas-reliefs on the arch rose thirty feet above a busy road, with a coppersmith's shop behind it and a tram passing less than yards from the carved legionnaires. The old town was a maze of winding streets. Behind it climbed the terraces of the Turkish quarter, alleys clinging to the sides of a hill, low houses with balconies, enclosed with lattice work and a teeming population who would squat suspiciously on the uneven pavement. Near the port, which had been modernized at the turn of the century, there were cobbled streets divided by tram lines, garish cafés on the pavement, and drab hotels with grandiloquent names ("Splendid," 24 rooms; "Olympus," 20). It was a dusty and a noisy town; and it was an overcrowded one, for Salonika had not yet absorbed the refugees thrust down upon it by the upheavals of the Balkan Wars. There were Greeks and Slavs and Albanians and Turks on the streets. Above all were the

Jews of Spanish origin, sixty thousand of them proudly
independent even after centuries of exile, and forming one
third of the population. Salonika, once the second city of
Byzantium, was by now a bustling, clanging, strident city on
the make.[14]

   The city was encircled by supply depots, military hospitals, machine shops, army camps, construction sites, and other features that
one would expect to find in a place overrun by troops. More varied
and numerous modes of transportation brought a mobile population in
touch with the school. The campus was surrounded by British, French,
and Serbian installations. On the same side of the city as the school, a
huge hospital, signal corps depot, machine shops, convalescent
centers, transportation points, prisons, and police barracks were
hastily erected during the fall and winter of 1915–1916. Activity of
this sort provided work for hundreds, decreasing to a large extent the
habitual poverty. Visible from the farm school was the railroad built
out from the city to the point on the gulf where French troops
landed. For the boys, the atmosphere was electrifying. They found
themselves catapulted into a modern, technological environment the
dimensions of which had never been seen in this corner of Europe.
Thousands of troops with packs on their backs, huge cannons on
wheels, and columns of trucks rumbled past the school continuously
on their way to the front.[15] The airplanes particularly made a strong
impression on the Houses as well as the boys. Mother House revealed
her amazement by writing, "It is a wonderful world we live in. We are
getting so used to seeing men flying about us that it is growing quite
commonplace."[16] A whole new world of mechanization and
organization on a grand scale opened to the boys, whose sense of
organization was limited to the smallest and most elemental
combinations. Imagine, then, their reaction to the following situation
as related by Father House:

---

[14] Palmer, *Gardeners*, 13–14. For another view of Thessaloniki, see Charles
Packer, *Return to Salonika* (London: Cassell, 1964) 1–15.

[15] John Henry House, "An American Outpost on the Balkan Front," 2, n.d., AFS
Archives.

[16] "War Time," 18, AFS Archives.

When an offensive was on, the continuous roll of heavy guns gave a weird sense of the meaning of all of this stir and preparation. Enemy or friendly airplanes were daily in evidence in the air above. Spent shells of anti-airplane guns fell upon the fields around, sometimes fearfully near. Once the very flash of airplane guns of a battle in the air was visible. Again the burning of a great zeppelin illuminated the night air while still again the evening sky over the bay was lighted up by the bursting torpedo of a submarine that was sinking a transport in the bay beyond the point.[17]

The machinery used by the British and French made a deep impression upon the minds of the young boys. It was then and there that conservative peasants witnessed the efficacy of gasoline propelled equipment. Before the war, such modern equipment was regarded with skepticism, but the war's advent familiarized all classes, urban as well as rural, with the practicability of machinery. Local laborers hired by the Allies to construct buildings, bridges, and fortifications were given sophisticated engines with which to work. Peasants were taught how to use and repair the Allies' vehicles. Small boys watching the scene with wonder were quick to enthuse over and readily grasp the unlimited possibilities of engine-driven contraptions. The ready acceptance of machinery by Macedonian farmers after the war is directly attributable to the Allied use of machinery during the war.[18] In 1957, for instance, about 43 percent of all tractors in Greece were found in northern Greece.[19]

Still another advantage of being brought into the mainstream was the personal contact the youngsters had with individuals outside their own culture. The interrelationship of the students and the foreign community was of course a two-way movement. Each group had its cultural horizon extended by intermingling. The school became a center particularly for members of the British expeditionary forces. These men joined the boys and staff every Sunday for church services,

---

[17] John Henry House, "Balkan Front," 2, AFS Archives.

[18] Charles Lucius House, "The Introduction of Farm Machinery in Greece," AFS Archives.

[19] Paul Vouras, *The Changing Economy of Northern Greece Since World War II* (Thessaloniki: Institute for Balkan Studies, 1962) 37.

walking in some cases for miles through cloying mud. At times the number of soldiers and officers reached eighty. Tea was served in the afternoon so that those who wished could stay, for the trek back into the city for something to eat and back out to the school would have been impossible. Sunday evenings the men formed duets and quartets, accompanying each other with musical instruments as they sang hymns and other music familiar to them. Macedonian boys, used to the herdsman's pipe that whistled the age-old tunes and the small drum that banged out ancient rhythms, were entranced by Western harmony. The youngsters, coming into such close contact with the Westerners for better or for worse, became more acculturated to Anglo-Saxon mores, since the bulk of the visitors were from the United Kingdom and Canada, although to be sure other groups, especially the French, visited the school. The British became so intimate with the school that they formed the Friendship Association of the Thessalonika Agricultural and Industrial Institute whose function, as stated in the association's constitution, was "[t]o meet the desires of many members of the British Salonika Force, visitors to the Mission who feel that they will want to know how the work is going when they are back in their homes. The Association Secretary will receive periodic reports from Dr. House and other missionaries and will circulate them to the members." The new association reported, "Spiritually it is harder to define what has been accomplished, but the school staff has not failed to notice in the attitude of the boys and in the spirit of the school that the prayers and lives of these soldiers of Christ reach their mark."[20]

The youngsters also profited from the presence of Allied troops in other ways. Allied doctors and nurses gave unstinting service to ill students. This medical assistance not only saved the school great expense; in at least five cases, boys were taken into army hospitals where delicate life-saving operations were performed. The caring for sick students had always been a great drain on the staff's moral and physical stamina and particularly on Ruth House's, who regularly took it upon herself to nurse the sick. The services of a well-trained and generous medical corps released the staff from one of its most serious

---

[20] "Annual Report, 1913–1916," 3, AFS Archives.

responsibilities, leaving it free to focus its attention more squarely on the curriculum.

Whatever the school received by way of good works, it repaid in full by reaching out in small ways to the community outside its doors. For instance it started a circulating library of English language books that became very popular with anglophone troops. Chaplains and theology students were always pleased with the selection; even those interested in fiction or textbooks were gratified to find a library in this outpost of Europe. In addition, British and French soldiers often came for seeds and to ask advice on cultivation of land around their camps. At one time the school lent a seeder to one of the camps. Using the proper machinery, the soldiers were able to sow 150 acres of grain for their horses.

Relations with the Greek government, too, especially the segment interested in agriculture, were cemented during the war period, setting a pattern for cooperative efforts when the war ended. For example, an inspector general for vineyards noted the vines of American stock that had been grafted to local vines to make them resistant to phylloxera, as well as certain other American stock that seemed to thrive in the Macedonian soil and climate. The report he submitted to his department called attention to the positive contribution the school was making to the development of Greek agriculture.[21] Seemingly minor details, such as the fact that the school possessed shade trees, a unique phenomenon in the area, enabled it to provide comfort to soldiers parched and exhausted from the blistering Macedonian heat. They were often allowed to break ranks at the school in order to rest near the well and fill their canteens from its cool water. Sometimes the boys pumped the well long beyond midnight when there was no wind, so that no soldier would proceed without water. Dr. House, a missionary beyond all else, did not lose the opportunity to spread the Word. Bibles printed in Balkan languages were distributed by the boys as the soldiers lingered by the well. If the soldiers were not allowed to linger, the boys trudged alongside, often confusing nationalities so that Serbs received Bibles in Bulgarian, or Greeks received Bibles in Turkish.

---

[21] House, "Balkan Front," 5.

In other instances, the institution reached out in truly gigantic ways. The following are cases in point testifying to the school's underlying spirit, which found continual expression in acts of human kindness. In one example, two starving Greek boys separated from their parents in the chaos of war were brought to Thessaloniki. Dr. House had the children taken to the school, where they were put in the dispensary. After a year of almost constant nursing they were brought back to health. Also, a Jewish child from Monastir whose mother had been killed by a bomb and who had no means of support was brought to the school by a French airman. There the child stayed for three months, supported by his French friend until other accommodation could be found. Two emaciated peasants who stumbled onto the campus were cared for and given work. They became healthy and trusted herdsmen at the school.[22]

The war brought a wide variety of personalities to Thessaloniki. Not only the military, but missionaries, Red Cross and other relief workers, and government officials came to the city. At one point within a few weeks, visitors representing sixteen nationalities came to the campus. After the static years of social intercourse during the Ottoman occupation, this was suddenly a world of motion. The Houses' spirit of hospitality, as well as their keen awareness of public relations, served them well during this time. In fact, missionaries who depended for their good works on the generosity of donors had a highly developed sense of "image" and how to project it long before New York firms made public relations into a science. Mother House, who presided over tea, dispensed hospitality with Homeric grace. According to her daughter-in-law, she was the center of any social gathering: "She was an aristocrat, a leader, and took her role naturally and unconsciously as hostess, in complete charge."[23] No wonder the farm school became a gathering place for all sorts of personalities who journeyed to the Balkans.

As might be expected, King Alexander's visit in 1917 was very important in terms of gaining recognition from the host country. But the occasion was also memorable (and embarrassing) for Mother House: it marked her first public faux pas. The Houses were entertaining the king and several of his retinue in their dining room,

[22] "Annual Report, 1913–1916," 20, AFS Archives.
[23] Ann Kellogg House, notes to the author, 15 November 1972, AFS Archives.

then in James Hall. Mother House, aware of the room's dark and unattractive appearance, wanted to move upstairs to the living room, which, situated on the main floor, was bright and pleasant. The king, apparently enjoying the conversation, showed no inclination to move. Mrs. House twice made an effort to rise. The second time, one of the king's aides leaned over her chair and said, "Madame, we must remain seated until His Majesty is ready to move."[24] The initial visit by King Alexander set a pattern for future relations with the royal family. All of the kings since then visited the school and evinced an appreciation of what the institution was doing for the farmers of Greece.

It was during the school's great disaster that the hospitality it had extended to all visitors was generously returned in full measure. On 15 April 1916 a workman, without announcing his arrival, climbed to the roof of James Hall in order to begin his job of repairing the eaves. A powerful gale—perhaps the Vardaris, which had been howling all day—turned the coals the worker had brought up to the roof into flying sparks that set fire to the building. Soon, the high winds fanned the flames into an inferno. With the primitive means at hand, it was impossible to contain the fire, which in less than an hour reduced James Hall to a hollow shell. In 1916, this structure, which had taken over ten years to complete, covered 796 square yards and stood at the very core of the campus complex. It had contained Hampton Hall, which served as an all-purpose room, in addition to a dining room, recitation rooms, dormitory space for fifty people, and an apartment for the House family. Without James Hall, the school was crippled. The remaining facilities consisted only of the first building, Haskell Cottage, with a room overhead then used as a carpenter's shop; a bakery consisting of a lower story topped by a two-room apartment; and a connecting laundry. Also, there remained the original granary and carriage house with an attached building for poultry and a new barn with a hayloft that stabled the cattle and horses. Even with a resident population shrunk by the war to forty-five, the existing structures were inadequate.

From the first, friends of the school stepped in not only to fight the fire but also to provide shelter and other immediate essentials.

---

[24] Ibid.

When the fire broke out, French soldiers, spotting the flames from their neighboring camp, ran across the fields to lend a helping hand. A large part of what was saved was thanks to the quick and efficient help of these *poilus* (French soldiers) and also to the reaction of the students, who, keeping their heads, dragged down the steps whatever they could or threw things from the windows before abandoning the building. While the fire was still raging, one of the nearby British camps sent an ambulance with food that had been collected from several British installations to keep the school supplied for a few days. The Scottish Women's Hospital sent bedding, medicines, and bandages to the boys, who had lost nearly everything in the fire. Soon nursing units, troops, and foundations such as the BLMAS sent money for clothes and books accompanied by messages of support and sympathy. The contribution of 50£ from the BLMAS was particularly important, for the fire had taken place just at the season when the care of silk worms, garden, and harvest taxed the labor of the school to its utmost.[25]

After the flames were extinguished, the immediate imperative was finding shelter for the boys and staff. The first few nights they crowded into the barns and other buildings while the House family, refusing the many invitations proffered by friends in the city, moved into Haskell Cottage. Before the week was over, the British quartermaster general donated a large tent to be used for dormitory and classroom space. Thankfully, it was mild spring weather so that classes and activities could be held outdoors. Most astounding was the decision of Dr. and Mrs. House to continue the work under the grueling conditions and against the almost insurmountable odds. All evidence points to the fact that neither of them vacillated in their decision to rebuild; if defeatism had crept into their outlook, it was not reflected in their reports and correspondence. Instead, the main emphasis in documents written at this time was on how to go about reconstruction:

> Very soon after the School burned the question of rebuilding came up. With the armies all about us using so much of the materials and with prices so exorbitant both for

---

[25] "Annual Report, 1913–1916," 16, AFS Archives.

material and workmen, it might seem that we were unwise to attempt it, but on the other hand there were three important reasons for our building. We all felt it would be very unfortunate for the standing of the School to close it even for a year. Besides we had about twelve boys who because of closed boundaries could not reach their homes and were too young to be sent out to work, so we could not break up this their only home. And lastly, the walls that were standing were pronounced safe by British Army architects so, as it would be unwise to leave them unprotected through the storms of winter, it was decided to rebuild at once.[26]

It bears noting that on Saturday the building was gutted and on Tuesday Dr. House reported: "A meeting of the executive committee [decided] that the burned building be rebuilt at once."[27] By Thursday, Dr. Ojavo, a Thessaloniki businessman, made his bid to rebuild James Hall using materials he offered at cost price and an interest-free loan of $2,000. By Friday the boys began hauling material for construction. Mr. Baxter, a Scottish engineer who had volunteered his services to the school, began work on a tool shed to be used also as a temporary kitchen and dining room for the students. Contributions were substantial enough to provide the difference between the insurance settlement and the cost of rebuilding; thus it was necessary to borrow only a limited amount to complete the job.

Exactly six months after the catastrophe, the doors of James Hall opened in time for the fall semester of 1916. The school had continued, closing only for the summer vacation. The reconstruction is extraordinary when viewed in the light of the pressures of war, the internal upheaval in the host country, and the advanced age (seventy-one) of the director. Perhaps nothing better exemplifies the school's resolute spirit and its determination to survive than the rebuilding of James Hall.

The very next year, a great fire in Thessaloniki cut a swath through the city destroying the core. Because of the exigencies of war and the ensuing refugee problem, no work was done to reconstruct the burned buildings until the mid-1920s.

---

[26] Ibid.
[27] Ibid., 32–33.

At last the war was moving into its final phase. By mid-1918, to support the Allies the Greeks had 90 battalions in Macedonia consisting of at least 250,000 men. On 15 September 1918, troops on the Macedonian front began their final campaign. The Serbian army, accompanied by a French division, assaulted an enemy position at Dobropolji (Monastir), resulting in a complete breakthrough for the Allies. In a few days, in a lightning advance, the Allies separated enemy forces in the Vardar valley from those in Monastir. Three days later, combined British and Greek divisions began to attack on the Doirani front. Allied pressure on the German and Bulgarian armies was so irresistible that they went into full retreat, asking the Allies on 26 September for a suspension of hostilities. By 30 September all was quiet on the Macedonian front. The Allies were racing eastward into Bulgaria. The much ridiculed "gardeners of Salonika," with their Greek comrades, had delivered in a nonstop offensive the first staggering blow to the enemy. Since the Greek army was present during this successful drive and absent during previous unsuccessful attempts, some historians feel that the victory was made possible by the Greeks' contribution.[28]

From a military point of view, the achievement was considerable. As a result, Turkey was cut off from the Central Powers, Bulgaria was neutralized, and Austria-Hungary faced the Italians with little courage since her southeast borders were now insecure. Von Hindenburg himself admitted that the collapse in Macedonia excluded all hope of forcing the Entente to make peace.[29] Six weeks later, on 11 November 1918, the Germans signed an armistice with the Allies, bringing the Great War to a conclusion. The boys at the farm school heard the happy news with an ecstatic roar that rivaled in volume "the guns and mines of the first night of the big offensive." Father House assembled them for prayers and then dismissed them from their usual study hour to watch the firework celebrations. He studied the faces of the boys whom he knew and loved so well: "It struck me as I looked into the different boys' faces what a mixture of feelings they represented. One boy's father had been shot as a spy, the parents of two of them had starved to death, some dozen of them looked

[28] See Palmer's conclusions in *Gardeners*, 237–43.
[29] Edward S. Forster, *A Short History of Modern Greece* (London: Methuen & Co., 1960) 134.

forward to seeing parents they had left three or four years before, others going to homes they had fled, and still others to better food, more clothes and other comforts." In trying to find some evaluation for his own life's work, he ended on this note: "I could not but feel a great hope and faith that perhaps many of that group of boys would be strong leaders in the New World which could date its beginning from that day, and also if I could be in the least way instrumental in such an end, how tremendously worthwhile every effort would be."[30]

For Greece, at least, there was to be no "new world" in the immediate future. The havoc wrought just in Eastern Macedonia by the Bulgarians and Turks was horrendous. For example, the three hundred square miles of the Pangaeon area—bounded on the west by the Struma valley, on the southeast by the Musthenie valley, and on the south by the Aegean Sea—had been occupied since 1917. All books, schools, churches, and monasteries were destroyed; priests and teachers were murdered on the spot. The population living on the meager ration allotted to them by the Bulgarians was decimated, 5,196 people perishing from starvation. All males from the ages of fifteen to seventy were rounded up and sent to collection points in Drama and Kavala and then transshipped to Bulgaria. The remaining villagers were forced to work on road construction.[31]

The survivors who had been detained in Bulgaria were straggling back to their homes. Refugees still unsettled owing to the Balkan Wars were joined by the new flood dislocated by World War I. The worst human upheaval was still to come; in 1918 when Dr. House spoke of the "new world" he could never have foreseen the Asia Minor disaster of 1922.

---

[30] As quoted from a letter by John Henry House, in "Glimpses of the American Farm School," *The Sower: Sixty Year Issue*, 1964–1965, AFS Archives.

[31] Berry, "Relief Work," AFS Archives.

# CHAPTER 7

## CHARLES LUCIUS HOUSE:
## THE SECOND DIRECTOR

Of the four characters who figure most heavily in the spirit and survival of the American Farm School in the first five decades, Charles Lucius House is by far the most complex. A maverick by any measure, he possessed a defiantly independent nature, replete with paradox. If, by today's norm, his father was an old-fashioned missionary and humanitarian, then Charlie was an educator and humanist. In one way at least he was a humanist in the sixteenth-century sense: his outlook reaffirmed the concept that the human being's life on earth should be the object of God's and man's concern. He wrote for his alma mater, Princeton: "I believe that my friendships and my faith in a Kingdom of God *on earth* have been my greatest help in life" (italics added).[1] He was earthbound—of supreme importance to him was what happened in the here and now.

There was little remarkable in his face or carriage. While his father carried himself erectly, Charlie was careless of posture and dress. There is an anecdote about him that typifies his attitude toward clothes. A dapper friend flashed the label inside a suit jacket he was wearing, reading aloud, "Hart Shaffner and Marx." Charlie responded by exhibiting the inside of his own jacket, which was without a label, and mockingly read "Salvation Army."

Of medium build, but wiry, he remained lean all of his life. His 1909 class picture from Princeton is the last one in which we see him

---

[1] Charles Lucius House, *Philosophy of Life: 20th Year Record Book* (Princeton NJ: Princeton University, 1939).

clean shaven; after that he sported a small box mustache over his upper lip. In his bachelor days, although he wore his hair combed back, a loose curl often hung rakishly over his forehead. In his pictures the smile that plays about his lips is tinged with something akin to impudence or mockery, yet the eyes brim with warmth and a zest for life.

From his parents he caught the passion for service, devotion to Christian ideals, determination and courage that went well beyond the normal range of what most men consider duty. His most effective qualities were boldness, intuition, leadership and, more than anything, a sense of opportunity. The House spirit that had animated the school under his father's directorship was given impetus and broader application under Charlie's guidance. From the moment he arrived as associate principal in fall 1917, the school was stamped with the impress of this highly spirited individual.

In summer 1917, in the midst of World War I, Charlie House set out from the United States through submarine infested waters, headed for Thessaloniki. Steeled for life by his childhood in the Balkans, he was not one to delay this trip for safer times. Moreover, he had just come through a darker odyssey: a moral crisis that had strengthened his innate courage. He had decided to turn his back on a satisfactory eight-year career in the United States as a civil engineer to dedicate his life to a small institution located in the Balkans where he had spent his boyhood. As a result of this crisis, Charlie had settled himself on a new life course. He was a study in contradiction: an "anarchist," scrapper, conscientious objector with a strong belief in God, an individualist and rebel with an infinite capacity for love and life. Answering what he considered a "call" in the religious sense, he was coming to Greece "to help Father and ultimately take his place."[2] At first Charlie had volunteered to go out only in the position of tutor, and for an undesignated period, but Dr. House—aware of the turning point that his son had reached—requested that the board of trustees consider Charlie as the next director. The fire that had destroyed James Hall had taken its toll on the older House, who realized that it was time to groom his successor.

---

[2] Charles Lucius House to Lucius Beers, 12 April 1917, AFS Archives.

The proposal that he go to Greece was tendered to Charlie by his uncle, Lucius Beers, then serving as vice president of the board of trustees. Considering it, Charlie wrote:

> This leads to the decision which I must make, that of whether I feel called to this work and could go with my heart entirely in it and with the true spirit which alone can make my work count. At this point I have tried to consider every sphere of usefulness that I can foresee in my present course and although I can see real and worthy spheres along these lines and feel a strong regret at turning my back and breaking associations, I feel sure that in this call lay my duty even if it were to some other place than where Father and Mother are.[3]

In conclusion he stated, "I have delayed this letter in order that I might test my decision with the different moods that time brings and still feel unshaken in purpose."[4]

Not simply a man changing his career or selecting a new job, he was instead being called to a mission, embarking on a sacred lifetime commitment. The key phrases—"whether I feel called to this work," "in this call lay my duty," "with the true spirit"—are telling when analyzed within the framework of the personal crisis through which he was passing. He was at that moment searching for some outlet for his spiritual forces, some place where he could put into practice a philosophy of life that had taken thirty tempestuous years to gel. The American Farm School, which he had never seen, as he had left the Balkans in 1902, was the sort of place in which he thought he could be spiritually at home. The fact that his parents were there was a happy coincidence, but incidental to the main components that shaped his decision.

His crisis was precipitated by America's entry into the Great War. Right after his twenty-ninth birthday he came "to a dividing of the ways"[5] as the United States came closer to involvement in the war. "I was convinced," he wrote, "probably as a result of my birth

---

[3] Ibid.

[4] Ibid.

[5] Charles Lucius House, *Philosophy of Life: 45th Year Record Book* (Princeton University, 1954).

and childhood experiences in the Balkans, that the interests of my country and ultimate peace in the world would be retarded by our entry into the war."[6] At the same time, he was seeking a positive route in order to contribute to a better world. He doubted "whether there were any real moral issues in the war."[7] He was gradually but surely coming to the conclusion that the example of his parents and his own Christian training offered an alternative to participating in the war. "After a month of agonizing indecision, I decided in November 1916 to drop any philosophy of life as I saw it...."[8] That year he had come under the influence of the well-known Quaker, Rufus Jones, an articulate pacifist whose concepts of religious pacifism were expounded in his influential the *Later Periods of Quakerism*; there is no doubt that Jones was instrumental in Charlie's resolve to find a life of service that would be an expression of the forces that he felt taking form within him. But his crisis, although perhaps deepened by the war, was a more profound matter than simply confusion over America's foreign policy. He found it "hard for me to put into a few words my spiritual life in the last year or so, if I could do it at all. It has been very troubled. I gave up going to church because I could not follow Christ and live in the world as it is today, and because a Sunday Christianity and weekday respectability and compromise seemed like mental slavery to me."[9] By 1917 his stand had become clearly that of conscientious objector. It was prompted, among other things, by his view of Christianity, which was bound up inextricably with notions of service and mission.

If the moral crisis was aggravated by the necessity of committing an act of civil disobedience, then it was made almost unbearable by pressure from family and friends. His brother, John Henry House, Jr., who was very patriotically motivated, said that in actuality his brother was a traitor. A photograph taken of the two brothers is a striking portrayal of the two opposed polarities. The older brother,

---

[6] Ibid.

[7] Ibid. For a perceptive analysis of pacifists, their philosophy and their place in American society during World War I, see Charles Chatfield, "World War I and the Liberal Pacifist in the United States," *American Historical Review* 75/7 (December 1970): 1920–37.

[8] Ibid.

[9] Charles Lucius House to Susan Adeline House, 25 December 1916, AFS Archives.

posing in stiff military bearing, his officer's hat crooked stiffly on his arm, his legs tightly bound in gaiters, stares stonily into the camera. Immediately beside him, Charlie, in rumpled shirt half tucked into his unpressed, baggy trousers, grins playfully at the camera. The nonchalant posture and deliberately devil-may-care expression masked a bravery and loneliness that few could appreciate. Many of his closest friends chose not to talk to him; others begged him to sign up with the hope that he would be assigned to the engineering corps, thus avoiding active participation in combat; the most persistent tried to change his mind by pleading or argument.[10] His parents, although not conscientious objectors, respected his commitment without argument, fearing only that their son might be put in jail.

In May 1917, six months after he had decided definitely on his objector's stand and a month after he had received the proposal that he come to Greece, the selective service conscription bill was enacted. It called for the registration of all men between the ages of twenty-one and thirty. Charlie, who was to turn thirty in October, was in the first draft. His position had softened to the point where he would obey all orders short of bearing arms; otherwise he was prepared to go to prison. At the physical exam he was pronounced fit by the physician. However, a second doctor, who knew him, beckoned as he saw him leaving, ripped up the acceptance forms and signed others, declaring that he was physically unfit for military service owing to a weakness in his lungs caused by pleurisy. In this way the crisis of his life was solved, freeing him to pursue what he believed was the divine destiny of his life.

The circumstances under which he was rejected for service reinforced his intuition that he was on the correct path. He had expected to be accepted when called; therefore he had also expected, when he refused, to go to jail. He wanted to pay in some way—to suffer—for his convictions, for he loved his country and wanted to make some sacrifice. Thus when he was considered qualified for service all was in line with his feelings of right and justice. He was looking forward to the next step, which would end the affair. The second doctor, however, realized that no matter how Charlie was punished he would not change his convictions. The doctor decided

---

[10] Ann Kellogg House, "Memoirs," 31, AFS Archives.

that he could not allow Charlie to experience the dire consequences of his stubbornness. At first Charlie agonized over the second doctor's evaluation. Should he denounce him? Should he complain to the draft board that the doctor had lied? On the other hand, the second doctor was in higher authority than the first. Beyond that, he had no proof that the first doctor was right and the second wrong.[11] In a real sense, other forces had taken control of the situation.

Charlie was an extremely complex individual. Seemingly irresolvable contradictions existed side by side within his character. In an analysis it is almost impossible in many cases to suggest one behavioral pattern without immediately contrasting it with another. For instance, the conscientious objector totally opposed to violence in any form was as a youth pugnacious and belligerent. Granted a scholarship in 1902 to Blair Academy at the age of fifteen, he found himself, fresh from Thessaloniki, a foreigner in his own land. His childhood in the Balkans had made him culturally different from his classmates, so they teased him. His method of adjustment is significant. Rather than conforming, he took an antagonistic tack. Donning the Turkish fez that he had brought from Thessaloniki, he flaunted his foreignness. His schoolmates took the fez as a challenge and pounced on him. Slight of build (his adult weight hovered around 145) but wiry and strong, Charlie took on all comers with such fury and gusto that he was dubbed the "Terrible Turk."[12] The epithet, which was shortened to "Turk," stayed with him, and his classmates in the 1909 class at Princeton referred to him all through his life as Turk House.

Yet running crosscurrent to this testiness was a spirit of sympathy and understanding so broad that it spilled over from his own personality, affecting the outlook of others. Judge Harold Medina, a classmate of his, illustrates this further: "We were all sitting around our reunion tent one evening and one of the men said, 'Oh I hate that man'...and after a while Turk came up to that man and he put his arm around his shoulder and he said, 'You don't really mean that. You know we mustn't hate people.'"[13] Judge Medina admitted, "I was that

---

[11] Ann Kellogg House, notes to the author, 16 August 1973, AFS Archives.

[12] Ann Kellogg House, "Memoirs," 13, AFS Archives.

[13] Harold Medina, memorial service for Charles House, New York NY, 9 November 1961, AFS Archives.

man. I will never forget it. I will never forget it if I live to be a million years old. It just burned into my soul."[14]

As he grew to manhood and began to see violence as an evil, his aggressiveness was channeled into verbal argument. All who knew him, from the most casual acquaintance to the most intimate, stressed that Charlie loved an argument. He would often argue for the pure sake of debate, taking a contrary thesis to the one being discussed and exposing his somewhat radical views. Frequently, the topic was political; yet later as director of the American Farm School he managed somehow in the Greek milieu of troubled governments to keep the farm school and his person free from political involvement.

One of his most controversial positions concerned established government. In early manhood he was drawn to the cause of anarchism although he never had contact with other anarchists and did not approve of violence as a means of overthrowing the established authority.[15] He did believe that all governments exploit their people. His general philosophy suggests a direct kinship with Henry David Thoreau, and the farm school was his version of Walden. But his actions were not as consistently hostile or defiant to government as Thoreau's. Once he assumed the directorship of the farm school he was circumspect in his dealings with government officials, both Greek and American. The rapid reconstruction of the school after the devastation of World War II was directly attributable to Charlie's facility in cooperating with United States officials so that optimum advantage could be taken of various aid programs.

Yet in company he made no secret of his espousal of anarchism. His wife, Ann, thought he "rather enjoyed shocking people, by making extreme statements, especially on the subject of anarchy and the way governments exploit their citizens."[16] In casting around for a proper category in which to fit Charlie's particular brand of anarchy, she concluded that he was certainly not the characteristic type. Herbert P. Lansdale, Jr., his longtime friend and father of the future director, Bruce Lansdale, labeled him a "philosophical anarchist."[17]

---

[14] Ibid.

[15] Ann Kellogg House, notes to the author, 13 March 1973, AFS Archives.

[16] Ibid.

[17] Herbert P. Lansdale, Jr., interview with author, Athens, 22 May 1973, written, AFS Archives.

Indeed all of his more radical opinions became philosophical, bridled as they were by the realities of maintaining the American Farm School on a steady course. The safety of the school, its success and worth, became the passion of his life to which all other considerations were subordinated.

As suggested, there were so many contradictions in his anarchism that one wonders if he were aware of all the implications of the cause to which he claimed to be devoted. After 1935, for instance, he was a close friend of Greece's King George II (reigned 1923–1924 and 1935–1947). Both men shared a love of carpentry and farm life, the model farm at Tatoi being the king's pet project. At the king's request, Charlie went to Athens to install American-style counterhung windows at the palace. The king followed him around, supporting him as he climbed ladders and stood on windowsills. He once remarked to Charlie, "What a scandal it would be if you fell out of the palace windows."[18]

In the evening they would sit by the fire chatting, covering a wide range of subjects. Whether or not Charlie discussed reforms with this king, who presided over a repressive dictatorship in partnership with dictator Ioannis Metaxas (in power 1936–1941), we do not know. We do know from his astute analysis of conditions, recorded during the Greek civil war, that Charlie was acutely aware of the oppressions generated by this government, which created or reinforced divisions within Greek society.

How, then, can we explain his friendship with the king? Probably Charlie's love of people, a dominant theme in his life, overruled his philosophic considerations and allowed him to have close personal relationships with individuals who personified the very political outlook that he could not intellectually abide. Also, Charlie recognized that his personal relationships with government officials strengthened the school's position. His later friendship with German officers during the occupation can also be analyzed from this point of view, for his singular devotion to the school was the monolith upon which he placed the whole structure of his life.

There is reason to believe that Charlie House himself was never able to resolve all the contradictions in his personality. The resultant

---

[18] Ann Kellogg House, notes to the author, 13 March 1973, AFS Archives.

tension might well have been the cause of his frequent and severe migraine attacks but it was also, conceivably, the very dynamic that propelled Charlie to act with both extraordinary persistence and courage.

His courage was supported by a profound, lifelong faith in God. Once again his formal religious ties, although not contradictory to other values, lack facile definition. To paraphrase T. S. Eliot, we cannot fix him in a formulated phrase and pin him to the wall for examination. He remains elusive and individual. That he was a Christian is obvious. When he was sixty-five he wrote, "There is no way under Heaven whereby man can be saved except through Christ and his way of life."[19] That he felt himself Protestant is doubtful. Certain notions—his response to "call," his passion for service—reflect the Protestant conditioning of his family, but there his sense of particularism stops. His religion was all-embracing, universal, and shot through with humanism. He never found a Protestant denomination that suited him. At one point he was even attracted to Catholicism.[20] Quaker testimonies such as pacifism, abhorrence of rigid structure, and emphasis on social service are all present in Charlie's world outlook. He fit, if we must, more easily into the Quaker mold than any other.

His extreme courage was exhibited when his friend, author Sidney Loch, died in Pyrgos, at that time a remote area bordering Mount Athos in the wilds of Macedonia. Knowing that Loch's wife, Joice, passionately wanted her husband to be buried at the farm school, Charlie set out through a heavy snowstorm with two companions to collect his friend's body. The men literally had to dig their way through snowdrift after snowdrift while hungry wolves eyed their progress. At one point it took them nine hours to cover 11 miles. They did succeed in making the return trip; consequently, Loch is buried in the school's cemetery. Charlie attributed the success of the trip to the hand of God. "Humanly speaking," he allowed, "it was

---

[19] Charles Lucius House, *45th Year Record Book.*
[20] Ann Kellogg House, interview with author, Athens, 25 May 1972, AFS Archives.

impossible."[21] This was in 1954, when House was sixty-seven years old—seven years before his death.

If almost blind faith was a heavy weight on the scale of his personality, then pragmatism acted as the counterweight. These two balancing elements were symbolically expressed by the two objects that Charlie always carried with him: Oldham's *Devotional Diary* and a slide rule. President Emeritus of Anatolia College Carl Compton said, "I've seen Charlie at a gathering of elegantly dressed guests at some official reception with his slide rule protruding from his hip pocket."[22] Of the devotional he said, "Every morning he took out his dog-eared copy and read from that diary."[23]

If one can say that contradictions give rise to irony, then Charlie House's life proves that axiom. Denounced and cursed by most of his friends and members of his family for being a conscientious objector during World War I, in 1957 he received the highest honor bestowed on Princeton alumni, the Woodrow Wilson Award, granted in Charlie's case, ironically, for achievement in the nation's service. At the award's ceremony, Princeton's president, Robert F. Goheen, lauded him for being teacher, linguist, school executive, government adviser, unofficial diplomat, and "exemplar of the Christian gospel. When he graduated from Princeton in 1909 after a stormy college career as rebel, prankster, and near flunk-out, Woodrow Wilson, then president of the college, had delivered his well known baccalaureate address based on the tenth verse of the seventeenth chapter of Luke: "unprofitable servants; we have done what it was our duty to do." By centering on this chapter, Wilson hoped to impress upon the class the need to do service unstintingly, for the unprofitable servant, the one who does only what is absolutely necessary, was a slacker by Wilson's standards. With Charlie's poor reputation both for academics and behavior, he seemed on that graduation day in 1909 the least likely of all the graduates ever to fit Wilson's image of "profitable servant." Yet President Goheen, upon presenting the award, observed, "in thirty-seven years of his half century of service since his graduation,

---

[21] Craig Smith, Memorial Service for Charles House, New York City, 9 November 1962, AFS Archives.

[22] Carl Compton, Memorial Service for Charles House, New York City, 9 November 1962, AFS Archives.

[23] Ibid.

Charles House, at a distance of six thousand miles, has exemplified Princeton in the nation's service."[24]

Perhaps he might just as well have exemplified Princeton in Romania, China, or Hawaii. Yet it was from Greece that he derived his special impetus. If he had a large influence on the Greeks, he was influenced in equal measure by them. His Greek friends and associates generated within him enormous energy. Bruce Lansdale has written, relative to this point: "I remember…seeing him in the United States on two occasions, once when he was in the bureaucracy in Washington prior to his return to Greece after the war, and once when he visited us in Rochester. I remember thinking how rather small and unprepossessing he seemed compared to the dynamic, energetic and innovative person that I had known in Greece."[25] Some forty years before receiving this award, Charlie had been regarded as a maverick in some circles, especially for his anarchist and pacifist views. Apparently, the American Farm School's board of trustees was not worried by his radical attitudes at the time of his appointment in 1917. It was thanks to a tolerant, liberal strain in the members of the board that Charlie was selected. Their wisdom to see beyond the flashier aspects of his personality to the capable, strong, and affectionate core of his character was unusual when one considers that for the most part they were men with solid business interests, patriotically motivated at a time when patriotism ran high. Some members, however, were ministers or Quakers, or themselves connected with educational institutions. It was not unusual at this time to find pacifists in the ministry and in education, and the Quakers were the leaders of pacifism in American thought.

They were also members of the establishment, so to speak, presiding over a country in social transition. Since Charlie numbered among his friends such "radicals" as socialist-pacifist Presbyterian minister Norman Thomas, whose socialist approach to problems seemed a threat to the country's present order, the case against Charlie's appointment could have been strongly made.

His uncle, Lucius Beers, with whom he stayed frequently during his years at Blair and Princeton, probably knew him better than

---

[24] Princeton University, Woodrow Wilson Award, Charles Lucius House file, 1957.
[25] Bruce M. Lansdale, notes to the author, 31 August 1973, AFS Archives.

anyone. Beers was naturally very fond of his sister's son; he was also a heavy contributor to and constant supporter of the school since its inception. A lawyer by profession, he was not apt to allow affection to override clear thinking when it came to his vested interests. He wrote that while Charlie was "not an expert farmer, he has a good knowledge of modern farm requirements and I think is well qualified to take the same oversight of work of this kind which his father has while I feel that with study and experience he would be an efficient expert along these lines."[26] Beers felt that Charlie, having practiced his profession as civil engineer doing development work, road construction, bridge building, and additionally working as engineer and adviser for a large farm on Long Island, had an excellent foundation.

Beers's main hope was that "if Charlie can work with his father for a few years we can feel assured that there will be this continuity of work."[27] Beers grasped the concept that, to sustain the institution, it was necessary to fill the position of director with an individual who was attuned to the spirit of the school. Another board member, Hollingsworth Wood, a Quaker, expressed enthusiastic support for Charlie's appointment: "It seems to me one of the great opportunities for a 'conscientious objector' to be of real constructive international service and your response to it does me a lot of good."[28] The one negative opinion regarding his selection was written by his father: "There is only one thing that troubles me and that is his smoking. I cannot feel so well about the Principal of this School using tobacco."[29] He goes on to console himself and then torment himself again with his son's foible: "But there are worse things than that, tho' I do not see how it could be otherwise than that the boys would follow the example and smoke."[30] However, Charlie never smoked on campus in deference to his father, even after the latter's death. Had the elder House known that Charlie also enjoyed a drink, the whole history of the farm school might have been different.

---

[26] Lucius Beers to the Reverend Irving Metcalf, 17 April 1917, AFS Archives.

[27] Ibid.

[28] Hollingsworth Wood to Charles House, New York, 15 April 1917, AFS Archives.

[29] John Henry House, letter, addressee unknown, Thessaloniki, 20 April 1917, AFS Archives.

[30] Ibid.

Still, it is certain that he wanted his son to succeed him. After the burning of James Hall and the travail connected with that event, the aged House was overjoyed to learn that his son had felt called to the path of service. Thinking as he did that God personally planned all earthly operations, Father House could only be convinced that his son's coming was part of a divine plan.

Clearly no one could have had a background more suitable for the position of the American Farm School's director, which required a person who knew many languages and understood the Greek people's mentality. Charlie had lived from birth until the age of fifteen among the peoples of the Balkans. Born in Samakov, Bulgaria, in 1887, he was the second boy in a family of seven children, with four sisters intervening between him and his grown brother, John Henry, Jr. His early childhood was spent in Bulgaria. At the age of seven he moved with his family to Turkish-occupied Thessaloniki, where the House family lived in an old Turkish house on a narrow cobblestoned street behind the church of Aghia Paraskevi among a polyglot population of Greeks, Jews, Turks, and Slavic-speaking peoples.

As a child, he learned the ways of the city as he wandered with his father and sisters Ruth and Gladys through the market crammed with immense tubs of feta cheese, barrels of black olives, and trays stacked with baklava. Meat markets hung the carcasses of goats and sheep festooned with colored paper; fruit and vegetable stands displayed the peaches and apples of Macedonia. Life thronged through the coppersmiths' quarter whose landmark was (and still is) the coppersmiths' church. There, smiths with hammer and heat forged vases, kettles, and pots. Women clopped by on wooden clogs carrying on their heads large trays of bread to be baked in the public ovens. Camels loaded with charcoal or wheat lumbered by. Young House observed the men sitting in the open coffee shops sipping tiny cups of muddy Turkish coffee. In the background the discs and dice of *távli* (backgammon) clicked interminably. Dr. House would buy the children rice pudding and a glass of lemonade, or candy on a stick.[31]

When the parents had time, sometimes they would take the children to Allatini by horse car. Although the children were disturbed by the sight of the scrawny horses, they loved the ride. At the end of

---

[31] Ann Kellogg House, "Memoirs," 6, AFS Archives.

the tram line they would walk toward the sea to gather shells or watch fishermen repairing their nets. The depot at Allatini was near the land that one day would be the American Farm School.

Later the family moved to Rue Franque near Vardar Square. Here Charlie could wander along the wharves watching the bobbing *caíques* (small boats) unloaded by men with a saddle strapped to their backs. In his wanderings he picked up the languages of the polyglot city. Owing to his linguistic ability, he managed at a tender age to get packages cleared through the labyrinth of Turkish bureaucracy when his father often failed. His varied experiences during his childhood were equipping him for his future life.

The years of his boyhood were those of the Macedonian Struggle. Young Charlie witnessed his father's comings and goings during this time of terror. In spite of the constant danger to his life, his father never carried a gun, his rationale being that the safest place for a man is on the path of his duty. If courage can be developed by dint of example, then Charlie had every reason to mature into a courageous adult.

He was aware of all of the details of the kidnapping of Ellen Stone and Katerina Tsilka and was present at their release. When word came that the two women had been freed by the IMRO revolutionaries, he begged his father to take him along to greet the victims. As they reached the place where the women had taken refuge to await Dr. House, Charlie ran ahead of the others. During the months of their captivity he had grown so tall that he had finally discarded short pants for regular trousers. Stone, still apprehensive from her nerve-wracking ordeal, failed to recognize him, thinking that he was a reporter, and refused to see him.[32] By the time he left the Balkans, he had been seasoned by danger. In later years, the German occupation and the civil war seemed to him only an extension of his childhood experiences.

His formal education in his early years was haphazard. At first he attended a school run by Peter Donaldson, a Scottish missionary. Then he attended the German School in Thessaloniki for two or three years, adding German to his list of languages.

---

[32] Ibid., 12.

In 1902 during the height of the Macedonian Struggle, Dr. and Mrs. House felt it was time for their younger son to leave Macedonia in order to gain a proper education among his own countrymen. He was fifteen years old and would not see his parents again until he was twenty-four. Granted a scholarship to Blair Academy in Blairstown, New Jersey, he graduated in 1905 and subsequently entered Princeton College, the institution of his choice, a selection he had made in early boyhood.

Princeton was a molding force in his development. Bereft of the stabilizing presence of his family during these formative years, he formed intense, lifelong friendships during the four years at college that substituted for family relationships. His wife Ann felt that "those four years...had a greater influence on his life than any other period. I did not know him then, but all through our years together, although we lived several thousand miles away, Princeton was an almost daily part of our lives."[33]

The Princeton of that time was a rather closed society, one in which certain prescribed ideas were perpetuated and fostered by the leadership of the college and the families of the students. For example, Woodrow Wilson, president of the college during Charlie's years, was the first non-clergyman elected by the board of trustees to hold the presidency of Princeton. But Wilson's father and maternal grandfather had both been Presbyterian ministers. Wilson's world outlook included a heavy dose of religion. "Protestantism provided for Wilson a cosmic framework, helping the 'higher realism,' and the emphasis on principles which produced reform during each of his executive positions. [He]...accepted the pietic evangelical ethos of their day."[34] It is valid to say that Wilson's view of foreign policy and political structure was highly colored by his religious convictions. These principles, although still formative in the years he served as Princeton's president, were absorbed by the small, intimate student body.

The jobs Charlie held as civil engineer after graduation were stimulating for a young man and the experience he gained was

---

[33] Ibid., 19.

[34] Joseph Grabill, *Protestant Diplomacy and the Near East: Missionary Influence on American Foreign Policy, 1810–1927* (Minneapolis: University of Minnesota Press, 1971) 87.

invaluable for the situations he would face later in Greece. His first job took him to a mining town near Gwynn, Michigan, where he helped lay out a new town site. The farm school, which he later consciously developed according to the pattern of a Greek community, undoubtedly harked back to this Michigan experience. His next assignment was with a team of engineers organized to survey the lower reaches of the Wallenpaupack River in Pennsylvania for the purpose of building a power plant. When the survey was finished he stayed on to relocate roads that were to be flooded with the construction of the dam. This stint was a rich human experience. Here he learned to handle rough, undisciplined workers. One of the men was particularly unruly and refused to obey orders. Charlie, still the tough scrapper, finally decided he would have to challenge him to a fight in order to secure dominance. A Princeton classmate, Harold Dolph, who was working with him, began dropping hints as to Charlie's agility and strength in wrestling. He exaggerated so successfully that when Charlie challenged the workman the latter refused, saying, "I know you, you're a professional prize fighter."[35] Charlie had no further trouble with him. His most instructive jobs were on the Long Island Motor Parkway, the Pennsylvania Railroad, and the Holland Tunnel.

Charlie arrived at the farm school on 27 October 1917, six weeks after leaving New York. The Thessaloniki he found was hardly the same quasi-oriental city he had left in 1902. Besides the ubiquitous presence of Allied troops who lent to the city a pulsating air, the great fire of August 1917 had cut a swath through its heart. The whole area along the shore and up the hill toward Aghios Dimitris had been swept away; rubble still lay scattered in the streets.

Upon his arrival, Charlie began work immediately. Using Father House's original concepts as a foundation, he set about building the school into a modern institution. He knew how to actualize all the processes necessary for genuine village life. By the middle of the 1930s he had made real craftsmen out of those who had come to him as raw Greek peasants. He installed them in the farm school, transforming it into a coordinated, more efficient farming and

---

[35] Ann Kellogg House, "Memoirs," 27, AFS Archives.

vocational operation.[36] Father House, who had visualized such an operation, lacked the technical training to put it into effect. He wrote years later, "We have attempted to develop the Farm School more on the lines of a community than an institution without sacrificing the essential features of an educational institution."[37]

The swift injection of Charlie's young, impetuous, aggressive, hard-driving personality must have been jolting to the older House, yet from all accounts father and son adjusted immediately to each other's tempo. The transitional period between Charlie's arrival and his official appointment as director in 1929 was smooth. The full burden of responsibility shifted rapidly but smoothly from father to son. Charlie built on his facility for language, improving the Greek he had learned as a boy, thus enabling him to represent the farm school to the Greeks. Father House, talking to them in Bulgarian or Turkish or through an interpreter, never established the intimate rapport that Charlie was able to achieve by speaking Greek. By 1923 all important matters having to do with the operation of the school had become Charlie's responsibility. In 1927 Father House offered his resignation, but on 11 November of that year the board of trustees tabled the issue until May 1929, when the resignation was finally accepted.[38]

The working relationship between father and son remained close. Interchanges were constant as they spoke across the desks in the office they shared in Kinnaird House. Charlie would consult with his father regarding subjects of interest, but the worrisome things he kept to himself, and he made the difficult decisions by himself. Father House spent most of his time working with the boys in the garden—grafting, pruning, sharing with the students his love of growing things, or "working with God," as he called it.

Although the older man's moment of leadership had passed, his continuing residence at the school until his death in 1936 continued to serve a real purpose. Dorothea Hughes Simmons, teacher and patron of the school, caught the essence of his old age and its meaning for the boys:

---

[36] Herbert P. Lansdale, Jr., interview with author, Athens, 22 May 1973, written, AFS Archives.

[37] Charles Lucius House to Miss Backhouse, Thessaloniki, 16 July 1953, AFS Archives.

[38] Minutes, Board of Trustees, 11 November 1927; 24 May 1929, AFS Archives.

You, Dr. House, remark that you look on yourself as a sort of useless person at the School now. I suppose to an active minded person, the position as a sort of patron saint does not appeal; yet I maintain that seeing you and hearing you makes the boys feel a unity with the past of the School and with your dreams of its future. The fact that you do not work shoulder to shoulder with them, as you used to, means only a shift in your influence and not a diminution, if I may judge from personal observation.... Those refugee boys have been so torn up, they had walked on such shifting sand, and even the Macedonians lacked any feeling that the present order had come to stay! Exile or revolution or war was to be expected. The Greek Church alone stood the shocks and held their reverence. When they came to the Farm School they saw an old man who had been there in the days of the Turks through wars and revolutions. His presence gave testimony of permanence, the permanence of the School he had founded, and behind that, the permanence of an ideal. I believe that every year you live is of infinite value to the School.[39]

Casting his eye around the primitive facilities that existed in 1917, Charlie set to work first on the pressing problems, chief among them the water supply. He immediately saw the possibility of drawing on British help—of buying old and spare parts that the British would not take with them when the war ended. Scavenging through their stockpiles, he found an assortment of hardware that imagination and ingenuity would transform into workable mechanisms. The quartermaster general of the British forces sent to the school one of the American experts, who was under British contract. Using an American machine for drilling oil wells, the American sank a 280-foot well that was capable of yielding some 800 gallons an hour. The total cost of sinking the well was $1400. The annual report reflected the significance that such a water supply had for the expansion of the school. It stated that the success of this well was an "event in the history of the School.... No such machine [the American oil well

---

[39] Dorothea Hughes Simmons to John Henry House, Boston, 24 February 1923.

digger] has been available to us before the war and may not be again as the machine was removed even before the British left."[40] The report added wistfully, "We are still waiting for some good friend to cancel the debt thus incurred of about $1400 and we should be glad to allow such a one to name the well as a memorial."[41]

The expenses of purchasing pieces of surplus British equipment weighed heavily on the leadership of the school and the board of trustees. Charlie, less conservative than his father, pressed for their acquisition, thereby incurring a sizeable debt, a situation that the older House had traditionally sought to avoid. Vast accumulations such as cement, building materials, motors, and scrap parts from vehicles were bought cheaply. A building dubbed "the hut," a 25 by 80 foot structure whose original cost was $5,000, was purchased for $950 for use as a dormitory.

Charlie then started to work on improving the washhouse and the lavatory used by the boys. The students worked with him as part of their training while he, side by side with them every day, was becoming more intimate with their personalities, problems, and hopes. In addition, he erected a blacksmith shop fairly well equipped with forge, anvil, and necessary tools. The forge was fed air from huge Turkish bellows that gave off tremendous power but were basically a primitive design meant to be easily copied by the boys and applied in their own villages. As an alternative, he simultaneously built a more sophisticated method using compressed air that was funneled in from a nearby machine shop. The students were exposed to both methods so that when they returned to their villages they could adapt to the most primitive circumstances yet at the same time be prepared to introduce, if possible, more complex and effective systems.[42] Charlie's early realization that the boys should be able to operate on many levels, and his ability to instruct them with a multitude of variables produced, in effect, scores of improvising, flexible technicians for Greece. As a result, boys going back to their villages often became the primary—or only—people trained in technology.

---

[40] "Thessalonika Agricultural and Industrial Institute Annual Report, 1916–1919," 17, AFS Archives.

[41] Ibid.

[42] Herbert P. Lansdale, Jr., interview with author, Athens, 22 May 1973, written, AFS Archives.

As a matter of fact, farm school graduates were among the few people capable of immediately harnessing electrification when it first appeared in northern Greece. Although from a short range point of view all innovations and improvements put the school into a financial bind, Charlie was actually on solid ground, for in the long run he was increasing the capital value of the school's property with each improvement as well as developing its productive output.

A year after his arrival, the school was put on a more secure legal footing. On 3 January 1918, the Greek government recognized its unique contribution to Greek life by publishing a royal decree in the name of King Alexander. This recognition placed the school under the jurisdiction of the ministry of agriculture.[43] At the same time, the director was informed that the same royal decree remitted a severe tax that had been levied on the school by the provisional government established in Thessaloniki by Eleftherios Venizelos after he had split with Alexander's father over Greece's entry into World War I.

Since Sultan Abdul Hamid's *irade* in 1907, the school had run tax-exempt, paying only a nominal fixed tax of $3.52 on the original 53 acres. After the Balkan Wars, when Turkish territory fell into Greek hands, the Greek government was treaty-bound to acknowledge all the rights of foreign property. Yet because of the chaos of continued war and the ensuing upheavals, the royal decree of 1918 was the first legal document promulgated by any Greek government that legally established the school under Greek law.

During and immediately after the First World War, the school became known to all in distress as a haven where a hungry man could get a piece of bread, a traveler could obtain shelter, a wounded person could have his injury dressed, a farmer could borrow a tool or learn a new method of planting. Owing to all of these attentions focussed on the outside world, the students learned the spirit of cooperation and responsibility for their fellow human beings. Whatever was preached was lived.

The 1920–1921 annual report, the last one before the refugee exchange wreaked new havoc upon the land, reflects dynamic growth in every aspect of the school's activities. From the scanty wartime enrollment the number of boys jumped to fifty-two. The harmony of

---

[43] "Government Gazette of the Kingdom of Greece," No. 7A, 4 January 1918, Vivliothiki Voulis, Athens.

nationalities that Dr. House longed for was operative at the school, which now boasted a student body composed of Greeks, Albanians, Serbians, Armenians, Maltese, Bulgarians, Vlachs, and Russians escaping from the Russian civil war.

There was still no systematic recruitment program or selection process for acceptance into the school. Children—some of them orphans, many of them refugees, others whose parents thought their children needed character strengthening, and some who wanted a good education—simply arrived at the school's doors. It was a measure of Greek tolerance that religiously oriented Greek families sought out a school that they knew was run by Protestants and in which religious services were offered only in the Protestant form. It is also a measure of Dr. House's tolerance, given the fact that he began life as a missionary, that he never alienated his Orthodox students or their families by insisting on practices that would be repulsive to them. The mutual respect existing between these tenaciously loyal religionists stands out as an admirable quality that fostered a rare relationship.

# CHAPTER 8

# ASIA MINOR AND THE
# POPULATION EXCHANGE

In 1954 Adlai Stevenson wrote, "Gentle grace has not been our lot in the twentieth century. Instead the first and second planetary wars have helped to make of this half century the most barbaric interval of the Christian era."[1] Professor William Vogel underscores Stevenson's opinion by singling out the population exchange among Greece, Turkey, and Bulgaria at the end of the First World War as one of the most barbaric actions perpetrated in our time.[2] The events leading up to this exchange, the actual exchange itself, and its aftermath created one of the largest if not the sorest refugee upheavals in modern European history.

Greece, a nation already impoverished, received the greatest number of dislocated individuals. Over 1,250,000 souls swarmed over its borders within a decade, the bulk of whom (60 percent) settled in Macedonia and Thrace. More than a few of the farm school students have been descendants of those uprooted, and many of the staff descend from families who made their way from Asia Minor as part of that dislocated throng. Since the school had managed to extend its services far beyond its own boundaries during earlier catastrophes and also to survive this new disaster, it was not surprising that it should become a nucleus for relief and resettlement work.

---

[1] Adlai Stevenson, *Call to Greatness* (New York: Harper, 1954) 2.

[2] William Vogel, *Lectures in Twentieth Century European History*, University of Cincinnati, 1968, author's possession.

Its willingness to extend itself was a crucial factor in the school's acceptance by some reluctant sectors of Greek society. After the absorption of Macedonia into Greece, ethnic sensibilities in the country were as strong if not stronger than before. Many Greeks felt that foreign schools were undermining the ethnic and religious foundations of the youngsters; some schools were suspect for having been pro-Bulgarian or pro-Slavic during the Macedonian Struggle and were stamped with a stigma. The school had not entirely escaped criticism. Father House's affection for the Bulgarian people was genuine and lasting. His ability to speak and preach in Bulgarian,[3] which he unwisely continued to do during the Great War, made him suspect in some Greek and even Allied circles. Yet to confuse his love of the Bulgarian people among whom he had lived for thirty years with a militant or even active pro-Bulgarian stand is not to understand John Henry House. Documents support the conclusion that he categorically condemned IMRO's activities in Macedonia. In citing both its violent tactics and harmful strategy, he wrote:

> The cry of late among the revolutionists has been, "Macedonia for the Macedonians," but what is the bond that will unite these conflicting elements and enable them to establish a stable government? Russia doubtless has her answer to that question, and when the time comes she will not be slow in announcing it…. Meanwhile, Macedonia is in a constant ferment caused by the oppression of the Turks and the agitation of lawless revolutionists who call themselves "patriots," but who in reality hate the Turk more than they love their country, and care more for freeing Macedonia from Turkish rule than they care for the real progress and development of their land.[4]

---

[3] For one version of how Dr. House's insistence on preaching in Bulgarian brought him into a collision with some Greek authorities, see George Horton, *Recollections Grave and Gay* (Indianapolis: Bobbs-Merrill, 1927) 293–96. For documents concerning the incident, see "Dr. John Henry House's Arrest" folder, 1919, AFS Archives.

[4] John Henry House, untitled article, *Southern Workman* (June 1903). This article was printed anonymously, for it was during the height of the Macedonian Struggle. Such was House's method of expressing a political opinion on an extremely incendiary issue without exposing his identity.

Nor was he interested in the political aspects of the struggle. His interests remained rooted in his mission, his concerns purely humanitarian and religious.

As one of the co-founders, Reverend Edward Haskell and his activities during the war contributed to the suspicion. Younger and less detached than Father House, Haskell went to Washington at an emotional moment in history to urge his government to refrain from making any hostile moves against Bulgaria. He spoke up at the hour when a resolution to declare war on Bulgaria was before the United States Senate: he was bitterly assailed for his position and was labeled pro-Central Powers and anti-American.[5] He had left Greece some years before, at the close of the Balkan Wars, and thus had no official connection with the school. Missionary newspapers, too, were expressing the opinion that the United States government should have no quarrel with the Bulgarians, who felt warmly toward Americans. Such sentiments irritated the large mass of Americans, who were anti-Central Powers. Once Greece entered the war on the Allied side these statements were used by some Greeks to show the pro-Bulgarian colors of the missionaries.

Most Greeks who knew the Houses respected them for their dignity and honesty, yet later in the immediate postwar age it seemed to some that Father and Mother House with their tendency toward religiosity and moral absoluteness were out of rhythm with the Greeks' zest for life and less inhibited behavior. Charlie, a fresh personality without residual connotations and with obvious affection for the Greek people, attracted elements of society that hitherto had remained aloof from the farm school. Among the more prominent liberal leaders in Greek civic affairs were two brothers, Alexandros and Konstantinos (Kostas) Zannas. These men of good will, conscious of their community responsibilities, sought ways to improve the quality of Greek life. With their sense of social mission it was not unnatural that they would quickly seek out and become close friends with Charlie House. Alexandros Zannas was an intimate of Prime Minister Venizelos's, serving in his government as deputy air minister from

---

[5] W. W. Hall, *Puritans in the Balkans, The American Board of Mission in Bulgaria: A Study in Purpose and Procedure* (Sofia: Cultura Printing House, 1938) 258–60.

1928 to 1933. There is little doubt that Venizelos's deep interest in the school was first cultivated through Alexandros Zannas.[6] His brother, Kostas, a lawyer, became the first Greek member of the school's board of trustees and remained Charlie's lifelong friend. When we speak of the steps in the Hellenization of the American Farm School, the process whereby the school began to stress and reflect the Greek mentality as well as the American, Kostas Zannas must be recognized as fundamental in that process. He and his brother recognized the value of the school and joined with Charlie in an effort to broaden its services.

The Zannas family stemmed from old Greek stock and had played a leading role in the Macedonian Struggle.[7] Endorsement by this influential family and its direct participation in school activities formed an anchor connecting the school to a broader Greek social spectrum. Kostas from the first offered his legal services to Charlie and served also as an invaluable guide to him, leading him through the labyrinth of government bureaucracy. Zannas's participation in the founding of a refugee hospital, a school for the blind, and a seaside center for underprivileged children placed him at the very center of philanthropic activity in Northern Greece. Besides, he possessed a talent for public relations and knew all of the other civic leaders. Beyond that, his interest in farming (the family always owned land and worked it) brought him into direct communication with peasants. Few men understood better than Kostas Zannas the needs and aspirations of Greece's rural communities. His relationship with the farm school remained close until his death in 1966.[8]

His son Dimitris, also a lawyer, assumed many of his father's community responsibilities, chief among them the position of legal adviser to the school and membership on its board of trustees. If his father had been indispensable to Charlie, Dimitris was equally important to Bruce Lansdale. Since family continuity has been a

---

[6] Virginia Zannas, interview with author, Athens, 16 November 1973, written, AFS Archives.

[7] For references to the Zannas family's participation in the Macedonian struggle, see Douglas Dakin, *The Greek Struggle in Macedonia: 1897–1913* (Thessaloniki: Institute for Balkan Studies, 1966) 79, 198, 204, 210, 211, 212, 381, 449.

[8] The information on Kostas Zannas is based principally on three interviews with author: Dimitris Zannas, Thessaloniki, 8 January 1974; Herbert Lansdale, Athens, 12 March 1974; and Harry Theoharides, Thessaloniki, 8 January 1974, AFS Archives.

fulcrum upon which the stability of the school has traditionally rested, the Zannas family is part of the pattern.

Other Greeks also turned their attention to the school. Among them was Professor Yiorgos Soteriadis, first rector of the Aristotle University of Thessaloniki. Understanding the cultural synthesis that was taking place, he wrote: "The excellent American school situated a little outside of Thessaloniki is one of the important things of which the Macedonian people may be proud of being of their own making, although those who helped the Macedonians create the school and give it breath of life were Americans, and these Americans are still continuing to direct, keep alive and feed the organization with the most wholesome and dynamic food."[9]

This ability to work in concert with Greek civic and political leaders enhanced the school's capacity to render service in this moment of almost unprecedented need in Greece. An understanding of the events that led to the population exchange will clarify the abysmal predicament of these Balkan peoples.

As the Greek army had performed admirably in 1917 when reunified Greece finally joined the Allied cause, the French and British agreed to support, at the end of the war, Greece's extensive irredentist claims in Asia Minor and other parts of the now defunct Ottoman Empire. At the Versailles Peace Conference in 1919, Prime Minister Venizelos, who was a proponent of the *Megáli Idéa* (a policy calling for the absorption into Greece of many territories outside of the Greek state that held sizable Greek populations), presented his case ably. In May the conference members extended to the Greek government a mandate to occupy Smyrna and its hinterland, a portion of Turkey that had been inhabited by Greeks (along with other peoples) for thousands of years. A year later at the conference at Spa, Greece was granted an additional mandate to occupy Eastern Thrace and the northwestern section of Asia Minor. The Treaty of Sèvres (drawn up but never ratified) put a somewhat official seal on these and other concessions made by the Allies to the Greeks.

Running crosscurrent to Greek ambitions in Asia Minor, meanwhile, was the rapidly growing power of Mustapha Kemal, who was determined to construct a modern Turkish nation-state. He had

---

[9] *Néa Alithea* [Thessaloniki], 1923, n.p., AFS Archives.

established his own government in Ankara, independent of the constitutional one in Constantinople. Riding at the head of the Turkish army, he challenged the Treaty of Sèvres by refusing to recognize the sultan's capitulation to the Allies that had allowed Greek occupation of Turkish lands.

With the verbal but not the military support of Great Britain and France and the approval of President Wilson, Greece had landed troops in Asia Minor. This doomed army, supported by a country exhausted by continual warfare, was split by political dissension and further weakened by the futile military strategy of its leaders. Possibly the most lethal blow came from its allies, who for various reasons of self-interest withdrew support and, in the specific case of France and Italy, actively turned to supporting the Turks.

By September 1922 Mustapha Kemal had broken the spine of the Greek forces, driving them back literally into the sea. A large segment of the routed army fell back into Smyrna, where thousands of Christians fleeing from the victorious Turks had gathered hoping to find some protection in this Greek center. The Turks entered the city on 11 September in a spirit of triumph and vengeance. They set fire to the Christian quarters, first to the Armenian and then to the Greek, while participating in a wholesale massacre of the terrorized Christians. Meanwhile, riding at anchor in the harbor, European ships remained passive as the Turks obliterated the city. However, boats of many nations and descriptions, from small fishing boats to battleships, offered their services in the evacuation of the fleeing Christians. The huge flow of escaping people cascaded into the city and onto the docks, finally overflowing into the sea, some of them to be fished out by waiting boats, others to drown. Of the 200,000 who reached the shores, over 25,000 died before boarding the vessels. Those who failed to be evacuated—mostly able-bodied men, for the Turks had allowed only very old men, women, and youngsters to escape—were driven by the Turks into the interior and in most cases were starved or beaten. Those who did manage to escape later from the trek into the interior turned up as human wrecks.[10]

Leland Morris, American consul general in Thessaloniki, was one of the first foreign officials to be contacted in Greece after the

---

[10] Harold Allen, *Come Over into Macedonia* (New Brunswick: Rutgers University Press, 1943) 19.

disaster. He was merely asked to meet an incoming American battleship. In those days of less-than-rapid communication he had no idea of the extent of the holocaust that had taken place in Asia Minor. To his astonishment he saw the ship's passengers, young hysterical women who had been plucked from the sea without shoes, standing in the clothes they were wearing when they plunged into the water. These were as far as we know the first of the Asia Minor refugees to arrive in Macedonia after the Smyrna inferno.[11]

Although many souls were saved, the city of Smyrna perished. Marjorie Housepian writes in her haunting and blistering account of that holocaust: "Smyrna died in 1922. Its character obliterated, its name erased from the maps of the world.... In its place grew Izmir. Izmir emerged from Smyrna's ashes to the accompaniment of official applause by the great world powers who had a short time before witnessed the city's destruction in deafening silence."[12] Although Housepian makes her stinging indictment against the world powers from a perspective of fifty years' hindsight, Charlie House, a contemporary observer, shared her view: "...and so the endless chain drags on until...we as nations refrain from exploiting the weaknesses and faults of smaller states. It seems to me that Turkey should be judged less harshly than France and Great Britain, as well as other Great Powers. She is acting according to her professions."[13] He was observing at first hand the effects of the Smyrna tragedy, whose reverberations were being felt in villages around the school: "Already one of the evil results in this short time showed itself in a Turkish village near us where I understand six people were killed by a band of these refugees who were taking revenge for what they had suffered at the hands of the Turks."[14] To witness human misery was by no means

---

[11] Marie Morris, interview with author, Athens, 20 October 1973, AFS Archives. Leland Morris served later as chargé d'affaires of the American embassy in Germany. It was to him that von Ribbentrop handed Germany's declaration of war on the United States in December 1941. Although Morris communicated the contents of the declaration by telephone, he did not actually hand the document to President Roosevelt until June 1942. He had for a large part of the interim been confined by the Nazis in Bad Nauheim, Germany, and was finally exchanged.

[12] Margorie Housepian, *The Smyrna Affair* (New York: Farber & Farber, 1970) 14.

[13] Charles Lucius House to Ann Kellogg Chapman, Thessaloniki, 14 October 1922, AFS Archives.

[14] Ibid.

a new experience at the school, but the sheer magnitude, after the Smyrna holocaust, of thousands of destitute people pouring like a tidal wave into Macedonia gave pause to even the most seasoned veteran. Charlie House, always sympathetic to the plight of the refugees, was aghast at the flood and wondered in a letter to his fiancée, Ann, how he would react to this test: "I dread becoming hardened to the suffering of these poor people, yet when you listen to them and have the same kind of story told you time and time again, and are in most cases compelled to refuse the help which may lead to bitter words from the one refused, it is difficult."[15]

Indeed to all observers it seemed that the hordes of refugees would be the country's undoing. Instead, the energetic reaction by the Hellenic people to solve the enormous problem fixed for the war-torn nation a new internal focus and a fresh set of national goals. The small country was aided by scores of British and American agencies such as the Red Cross, the Society of Friends, Save the Children's Fund, and Near East Relief. The most important organization was the Refugee Settlement Commission created by the League of Nations, whose goal was to settle the refugees on a self-supporting basis. The resettlement program in this sense remains a historic example of international cooperation. Although the United States was not a member of the League it did contribute significantly toward relief and settlement. The school joined directly or indirectly with most of these organizations in relief or resettlement work.

Charlie's letters in the few weeks after the massacres describe vividly the refugees' arrival into the city. His day-by-day observations express the staggering problems of the time and record the growing numbness, almost paralysis, on the part of agencies and individuals who wanted to help, as the number of refugees mounted.

> Our firsthand knowledge of what had gone on in Smyrna was brought to us by Mr. Getchel [a missionary], who got away from there with only the clothes on his back.... We are taking nine more orphans into the school. They escaped from Taurus last year when the French turned the place over to the Turks and now had to flee again. We have no visible funds for

---

[15] Charles Lucius House to Ann Kellogg Chapman, Thessaloniki, 25 September 1922, AFS Archives.

their support but I believe it will be found somehow. It is hard not to be able to gather more boys in, for this is such an ideal place for boys. One of the teachers from the American College at Smyrna is with us. He brought in on an American destroyer over 400 orphans for whom we are now trying to find places.[16]

The next day he reported the following situation in Thessaloniki: "The dazed refugees from Smyrna are crowding into the city by the thousands. Such a sorrowful lot, families divided with no knowledge as to where the others are. One of our students was here the other day. He had escaped with his life after living fourteen days in the field.... He had cut off his trousers to the knee to make himself appear like a younger boy, as the Turks were taking the men and boys they did not kill to put into their army."[17] Charlie mentioned that there were about 65,000 refugees in the city by 26 September and that a bread shortage was now apparent. New refugees were arriving every day although there was not enough shelter for those who had already come. Clothing, too, was an immediate problem, for all had arrived with only the clothes they were wearing.

A fortnight after the Turks entered Smyrna, Charlie wrote that the situation was growing more serious with each day. His evaluation was that there "seems to be a real desire on the part of the people to help but it has passed all bounds that could be coped with from private efforts and as yet there does not seem to be any adequate public body functioning. It is probably that unless grain can be bought very shortly the people will be without bread."[18] The school had an order with the government for one and one half tons of flour that Charlie felt would be forthcoming, but he added: "There is no satisfaction in thinking that you have provided for your own needs in the face of all this suffering but I am responsible to the large group gathered here."[19] On 14 October he learned that half a million refugees from Asia Minor

---

[16] Ibid.

[17] Charles Lucius House to Ann Kellogg Chapman, Thessaloniki, 26 September 1922, AFS Archives.

[18] Charles Lucius House to Ann Kellogg Chapman, Thessaloniki, 30 September 1922, AFS Archives.

[19] Ibid.

had arrived in Greece. He expressed alarm that the country was in no position to take care of them with winter coming on. Five days later he grew more hopeful with the news that international organizations were about to take action: "Fortunately there is a relief committee in the city but with the many thousands of refugees they cannot cope with all that come to them. We are hoping so much that the workers who are sent to us from the Red Cross and Near East Relief will not merely give relief which would tend to pauperize the people."[20] It was Charlie's wish that the refugees who had just poured into Greece could be used in building or repairing the primitive road network, in constructing houses or in any worthwhile projects through which the country as a whole could benefit and which would make the refugees economically independent.

By the first of November, with the onset of winter, the situation became nightmarish. A paragraph from Charlie's letter of that date bears quoting:

> I have been to town today to get flour without success. I am told that it will be a number of days before we can get it. I am only hoping that we shall be able to get bread. When bread is the main food for so large a part of our community it always makes me restless when we are without flour, especially when we have 100,000 more people here than usual with ships coming in all the time. Ten came in full yesterday. Many of these people are out in the open with no great hope of being able to find shelter before the cold weather comes.[21]

Desperation and futility were never part of the farm school ethos. A sense of the immensity of the Anatolian refugee catastrophe can be obtained from this extreme statement from one who rarely in life succumbed to pessimism:

> It is raining hard now and I cannot help thinking of the thousands that are out of doors with very little to cover them

---

[20] Charles Lucius House to Ann Kellogg Chapman, Thessaloniki, 20 October 1922, AFS Archives.
[21] Charles Lucius House to Ann Kellogg Chapman, Thessaloniki, 1 November 1922, AFS Archives.

and no shelter. I was talking with the representative of the League of Nations and he said that by spring fifty percent of these people would be dead. It may sound hard but humanly speaking under the circumstances I believe that in most cases it will be a blessing. Most of these thousands that have come are women and children and old men. There are many tiny babies and our women teachers are hard at work making garments for as many as possible.[22]

The school itself soon took on the aspect of a refugee camp as it consolidated its space to accommodate as many refugee boys as possible. Charlie happily wrote on 14 November that they had, by consolidating, found room for ten more boys than they ever thought possible. During that week the official statistics placed the number of refugees in the city at 140,000.

The school's regard for all people without distinction of race or nationality became even more obvious as a welter of multiethnic groups surged through Macedonia. A case in point is the following related by Charlie: "[A] few days ago we had an opportunity to help a Turk and his family who live a few miles away. They have a large prosperous farm but the refugees had gone out there and they had to leave. We have helped each other at different times and the young men from the farm often come here. The other day they brought forty head of cattle for us to keep a few days until they could sell them and not lose everything."[23]

On 5 December the school received a cable from the board of trustees that it had been given a donation of $2,000 to begin work on a dormitory to house the orphaned refugees. Word was sent immediately to refugee men having families that work was now available at the American Farm School. The next day the refugee laborers began hauling stone for the foundation of Princeton Hall. More and more, the school assumed the appearance of a refugee station as outbuildings were hastily thrown up to house the workers. The lack of the simplest necessities was appalling: "We have started

---

[22] Charles Lucius House to Ann Kellogg Chapman, Thessaloniki, 5 November 1922, AFS Archives.

[23] Charles Lucius House to Ann Kellogg Chapman, Thessaloniki, 25 November 1922, AFS Archives.

the foundations of a shed to house the refugees while they are at work on the larger building and I expect about 50 more refugees out tomorrow to help with the work. I do not know how we can get covering for the men. That is one of the things that worry us. We hate to think of their being cold at night. We shall have to try in some way to get a supply of quilts that will make these poor men more comfortable."[24] The Red Cross and Near East Relief, organizations that were supplying such articles, were restricted in their donations. They were able to give blankets only to the unemployed. The employed had to buy their own, which were prohibitively expensive, costing more than half a month's wages.

Understandably, the Asia Minor disaster formed a watershed in Greek-Turkish relations that culminated in the Treaty of Lausanne, signed on 24 July 1923. The treaty's preliminary, the Convention of Lausanne, specified that a compulsory exchange would take place between Turkish nationals of the Greek Orthodox religion living in Turkish territory and Greek nationals of the Muslim religion living in Greek territory. The only exceptions were those Greek Orthodox inhabitants of Constantinople who had been living there before 30 October 1918 and all Muslim inhabitants of Western Thrace. From 1923 to 1924, accordingly, Greek Orthodox individuals who had not fled in the massacres were now forcibly sent to Greece; thus over 188,000 more Greek souls augmented the flow within those two years and merged with those who had been dislocated previously. Under this same agreement 335,000 Muslims left Greece for Turkey. The treaty superseded the now defunct Treaty of Sèvres and forced Greece to return to Turkey all of Eastern Thrace and a small part of Western Thrace together with some small islands and all claim to regions in Asia Minor.[25]

Four years earlier, another population exchange that was a forerunner of that between Turkey and Greece had been stipulated in a

[24] Charles Lucius House to Ann Kellogg Chapman, Thessaloniki, 28 December 1922, AFS Archives.
[25] Harry Psomiades, *The Eastern Question: The Last Phase* (Thessaloniki: Institute for Balkan Studies, 1968) 66. S. Victor Papacosma, "Minority Questions and Problems in East European Diplomacy between the World Wars: The Case of Greece," paper read at the American Association for the Advancement of Slavic Studies, October 1978, author's possession.

convention on mutual emigration that accompanied the Treaty of Neuilly (1919). That treaty, which ended the war between Bulgaria and the Allies, inflated the migration problem, particularly in Macedonia, when about 50,000 Greeks were exchanged for some 100,000 Bulgarians. It was according to this treaty that Bulgaria relinquished its sovereignty over Western Thrace to the Allies who, a year later, handed that region over to Greece.[26]

In the case of Greek Macedonia particularly, the influx of Greeks substantially altered its ethnic composition. In 1912, the population of the part of Macedonia that was absorbed into Greece as a consequence of the Balkan Wars was 42.6 percent Greek, 39.4 percent Muslim, 9.9 percent Bulgarian, and 8.8 percent miscellaneous (including Thessaloniki's Jews). By 1926, the League of Nations Refugee Settlement Commission's map depicting the population shift reported Macedonian Greece to be 88.8 percent Greek, 0.1 percent Muslim, 5.1 percent Bulgarian and 6.0 percent miscellaneous.

As far as onlookers were concerned, the population migrations were not separable into the various time frames that we have just described. It was simply a continuous flow of peoples whose numbers swelled and whose needs by 1922 had become overwhelming.

In assessing its own ability to help, the school was aware that its slender resources could easily be dissipated in some temporary effort that, in a relative sense, would be of little help and might seriously injure the permanent work of the school. From the beginning it was decided to stretch school facilities to the utmost to accommodate boys. The first to be accommodated were from the International College at Smyrna, since Dr. House had colleagues there on the staff. Dr. House requested aid from the Bible Lands Mission Aid Society (BLMAS) in England for the first group of nine boys and received 100£ for their support. Still other waifs arrived, many of them with no funds for their support and no relatives responsible for their moral wellbeing. Most of them were picked up along the roadside by passing soldiers who deposited them at the school, paying something toward their keep for a time. Then the soldiers drifted off, leaving the students dependent on the school. But they were not turned away. Ann House told of the plight of one youngster on whom she tried her

---

[26] Ibid.

own shoes that her mother had just sent her: "The whole front of his shoe from the toe up was flapping in the breeze and his toes were sticking out. I took off these shoes of mine and tried them on him but decided that it would be cheaper to buy him shoes that would fit. Mother House is buying clothes for him."[27] The orphan situation was one of the saddest aspects of the disaster. Arriving totally without resources were thousands of bewildered children who had lost their parents in the confusion in Anatolia or whose parents had been killed. Some were adopted, others worked in shops, many were put into orphanages, farm boys went out on contract to farmers, girls went into homes as domestics, while others were installed in special schools.[28]

Very early during the influx of refugees the school saw the possibility of enlarging its plant through refugee labor while at the same time giving work to the unemployed. Yet for a small operation such as the American Farm School, working with refugee labor and student help had serious disadvantages. It meant that the boys had to be removed from their regular course of instruction while Charlie had the increased burden of supervising a large force of completely unskilled laborers. Charlie himself did more than supervise: "He is working to the limit of his strength out on the roof of James Hall, lifting iron rafters and beams and working right along with the men. They are slow and inexperienced and he is so afraid they'll get hurt in some way that he does the most difficult part of the building himself."[29] Correspondence during this time of building Princeton Hall and remodeling James Hall reflects great strain. With the expanded enrollment, the dormitory area was now completely inadequate; the requirement for increased space had reached crisis proportions. As usual the shortage of funds impeded efficient organization:

> Charlie has to cable Uncle Lu for more money for he has
> to borrow from the dormitory fund for current expenses and

---

[27] Ann Kellogg House to her family, Thessaloniki, 27 January 1923, AFS Archives.

[28] Allen, *Come Over into Macedonia*, 7–8.

[29] Ann Kellogg House to her family, Thessaloniki, 14 September 1924, AFS Archives.

we are down to rock bottom. It meant that he would have to discharge the men whom he had just gotten trained to do the work on the new building and could not go on with the repairs…. Unless he can get enough of the new building done so that the boys can move into it and out of James Hall we won't be able to remodel James Hall and put in the classrooms we need so much.[30]

The construction of Princeton Hall is itself a saga. About 1,000 refugees were employed in the building of it. In order that the wages might be spread throughout a wider sector, one group of workers would work for a few weeks, then another would take a turn. The walls of the entire building were made of limestone quarried from one of the school's fields. The construction was accomplished without the paid services of a contractor or architect (John Henry House, Jr., Charlie's older brother, an architect by profession, gave his advice and services gratis), since Charlie conceived of and executed the plans. The boys installed the plumbing and electricity. The structure had sleeping space for 250 boys, a kitchen, dining room, play room, assembly room, and quarters for the faculty. Its limestone steps bowed by the feet of hundreds of boys, it still stands—an object lesson of what can be done with local materials, labor trained on the job, and perseverance in the extreme.

The health of most of the laborers was pitiable. Many of them lived in the neighboring village of Sedes. Often when the workers did not appear for work the boys were sent to investigate: they frequently found whole families down with malaria.

The school was most interested in helping to eradicate this plague. Malaria was historically the gravest health menace in Macedonia. Both Father House and Charlie suffered chronically from it and Ann House was in Greece barely a year when she almost succumbed to it and its debilitating consequence, blackwater fever. The school infirmary always had a few boys down with an attack. The newly arrived refugees, already weakened by their ordeal, were a natural prey upon their arrival.

---

[30] Ann Kellogg House to her father, Thessaloniki, 10 April 1924, AFS Archives.

A many-pronged attack was launched by several aid teams to rid
Macedonia of this scourge once and for all. One of the first actions
taken by the boys was to distribute quinine to the newcomers.
Organized into teams, they went out from the school giving the
medicine to the victims, but it was recognized by all that the root
cause of malaria, the anopheles mosquito, had to be wiped out.

Dr. House had been interested for years in the gambusia fish,
which had proved effective against malaria in Italy. This small fish
eats the larvae of the anopheles mosquito. When some funds held by
the Quakers became available, the Society of Friends decided to
allocate the money for the purchase of these fish. Dimitris Georgiadis,
a member of the farm school staff, was sent to Rome to bring the fish
back to Greece. Most were put into a cistern that ran around James
Hall. The boys, well trained in malaria control but not notified of the
arrival of the fish, poured oil in the cistern. Luckily some fish had
been placed in other ponds and survived. Eventually, with their
extraordinary proliferation, these small fish sent their offspring into
the waters of the Middle East.[31]

It took fully two more decades to rid the area of malaria, yet the
school experienced fewer occurences of the disease in the 1920s. In
1924 an intensive campaign to stamp out malaria was launched on the
campus. Many methods were employed simultaneously such as the
oiling of ditches, the placement of the gambusia fish, screening of
windows, and the generous distribution of quinine. The results are
impressive.[32]

| Year | Enrollment | Malaria Cases |
| --- | --- | --- |
| 1924–1925 | 73 | 41 |
| 1925–1926 | 83 | 24 |
| 1926–1927 | 83 | 11 |

Any discussion of the malaria campaign should include the work
of Theodoros Litsas. This energetic, imaginative young man, who was
later to become a guiding force on the staff of the school, arrived
from Smyrna in 1922 with Near East Relief. He immediately grasped

[31] Joice Loch, "Dear Bruce Letters," n.d., AFS Archives. The Gambusia is a live
bearer and produces about 100 offspring at a single birth.
[32] *The Sower*, 1927.

the potential value of having scores of boys organized into disciplined, effective groups working on meaningful projects, primarily anti-malarial. With his arrival, the school Boy Scout troup, which had become inactive, was reorganized. At the school and in the villages he applied a philosophy that coincidentally ran parallel to Dr. House's: theoretical training combined with practical application, both directed toward community service. The groups learned to function as a team, oiling ditches, planting trees, mending roads, etc. The school's contingent became very active. At times whole classes would go out into villages, sometimes spending full days pulling out reeds from streams so that stagnant water could flow freely. In the neighboring village of Sedes the scouts copied the school's playground (playgrounds were not a feature of Macedonian village life) with a track and volleyball nets.[33] The scouting program did much to fill gaps in the boys' lives. In the summer, those who had no homes to which they could return spent the time at the Boy Scout camp on Mount Hortiatis.

Yet even before Theodoros Litsas came to organize the boys into scouting troops, the students had been active in the villages under the direction of Arthur Bertholf, the school's math teacher, who did much to organize relief programs. The students' first effort under Bertholf's direction, along with the distribution of quinine, was the loaning of wheat seed. In 1923, 76 percent of the peasants of Macedonia were engaged in growing cereals, chiefly wheat, for this crop provided the quickest return.

Since some of the relief work tended to pauperize the recipients as it consisted of give-away programs, the farm school had made a fast rule not to donate anything outright, but rather to provide work, especially during the no crop period, or to loan supplies with the understanding that the refugees would pay back the loan in kind. Thus the older boys distributed wheat seed that the refugee farmers repaid at harvest time. This returned seed was then loaned to others so that many could benefit from the seed loan. To house the grain, money was donated by BLMAS. From these funds a structure called the Ark was built with a grain storage area on the first floor, where rat-proof

---

[33] *The Sower*, June 1926. This first issue of the *Sower* contains contributions by the students and is touchingly amateurish. A periodical with the same name, printed by the New York office today, is of professional quality.

bins were installed. To accommodate other relief activities conducted by the farm school in the villages, the Ark was expanded to three floors. Its odd shape and the fact that it housed not only people but animals conjured up the vision and name of the Ark.[34]

By 1924 the school—in cooperation with the Society of Friends, the Commission of the League of Nations, and the BLMAS—was working in a radius of about 30 miles, using the campus as a center point. Fifty villages containing 300 families were encompassed in this radius. Farm school boys, besides contributing in the ways already mentioned, served as liaisons between the destitute families and the relief workers. For example, during the Christmas holiday of 1924, groups of students volunteered to survey the condition of these villages. The results by the second Christmas after the exchange were grim. They found that many refugees were still living in ruined Turkish stables with no roof and no bedding other than sacks and straw. In some dwellings there were just one or two blankets for five or six people, yet hundreds of other people were still arriving in a steady stream.[35] Making such a survey was no easy task for the boys and was done at considerable sacrifice to their own wellbeing. They had to travel many miles on donkeys over muddy or frozen ruts going from village to village carrying as much as they could of their own supplies. The poverty-stricken refugees were in most cases hard put to give these youngsters shelter, let alone food.

One of the school's greatest contributions to the refugee settlement was its graduates, a ready pool of resource personnel. For instance, the Near East Foundation, in an effort to initiate an effective extension program for the farmers, called on farm school graduate Demosthenes Economou. This young man, with no training other than his farm school education, was sent to Central Macedonia to direct an innovative course of classroom instruction for farmers in three villages. He visited the villagers to give advice to them in the fields.[36] Other graduates were used as survey men to go into the villages to assess needs. Another was hired by the government to

---

[34] Susan Adeline House to her children, Thessaloniki, 21 January 1923, AFS Archives. The Ark was torn down in 1960 when it became unsafe.

[35] Ann Kellogg House to "Dear Guilders," Thessaloniki, 13 March 1924, AFS Archives.

[36] Allen, *Come Over into Macedonia*, 54–55.

settle 500 refugees in villages, apportioning them land and looking after their needs. One graduate served on the staff of an orphanage, where he had charge of the practical work being taught to the children. These graduates had a revitalizing effect on the refugees, with whom they were able to become more intimate than foreign workers or even some indigenous agents. The graduates, having grown up in central Macedonia, were close to the problems the refugees were now facing.

As the school's primary goal was to train young men to be good farmers and community leaders, the fact that it was able to continue this mission was the essence of its contribution to Greece in the 1920s. Seen in the context of what agricultural education was available in Greece and balanced against the country's agricultural needs, the farm school's success acquires a substantial dimension.

To meet the crushing need to train youth as farmers, the Greek government made courageous attempts to introduce courses in agriculture into the state school system in rural areas. Simultaneously, there was a serious drive to establish solid agricultural schools throughout the country. Nine boarding schools were operating on the elementary level. The instructors were technically well trained but their knowledge was for the most part theoretical. Thus their understanding of the fundamental problems of the farmer's life was limited. The two state secondary level agricultural schools functioning in the country suffered from the same difficulties. Worst of all, many of their students had no intention of returning to the land. Most hoped to find positions in the governmental bureaucracy.[37]

In the farm school's case, what facilitated the flow of students back to the land was the school's emphasis on the practical application of farming and vocational principles, its underlying philosophy of training young men to be community leaders, and the attitude it instilled into the boys that farming was a worthy goal. The school's philosophy ran parallel to what Greek officials felt was most important for the rural sector of the country. In fact, Prime Minister Venizelos shared Dr. House's opinion that peace in the Balkans depended largely on "whether we can bring to a happy solution the

---

[37] Ibid., 59–60.

agrarian question."[38] He saw in Dr. House's institution a "fundamental contribution to the ends we so deeply desire."[39]

By 1925, 22 years after the first boys were taken in, 60 boys had graduated and 180 had passed through but had left before completing the 5-year course. The following breakdown of the placement of these students points to the fact that the majority went back into agriculture or chose some field that reflected their vocational training:

3 were receiving further education in Greek universities
7 were receiving further education in American universities
   or schools
11 were in business in America
15 were teaching agriculture or other subjects in schools and
   orphanages
3 were engaged in agriculture or silk culture
3 were employed by the Greek government as agricultural
   agents in refugee settlements
1 was serving in rural health
9 were employed by commercial organizations in various
   positions of trust
10 were in minor positions of ordinary employment
2 had died or been killed during the war
60 were in their own or other rural communities farming,
   doing carpentry, ...or participating in allied vocational
   fields[40]

The most important aspect of the resettlement program was the fact that the thrust in Macedonia was agricultural. The government's aim was to form a class of independent peasant proprietors who, as sole owners of their farms, would make every effort to increase production. In order to furnish refugees with enough arable land, large estates often run by absentee landlords were broken up, monasteries were forced to relinquish large holdings, and swamps were drained both

---

[38] Eleftherios Venizelos to Lucius Beers, Athens, 12 March 1931, AFS Archives. For a discussion of Venizelos's educational reforms, see John Campbell and Philip Sherrard, *Modern Greece* (London: Ernest Benn, 1969) 146–47.

[39] Ibid.

[40] *The Sower*, June 1926.

for reclamation and for elimination of mosquito breeding grounds. Fifteen hundred new villages were created in Macedonia after 1925; 2,000 had been built by 1930. The new farming population was settled into these villages; thus the influx of refugees and the government's attitude spelled the coup de grace to the formerly prevailing system of *tsifliks* (large estates) with tenant farmers. This agrarian reform started after the Balkan Wars had fundamentally altered the country's social and economic outlook.[41] The old system of the Turkish *tsifliks*, with the landlord occupying a two-story house and the tenant farmers living in huts round about, was replaced by that unit so compatible with Greek life: the village. New work attitudes among the recently settled Hellenic populations differed substantially from those that had prevailed under Turkish occupation. Although Greece has always been considered an agricultural country, in reality the people had traditionally preferred the sea or commerce. The emphasis was never truly on agriculture, probably because the Turkish exploitation of the peasant stifled any initiative and because the land in most places was non-arable. The new conditions closed the value gap between the school and the host country to a great extent. A sense of common goals and the unlimited possibility of cooperation united the American Farm School with the Greek people.

It would be hard to understand the school's role without knowing at the same time about the efforts of the Quakers. The school was used as a nucleus for members of the Society of Friends who, arriving in 1923, established their home and headquarters in the Ark. With their comings and goings into the villages, they created another link between the school and the rural population. By providing medical care, constructing clinics, hiring specialists, and establishing laboratories for research against disease, the Quakers contributed comprehensively to the rehabilitation of the refugees. They supported the Boy Scout movement spearheaded by Theodoros Litsas, providing funds and sending a scoutmaster from England to help. They soon took over many of the programs that had been conceived

---

[41] Dimitrios Pentzopoulos, *The Balkan Exchange of Minorities and Its Impact Upon Greece* (Paris: Mouton, 1962) 152–53.

and initiated by the boys and the farm school staff during the first days of the catastrophe.[42]

In each of these efforts there was an interaction between the school and the Friends. Interchange of personnel between the two groups was continual as teachers became relief workers and Quakers with teaching skills stepped into the classroom. The most significant exchange of roles concerned Arthur Bertholf and Dorothea Hughes. Hughes, who had come to Macedonia to do relief work among the refugees, became instead a teacher and nurse at the farm school. Arthur Bertholf, relieved of his teaching duties by Dorothea Hughes, was then free to devote all of his time to the rehabilitation projects that he, aided by the boys, had initiated in the villages upon the arrival of the first refugees from Smyrna. It was Dr. House's ardent wish that the boys would be relieved as soon as possible of the large burden of relief work so that they could devote themselves to their education. With the arrival of the Quakers and other organizations, the students returned to a more or less academic routine, spending only their free time in relief work.

Dorothea Hughes became one of the school's chief benefactors, with much of her generosity directed toward higher education for qualified graduates. She was keenly interested in having farm boys study in the United States so that they might return to their villages armed with a superior education and the ability to be effective community leaders. Returning to Greece with their leadership and know-how after four years' study in the United States at excellent universities, these young recipients contributed immeasurably to their struggling country. For instance, Patroklos Hrisopoulos, a 1927 farm school graduate, is typical of the group that benefited from her generosity. He feels that Hughes was the molding factor in his life. She not only provided him with a scholarship but gave him guidance throughout his whole educational process. He claims that her gift to him of *Robert's Rules of Parliamentary Procedure* in 1926 had a formative impact on his life, for it gave him insight into the workings

---

[42] Loch, "Dear Bruce Letters." For a brief background of Joice and Sydney Loch's important contribution to the refugees' welfare, see Alke Kyriakidou-Nestoros, "Folk Art in Greek Macedonia," *Balkan Studies* 4 (1963): 29–30.

of democratic procedure. Dr. Hrisopoulos (Ph.D, University of Athens) keeps a framed portrait of Hughes in his livingroom.[43]

The automobile she donated to enable relief workers to travel around the countryside was the first passenger vehicle housed at the school. At a time when the road network in Macedonia contained less than 3 miles of practical highway for every 100 square miles, no doubt its usefulness was limited.

The Quakers were a colorful and eccentric group. They entered into the work of the school with the same contagious enthusiasm that they exhibited regarding their "concern" for the refugees. In a sense, the institution was influenced as much by the Quakers during these years as it was by the refugees; they lent color, wit, and imagination, as well as both moral and financial support, to its entire operation. Their lifestyle differed considerably from the more serious tone set by the elder Houses. Their hijinks and personal idiosyncrasies were part of the postwar spirit that was abroad in Western Europe in expatriate circles. It was a spirit born of war, misery, upheaval, and death coupled impossibly with idealism, faith, derring-do, and humor. Yet in spite of the hot blood of these Quakers, the elder Houses still set the pace in the social life of the foreign community. The social Mecca for many foreigners was their living room at tea-time. The Houses were warm hospitable people whose dignity and sense of honesty squelched any proclivity for malicious gossip. As a result of the Houses' influence, Thessaloniki had perhaps the least scandalous reputation among foreign communities in the Balkans.[44]

By 1928 the rehabilitation program for the refugees was 75 percent complete. Ninety thousand families had been installed on farms in Macedonia and provided with implements, animals, seed, houses, and enough land for their self-support. Multiplied by 4 to represent a family unit, this figure equals over 300,000 people. Henry Morgenthau, who served as chairman of the Greek Refugee Settlement Commission of the League of Nations, articulated the socio-cultural implications of this settlement when he commented that the Greeks had been settled "in conformity with the traditions of their race,

---

[43] Patroklos Hrisopoulos, interview with author, Ekali, 4 October 1972, written, AFS Archives.

[44] For an amusing account of the Quakers at the Farm School, see Joice Loch, *A Fringe of Blue* (London: John Murray, 1968) 112.

gathered into communities of acquaintances, friends, and relatives, with their churches and schools and institutions of local self-government."[45] It was an astounding achievement.

In 1926 the Greek government awarded Charlie the Silver Cross of the Order of the Phoenix "for service rendered in rural and agricultural development of Greece." But the greatest recognition, one that was to have lasting consequences for the school, came from Prime Minister Venizelos. His visit to the school and subsequent action to support it with government funds cemented the institution formally into Greece's governmental structure. When the Refugee Settlement Commission of the League of Nations terminated its responsibilities in Greece, the Greek government recognized that the school could play a vital role in the reorganization of educational programs for young farmers. The prime minister's visit took place on 26 November 1929. He was accompanied by Robert Skinner, the American head of legation. Their arrival was fraught with embarrassment for the school because their small cavalcade, which finally arrived after dark owing to numerous delays en route, got stuck in the mud on the road leading into the school.

The prime minister had to step out of his vehicle and slosh through mud to get into another car. Later, he spoke to the boys with Dr. House standing beside him on the platform. The prime minister took the octogenarian's arm and said, "I came out here for two things: to see you boys in your school and to have the opportunity to express to your founder the appreciation of the Greek government for the service he is rendering to the people of Greece through this work."[46] Subsequently, he made a public statement saying that he wished that there were a dozen such schools in Greece. By so doing he threw the full weight of his prestigious personality behind the school. A few weeks later the school received a proposal from the

---

[45] Henry Morgenthau, *I Was Sent to Athens* (Garden City NJ: Doubleday, Doran, & Co., 1929) 273.

[46] Ann Kellogg House to her family, Thessaloniki, 26 November 1929, AFS Archives. It is the opinion of Dimitris Zannas that his father, Kostas, hoped that the prime minister would get stuck in the mud since this would be the only way to get a road constructed connecting the school to the city. Apparently Mr. Zannas was right, for Mother House wrote later that the day after Venizelos got stuck in the mud he told the governor of Macedonia that the farm school must have a road. She added that the road was well under way and would be finished in two or three weeks.

government to cooperate in a common endeavor to increase enrollment by taking in boys who would be given government scholarships.

Mr. Venizelos and the ministry of agriculture proposed that in fall 1930 the school should accept a maximum of 200 scholarship boys, that this plan would be converted into an agreement that would be binding on any subsequent government, and that the independence of the administration of the school would not be altered by the agreement. Charlie's analysis was that

> this action on the part of Mr. Venizelos seems to us to be the culmination of the policy which we have followed since the close of the war, of preparing the school to serve the definite needs of Greece and making it a real factor in the solution of the cultural problems of the country. This proposal constitutes a very real testimony on the part of the most important leader in the country to the capacity of the school to render the required service and it was made after consultation with people well informed on the work of the school.[47]

On 8 July 1930 the Greek legislature passed law 4810 authorizing a contractual arrangement for the scholarships. In May 1931 the school signed the contract to expire in the school year 1937–1938 in which it agreed to take twenty-five boys into each of its four classes for the next eight years.[48] Fully as important as the scholarship grant was the government's decision on the question of land for these needy youngsters. So that the graduates could start on their own land immediately after graduation, the ministry of agriculture allotted plots of reclaimed land that could be paid for via a long-term loan. These government measures provided a great incentive for the boys to return to agriculture. The youngsters were now full of hope and the school provided an education and a philosophy on which they could lean for the rest of their lives. In 1934 the first class of scholarship

---

[47] Charles Lucius House to Lucius Beers, Thessaloniki, 8 December 1929, AFS Archives.

[48] Catalogue entitled "By Which Law the Operation of the School Has Been Authorized," July 1930, AFS Archives.

boys was graduated. Twenty-two completed the course of study; one had died and two boys had dropped out. Law 4810 was an important milestone in cooperative efforts between the Greek government and the American Farm School.

# CHAPTER 9

# ANN HOUSE COMES TO
# THE AMERICAN FARM SCHOOL IN 1923

A trustee remarking then on the vitality and beauty of Ann Kellogg House in her ninetieth year wrote: "To look at Ann House is a beatification; such is the light that shines from her face."[1] Her longevity provided a link between past and present; although she and Charlie officially retired in 1955, her influence was still a strong element two decades later. In the early 1970s, Bruce Lansdale wrote to a friend, "More than any other person you or I know, she reflects the value that we both strive for, and because of this has influenced us both beyond anything she or we might know."[2]

To marry Charlie House, she came in 1923 to Greece—a country that she had heard of vaguely in school, an area that to her was so obscure that, when she first met Charlie a year before, she could not remember whether he lived in Macedonia or Mesopotamia. Eventually she had Greece's culture and geography firmly in mind and actually experienced some of its recent history, including the population exchange, the Second World War, and the civil war.

Although she came as a stranger, through the passage of years she became as deeply embedded in the tradition of the farm school as her in-laws. She caught the school's spirit, amplified it, and applied it. Essentially she changed the institution little; her importance lies first

---

[1] Iphigene Bettman, "Founders' Day Brings Historic Reunion," *The Sower*, summer 1972.

[2] Bruce Lansdale to Arthur Lanckton, quoted in Catherine Lanckton to Ann Kellogg House, New York, 18 March 1971, AFS Archives.

in the manner in which, with her husband as a partner, she kept the
school on its original course, and secondly in her conviction that the
survival of the school was as precious as life or freedom itself. This,
coupled with consummate love for her husband, gave her courage to
remain in Greece during the Italian and German invasions of World
War II and to stay on even after she faced certain internment when
her own country went to war against the Axis. Her importance lies
perhaps most significantly in the conduct of her own life—this is
what Bruce Lansdale had in mind when he wrote of the "value" she
reflected. In this sense she strikingly resembles another woman who
came of age at the same time and whose personality dominated many
of the years that Ann spent at the farm school—Eleanor Roosevelt.
Roosevelt's biographer, Joseph Lash, wrote of his subject, "She was a
woman with a deep sense of spiritual mission...of extravagant
tenderness and piety."[3]

Although Ann was thirty-five years old when she arrived in
1923, from all indications her personality had not reached its
potential. There was definitely a cross-fertilization at work between
her and the farm school during the first decade or so after her arrival.
She seemed to grow at the school as if something in the soil was
providing her sustenance. Whatever she gave to the school, it
rewarded her amply. Her abilities and confidence developed with the
demands placed on her. If she had come to Greece without the
extraordinary courage to endure, she had developed that courage by
1940; if perhaps she had come without a dogged sense of mission,
other growing convictions made her steadfast; if she had lacked
creativity, her intuition of what was possible compensated. She
answered the needs of the school so precisely that it would be difficult
to imagine the institution's history without her.

Of course Father House saw Charlie's marriage as part of the
divine plan to insure the continuity and spirit of the institution that
he had founded. This sentiment is illustrated by his first letter to Ann
after the engagement: "How happy I am to welcome you to our
household... [Y]ou certainly have had an excellent preparation for
this work, and I do believe that your coming to us and our work is of

---

[3] Joseph Lash, *Eleanor and Franklin* (New York: W.W. Norton & Co., 1971) 517.

God. I am old fashioned and believe in the covenant mercies of God which are handed down from parents to children."[4]

Anna Mary Kellogg was born in New York City on 23 September 1888 into a family of restricted means. She grew up in lower Manhattan on the wrong side of the tracks from what had been the neighborhood of her mother-in-law to be. Yet her family was not without quality. She was the progeny also of church-oriented people: her maternal grandfather was a Presbyterian minister and both her parents and grandparents played important roles as laymen. Her mother, Amy Robertson, was an educated woman, having completed at least her sophomore if not her junior year at Elmira College. It was due largely to her mother that Ann received an excellent education. She was hardly a promising student in high school, being deficient in Latin and French. Frustrated over her studies and eager to earn her own way, she begged her mother to allow her to drop out of school. But on the subject of education Mrs. Kellogg remained firm. She was determined that both her daughters should be educated. Since Ann was older, she would go on to four years at a liberal arts college while Helen the younger would attend two years at nearby Pratt Institute, for the family did not possess financial means to send both children away to a four-year college. Ann insisted that "Helen had a much better mind."[5] Under the patient guidance of her mother, Ann finished high school after five years.

Her father, William Kellogg, had before Ann's birth left his father's farm in upstate New York to seek his fortune in the city, but his fate turned out disappointingly to be a grueling job in a bakery in lower Manhattan. Their small apartment consisted of four rooms: one used for ice-box and storage; a small bedroom for her father, whose hours were so ungodly that he slept alone to avoid disturbing the rest of the family; a kitchen; and a fourth room that served as sleeping quarters for Ann, her sister and mother, doubling during the day as a sitting room. Out on the fire escape Mrs. Kellogg had fixed a play area for the little girls that became so vital a part of Ann's childhood that even in her ninth decade she could describe its arrangement in

---

[4] John Henry House to Ann Kellogg Chapman, Thessaloniki, 12 December 1922, AFS Archives.

[5] Ann Kellogg House, autobiographical outline compiled for the author, 21 September 1973, AFS Archives.

detail.[6] These difficult circumstances prepared her for the material hardships that awaited her in Macedonia: "I did not find any physical difficulties at the Farm, except for lack of transportation in the early years.... Life at home in America was rugged in my early years; we had very little money, lived in New York slums."[7]

Such financial deprivation led to social problems for the Kellogg children, born of quality but unable to mix with children whom they considered to be their peers. This situation was bound to be painful as they grew into self-conscious young women. Ann developed an inferiority complex that was only completely erased years later with her marriage to Charlie. But at the same time her childhood years left her with a streak of independence and hardiness that became much more the measure of the woman than the inferiority complex.

After graduating high school, she spent a year at Packer Collegiate Institute before entering Mount Holyoke College. She explained that Packer was "a 'rich girls' school" that she entered "frightened, self-conscious, and unhappy."[8] But there was a natural gregariousness and optimism about the young woman that carried her through the year. She emerged all the richer for the experience, having gained the self-confidence needed to enter into the more difficult program at Mount Holyoke.

At Holyoke her personality began to exhibit clear traits of persistence and independence. She refused to do the popular thing and join a sorority, but did take an interest in labor problems, which carried her outside the college into the grimy mill towns of western Massachusetts. She joined a club with factory workers and became intimate with their problems. In spite of a shaky high school career, she dug into her studies and finished a normal BA program in four years as an English major.

Throughout her growing up, the Kellogg family relationship was intimate and loving. Letters attest to an enduring family bond that encompassed the qualities of love, obligation, and responsibility. Her entrance into the tightly knit House family circle was eased by her own finely honed perception of kinship. The many accommodations that must have taken place between, on the one hand, a mother-in-

---

[6] Ibid.
[7] Ibid.
[8] Ibid.

law of iron will and bedrock conviction and, on the other, a somewhat pliant yet independent middle-aged daughter-in-law. It is to both women's credit and a measure of both their characters that a harmonious, respectful bond sprang up between them and lasted. Considering the years of forced intimacy, when they lived together and worked together in the narrow, isolated confines of the school, their relationship seems all the more uncommon. Family allegiance—the matrix of all Greek values and the key to understanding village life—was immediately grasped by Ann since her own mode of upbringing had rested on this basic value.

In July 1917, five years after graduating from Mount Holyoke, she had married a young physician, William Harmon Chapman. That year he joined the army, as he had a "deep religious conviction that the war was really a holy war, a war to end all wars."[9] Ann, however, had espoused an opposing cause: she had become a pacifist and was marching in protest parades. She wrote of the lack of conflict in their marriage even though they had moved to opposite poles: "I learned one great lesson during these years. Billie and I respected each other's conviction; of course we tried to convince each other, but we came to realize that we could love each other, respect each other's point of view and even help each other sustain the view which he felt to be right. It was an experience of growth and understanding for which I have always been grateful."[10]

In September 1918 the brief but happy marriage came to an end when Dr. Chapman died of influenza. With silent yet eloquent testimony to the love of that first marriage, Ann always wore both his and Charlie's wedding bands.

After graduating from Mount Holyoke in 1912 and during and after her first marriage, Ann had been teaching at the Ethical Culture School in New York. Any liberal tendencies that she had entertained as a young woman were strengthened by her contact with the staff of this open-minded institution. The principal, Anna Gilingham, who was a Quaker and an unshakable pacifist, had formulated and articulated a system of ethics that was already more than formative in Ann's mind when they met. Ann felt philosophically at home for the ten years she remained on the staff.

---

[9] Ibid.
[10] Ibid.

One incident that involved Ann while she was teaching at the Ethical Culture School illustrates both her character and the climate at that institution before World War I. The school was very progressive for its time, holding strong convictions on the equality of the races: it actually made a conscious effort to enroll intellectually qualified black children by advertising for them. At one point the school took the students on a trip to Washington, DC. Among them was one black child, Mary Mason, who, it was learned, would be refused accommodation with the rest of the students in the capital, and would have to sleep in the basement of the hotel with the service employees. Mary's mother balked at this arrangement and insisted that the girl be left at home. The school, however, was determined that Mary should have the opportunity to join her fellow students on the trip to the capital of her country and travel in dignity. The solution that seemed best was to have Mary Mason stay in a private black home. Ann offered to be the child's companion and remained with her in the private accommodations.[11] It was owing to the Mary Mason affair that Charlie House was first attracted to Ann.

While studying at Columbia University for her master's degree, which she received in 1919, Ann became close friends with Charlie's sister, Florence, who was on the faculty. In 1922, Florence, playing matchmaker, tried numerous times to introduce her brother, who was in the United States on farm school business, to her widowed friend. Charlie, still instinctively rebellious, dodged such family machinations and rejected his sister's invitation to dine with her and Ann. It was when he heard of the story of her chaperoning Mary Mason that he decided she might be a person well worth meeting.

For Charlie it was love at first sight. Certainly their church-oriented family backgrounds gave them a common reference. Ann's independent attitudes especially on war were bound to attract Charlie. The fact that she marched in parades, an unseemly action for any woman in 1914, must have fitted perfectly with his own ideas about civil disobedience.

Ann's reaction to his almost immediate proposal was negative. She felt that she "wanted to continue living in the past, with no

---

[11] Ann Kellogg House, interview with author, Athens, 25 May 1973, AFS Archives.

uprooting; although in those few months I had come to respect and admire Charlie, I urged that we break off our growing friendship."[12]

But Charlie was a persistent man. He bombarded her with letters when he returned to Greece less than two months after they had met. His letters reveal his inmost being. From them, more than from any other source, we learn the depth of his spiritual nature, his extreme sensitivity and the terrible self-knowledge of his own weaknesses that resulted in a compulsive drive pushing him to his utmost to correct his imperfections.[13] In September 1922, she finally accepted his marriage proposal through the mail.

By that year, Charlie had lost much of his earlier recklessness. Through his compulsive conscience, he had become chained, albeit voluntarily, to his onerous duties as son, brother, respectable citizen, leader of the community, school headmaster, Christian, and now fiancé. Yet in his letters he spared Ann none of his deficiencies. Her tolerant, accepting nature was large enough to accept these deficiencies and those "vices"—alcohol and tobacco—in which he indulged, although one of his letters containing pages about his feelings on "drink" indicates that she must have addressed him seriously on this point. The attraction on both sides was strong. Certainly he was lulled by her relaxed and measured personality. We cannot rule out the possibility that she, with her religious background, teaching experience, independent streak, and sense of family possessed all the prerequisites for the wife of the headmaster of the American Farm School. It is also true that, when he left for America that year, his mother had told her thirty-five year old son to find a wife.[14] Whatever the attraction, it was substantial. The marriage, in spite of a variety of unusual stresses placed upon it, was characterized by profound affection and uncommon devotion.

On 21 July 1923 Ann Kellogg Chapman sailed for Corfu to meet her fiancé and his parents and sister, Ruth. She knew that her future would be marked by adventure and peril, but she did not realize that calamity would be visited on her within the first few weeks of her

---

[12] Ann Kellogg House, autobiographical outline, AFS Archives.

[13] Tape of Charles Lucius House's correspondence to Ann Kellogg Chapman, 1922–1923, recorded by Ann Kellogg House, 1972, tape, AFS Archives.

[14] Ann Kellogg House, "Memoirs," 11.

arrival in Greece. She arrived immediately before the "Corfu Incident," Mussolini's first aggressive international adventure.

On 27 August 1923 an Italian general, chief of an international commission that was delimiting the Greek-Albanian border, was murdered by unknown assassins on Greek territory, along with four members of his staff. Mussolini, blaming the murders on the Greek government, sent a list of lengthy and severe demands to Greece. The Greek government denied the Italian allegation that it was responsible for the crime. However, since the crime was committed on Greek soil, it would honor certain of the demands. The more extreme demands were rejected by the Greeks.

In response to these rejections, Mussolini on 30 August ordered the Italian fleet to bombard the Greek island of Corfu. He did this unilaterally, without appealing to the League of Nations, a flagrant abuse on his part since Italy was a member. As a result of the bombardment, sixteen people were killed and thirty-five wounded. Italian troops then occupied Corfu and some neighboring Greek islands in the Ionian Sea. On 1 September Greece appealed to the League and on 17 September, as a result of international pressure, Italy withdrew its troops. Greece, under protest, was required to pay the indemnity of fifty million lire demanded by the Italians as prescribed by the Conference of Ambassadors that had been delegated the task of settling the problem by the League.[15] It was the first test case of the League of Nations and a failure that prefigured the subsequent Manchurian and Ethiopian tragedies.

Ann, who had barely gotten to know her new family and situation in a strange country, found herself in the midst of a serious and violent incident. Mother House gives an intimate picture of the family, the quintessence of composure, as "one shell...came so near the hotel that Ruth heard it whiz by."[16]

No one on 30 August, the day the family arrived, had heard that the Italian fleet was entering the harbor. They saw twelve ships, four of them battleships with decks cleared for action. The Houses assumed

---

[15] James Barros, *The Corfu Incident of 1923* (Princeton NJ: Princeton University Press, 1965) 40.

[16] Susan Adeline House to her children, Corfu, Greece, 7 September 1923, AFS Archives. When the British turned the island over to the Greeks in 1864 it was with the understanding that it would not be fortified.

it was merely a bluff on the part of the Italians to frighten the Greek authorities and inhabitants of the unfortified island.

Mother House wrote:

> The bombardment began about half past four o'clock. I was sewing and we all kept on with what we were doing, saying that the guns on the citadel were acknowledging the arrival of the fleet. When the airplanes began to circle around us we would jump up to see them as they came so close the noise was almost deafening. There were two very heavy shots that we could not mistake and what sounded like machine-gun fire, but we did not suggest to each other that it was an unusual way of firing a salute, and kept on with what we were doing.[17]

They heard from their window the screaming of women and children as they fled from the citadel. From a window in Charlie's bedroom they saw a white flag hauled up the flagpole and then they knew the island was no longer Greek. At six o'clock in the evening they went out for a walk and saw that the white flag had been replaced by the Italian. Several days later, Mother House added this note of interest to her experience: "A very strange and thrilling thing happened to the flag last Saturday. We had a storm that was quite heavy but only one crash sounded as if something had been struck. Later we heard that it was the flag staff and that the flag had fallen and blown into the sea."[18]

Still another calamity struck the House family. While the bombardment was going on, Ann was taken seriously ill with an attack of appendicitis that resulted in peritonitis so that throughout the Italian occupation and during subsequent months she lay critically ill on Corfu. By 25 November she was able to be moved to the farm school, emaciated, weak, and having lost all her hair.

Her first view of Thessaloniki, the boys, and the school, she describes in a letter home:

> We reached Salonika before six in the morning, before daylight, and when we got off the train, there was Mr.

---

[17] Ibid.

[18] Susan Adeline House to her children, 10 September 1923, AFS Archives.

Bertholf and about twenty of the boys, many of them in their scout uniforms. They sang a song of welcome and grabbed our baggage and carried it over to the auto that Mr. Irwin had driven down there for us. I couldn't see much for I was so bundled up.... Another crowd of boys were lined up half way up to the school all in scout uniforms and standing at attention.[19]

Such was Ann's arrival. However, her physical condition was hardly prophetic for her tenure. She later proved to be a dynamic leader, instrumental in setting the institution's tempo and at the same time making friends among the Greek and foreign community for its benefit. As a fresh personality in Northern Greece, she would enter Greek society with her husband without taint or label. She was not a missionary; she had not been there to sympathize with any group during the Macedonian Struggle. For her, Macedonia was Greek, the farm school was Greek, and she would attune her life to that orientation. The only language she sought to learn was Greek and this she did accomplish, although her accent never lost its American twang.

As she and Charlie settled into their own apartment in Kinnaird House after their marriage, it became a second home also to the boys: "When we reached here the last of November school had been opened running for some time of course and little irritations had grown up because of Charlie's not being here to smooth them over. The boys came to him and he brought them in small groups upstairs to our own living room and talked things over.... It is such a joy to have our own home to bring the boys to and they seem to enjoy coming up as much as we enjoy having them."[20]

In addition she invited teachers and staff in small groups. She had a natural talent for getting people to warm to her; there was hardly a boy or family on the campus whose troubles and happiness she did not follow very closely. With its expanding population and the arrival of the refugees, the school became a larger institution with both

---

[19] Ann Kellogg Chapman to her parents, Thessaloniki, 26 November 1923, AFS Archives.

[20] Ann Kellogg House to Florence House, Thessaloniki, 24 January 1924, AFS Archives.

transient and permanent residents. Through her sincere interest in people and her appreciation of family structure she welded what could have been diversified groups into a solid, unified farm school family. There was thus a unity of purpose, a common scale of values, and a sense of responsibility held among the school population in varying degrees. Certainly the key to understanding the school's unique spirit is this sense of community.

Ann Kellogg's marriage to Charles Lucius House, while she was still weak and thin, took place on 2 December 1923 in a simple ceremony performed by the Reverend John Henry House in the family living room, attended by a small group of friends. She was still convalescent and short walks around the farm were the limit of her activities until the middle of January.

On 17 January 1924 she took her first walk off the campus to a nearby refugee camp, accompanied by Dorothea Hughes. "It was a huge barracks containing about 100 rooms and had been used as a French hospital during the war. It had fallen into ruin and a number of refugee families were living in it.... They had boards, rags and paper to keep the wind and rain out, but the rooms were damp and cold and so many people were down suffering from malaria that they hadn't much ambition to better themselves."[21] She noticed pathetic attempts by the desperate Greeks, even in these surroundings, to decorate their hovels by putting newspapers with the edges torn in designs on some shelves and pictures of saints on the walls.

By February, Ann became strong enough to begin limited activities. She started in the academic department and slowly assumed the duties of principal in addition to other tasks. By April she was ordering books, having conferences with the boys who were getting poor grades, writing letters to parents, and working on next year's program.

But her main contribution remained in the area of relationships with people. She was most keen in judging people's capabilities and assets—a talent she applied in tapping the resources of individuals who could aid the school. She had an innate trait of leadership. She eventually began to share the Houses' dictum that the school was an end in itself and that all means should be directed toward this end. Still

---

[21] Ann Kellogg House to William Kellogg, Thessaloniki, 17 January 1924, AFS Archives.

her tenderness for people was deep enough that the means to that end were never harsh or exploitative.

The effects of her charm and warmth reached far beyond the school. In 1931, when Charlie went on business to Athens, the prime minister's wife, Helena Venizelou, invited Ann to spend the afternoon with her. Mrs. Venizelou took her for a ride in the country, where they talked for an hour and a half. She led the conversation, asking questions about the school and exploring possibilities for the opening of a girls' department. Exhibiting great interest and trust in this project, she then sent Lucius Beers $1,000 out of her personal funds.[22]

With indispensable help from her sister-in-law, Ruth, Ann was hostess between 1923 and 1928 to the chairman of the Refugee Settlement Commission, Henry Morgenthau, and to many other prominent officials connected with the League of Nations as well as with British and American relief agencies and the Greek government. The full responsibility of entertaining guests gradually shifted from Mother House to her daughter-in-law. As gregarious as Ann was, this task was still onerous: "I think the part of my life at the farm to which I found adjustment most difficult was that I could almost never be alone!... There were so many visitors that it seemed so strange to sit down at a meal alone, and I felt as if we should have a chaperone."[23]

The first decade of Ann and Charlie's marriage was one of unusual stress in terms of farm school development. The upheaval caused by the refugee crisis, the dramatically accelerated pace of life all over the world in the postwar period, and the rapid expansion of the school put unbearable pressure on key personnel. Charlie, with his compulsive work habits, labored day and night attending to the myriad details that went into running the school. Such absorption in his work and other factors put a gulf between him and his wife. Ann found that

---

[22] Helena Venizelou to Lucius Beers, Athens, 13 March 1931, AFS Archives. In this letter Venizelou wrote: "Anyone who is familiar with conditions would recognize the importance of the woman's part in the social and economic evolution of rural life, but other questions have absorbed the capacity and resources of our country to the extent that it has been impossible to take adequate steps in the training of village girls. The question is fundamental and perhaps American experience in this field is the widest."

[23] Ann Kellogg House, notes to the author, 3 March 1973, AFS Archives.

neither her faith nor her sense of mission was as strong as Charlie's or her in-laws'. When, for instance, there was not enough money to meet expenses, Mother House would rock trustfully in her chair and say, "If God wants the work to go on, He will provide."[24] But Ann's stomach would constrict with nervous tension.

By 1928, a physical, psychological, and finally a spiritual crisis had settled upon Ann. Her energy was always titanic. For such an individual to have been laid so low was not just Ann's personal response to the hardships of life at the farm school. It was more indicative of the general quality of life and the gigantic stamina that was required to survive at the school and retain a sense of purpose.

She was suffering from malaria, which includes severe depression among its pernicious effects. This had warped her total perspective. She was unsure any longer of her direction, wondering if she could maintain stability with a husband who was as devoted to the school as he was to her. Additionally, her physical and psychological strength was further eroded by another disease, directly connected to malaria, called blackwater fever, to which she very nearly succumbed. Overcome by all of these difficulties, her spiritual resources finally ebbed.

She was advised by her physician that she could no longer remain in Greece. She left for Switzerland. Once there, she regained her strength. But the real restructuring of her life was affected there by her contact with Eleanor Ford, an American woman involved in the Oxford movement, an organization that had started at Oxford University after World War I and spread to the United States. Its goal was to reinvigorate religion. Socially minded, it laid great emphasis on group prayer and Bible reading. Ford helped Ann to reestablish her psychological balance; she was responsible for her spiritual rebirth. Whatever happened to Ann in Switzerland was an experience having permanent effect. She returned to the school convinced that her place was beside Charlie; from that moment on her faith and convictions strengthened.

Eleanor Roosevelt enunciated an axiom when she wrote to an anxious wife: "All men who make a success of their work go through exactly the same kind of thing you describe and their wives in one

---

[24] Ann Kellogg House, tape to the author, 17 October 1972, AFS Archives.

way or another have to adjust themselves. If it is possible to enter into his work in some way, that is the ideal solution."[25]

Ann had discovered that axiom for herself long before Mrs. Roosevelt expressed it. She found that if she could be Charlie's secretary she could enter with complete understanding and involvement into the workings of the school and be with him more often. It was a modus vivendi that worked for them until their retirement. The school had never had a proper secretary; one of Charlie's failings as an administrator was his inability to cope with office work. Illustrative of her full absorption into the job of secretary, and also of her growing moral fiber, is the following incident told by Dimitris Hadjis:

> Charlie, Ann, and I found it necessary during the civil war to go to Athens. The government advised us to take a military escort but Charlie felt that might be even more dangerous as it would attract the guerrillas' attention. In the worst guerrilla-infested area we got a flat tire. Charlie mumbled, "Now we've had it!" Both of us began to change the flat tire. My hands were so shaky I could barely hold the wrench. Suddenly I heard a click, clack, click, clack beside the road. Ann had climbed out of the jeep, opened her typewriter, and imperturbably was taking the opportunity to catch up on her correspondence.[26]

Owing to the serious illnesses from which Ann suffered, and perhaps her age when she married Charlie, they never had children. The students and the children of the staff filled that void. Perhaps because of this the farm school boys received a quantity and quality of attention that otherwise would have been impossible. In a sense, though, Ann and Charlie did leave heirs in terms of the farm school's direction. Tad and Bruce Lansdale, although not kin to the Houses, and strong independent personalities in their own right, patterned their responses after Ann's and Charlie's. Bruce was appointed director in 1955; he and Tad had had five years to understudy their

---

[25] Eleanor Roosevelt to Mrs. McFadden, 16 September 1930, as cited in Lash, *Eleanor*, 352.

[26] Dimitris Hadjis, interview with author, Thessaloniki, 8 September 1972, written, AFS Archives.

predecessors before the Houses' retirement. This overlap was indispensable for the school's continuity.

Ann was the perfect complement to Charlie. With authority she gained stature, with experience expertise. "Despite any shortcomings either of them might have had, others recognized them as superior human beings."[27]

---

[27] Nike Myer, interview with author, Thessaloniki, 12 September 1973, written, AFS Archives.

# CHAPTER 10

# THE 1930s:
## DEPRESSION, DICTATORSHIP, AND
## APPROACHING WAR

The development of the American Farm School during the 1930s was played out against the somber backdrop of worldwide depression, political instability in Greece that resulted in dictatorship, and approaching war. At the beginning of the decade, the campus, set off against a still barren landscape, was a testament to the founders' skill and endurance. The institution had become a self-sustaining village of approximately 300 acres with its own water, sewerage, and electric power system. A modern dairy housed a herd of purebred Jersey and Guernsey cattle that supplied milk for the boys and a surplus sold to customers in the city. Sturdy sheepfolds enclosed flocks introduced into Macedonia from the island of Chios. A piggery sheltered the large British Black pigs imported into Greece by the school. The poultry barn was scientifically constructed so that the boys could learn to raise chickens and turkeys by modern methods. In many villages, even in the late 1970s, poultry foraged on the land like goats or sheep. Cubed bee boxes dotted the campus and vegetable gardens bore witness to the fact that tomatoes, onions, and cucumbers could be coaxed from the meanest soil as a part-time project. Twenty buildings in various stages of construction and of various styles—an architectural hodge-podge—housed the farm machinery, boys, staff, and animals. Flowerbeds banked the houses while rose vines, some grafted by Dr. House from three or four varieties, climbed the walls. From this inner core stretched grain and barley fields, vineyards and

orchards. In a more fertile countryside the school would have blended
with the landscape; in a more developed country the campus would
have been insignificant in many respects if not substandard. But,
considering the surroundings and the standard of living in this war-
torn, poverty stricken nation, its very appearance was striking.

In the early 1930s, the campus accommodated 127 students and
a teaching staff of 17. To be accepted, the boys had to have
completed the sixth year of primary school and be between fourteen
and sixteen years old. They followed a four-year curriculum in which
each day was divided evenly into practical and classroom work; it was
so structured that the students learned the theoretical explanation of
the work they were actually doing in the industrial arts department
and on the farm. By taking courses in history, language, mathematics,
science, and Bible study, they widened their cultural horizons.

The advent of worldwide depression did not alter the course of
study, but it did curtail the development of the physical plant. The
consequences of the economic catastrophe were felt immediately as
the flow of contributions from America slowed to a trickle. Although
nearly 75 percent of the school's income was raised in Greece by its
farm and vocational activities, almost all of the remaining 25 percent
came from donors in the United States. In 1930 the school requested
$34,000 from its board of trustees. Of this, endowments, trust funds,
and securities yielded $16,000 while about $11,000 was obtained from
contributions. The remaining $7,000 was gained by drawing on
capital, since contributions were inadequate.[1]

To stimulate donations, a fund raising campaign was initiated in
1931 in the United States with the unrealistic goal of $1,000,000, 60
percent of which was to be invested in endowments. Ann and Charlie
went to America to lecture, mainly in religious centers. Predictably,
the campaign failed drastically. Depression-ridden Americans
possessed neither the mood nor the funds to support a small school in
the Balkans.

---

[1] *Newsletter*, 15 May 1930. Almost always written by one of the Houses, the
*Newsletter* was an occasional letter addressed to friends of the school, conveying
school news and expressing opinions and reactions about both the school and the
world around them. When the *Sower* began to appear regularly in the 1960s, the
Newsletter was discontinued. The *Newsletter* collection is stored in the AFS
Archives.

Reeling under the impact of increased financial problems, the board of trustees advised the director to cut the salaries of all employees in 1931. In June of the following year, as economic problems became even more acute, salaries were again cut, this time a staggering 50 percent. Notification letters informed each employee that "the financial situation of the School has become increasingly precarious. Our income has been cut in half and, in spite of all our efforts at economy, we are facing a new school year with a heavy deficit and half of our usual resources."[2] The fact that the school retained the entire staff on half pay instead of dismissing some and attempting to close down certain departments was a strong force in keeping the loyalty of the staff and drawing them close together. To say merely that the staff remained steadfast in response to this radical salary cut is to understate the prevailing temper of the school. The following letter from a devoted teacher epitomizes the staff's reaction: "I am ready and willing to undergo even greater sacrifices if necessary and do all within my power to help the school emerge from its present precarious financial position and launch into its regular course and even greater prosperity."[3]

In preparing this book, I asked former staff members or their wives, "What was your attitude when your salaries were halved?" The answer was invariably "We did not mind."[4] Given the bleak conditions during this decade, one could argue that their steadfastness sprang from self-interest rather than loyalty since school personnel were still better off than countless others. To a great extent this argument has merit; yet a plethora of examples attests to the devotion, loyalty, and particularly to the spirit of selflessness in almost every member of the staff. To emphasize this point, a fellow staff member of Lee Meyer's remembers walking down the street in Thessaloniki with Meyer when they saw a poor refugee shivering in the Macedonian cold without a

---

[2] For one such letter, see Charles Lucius House to Theodoros Litsas, 15 June 1932, AFS Archives.

[3] Eleftherios Theoharides to Charles House, Thessaloniki, 25 June 1932, as quoted in Ann House, "Memoirs," 134.

[4] This is a sample response. See interviews by author: John Boudourouglou, Athens, 26 February 1974; Chrysanthi Litsas, Thessaloniki, 8 January 1973, written, AFS Archives.

coat. Lee Meyer slung his own coat over the refugee's shoulders and walked on.[5]

During this period, the House family set an example for sparse living. Charlie and Ann existed on a salary of $75.00 a month. They also received a small allotment from the farm school dairy that they often contributed to needy children.

Even with the salary cuts, the board of trustees felt that more measures had to be taken to combat the school's economic woes. The board suggested to Charlie that the school close temporarily or at least shut down some of its departments. He argued that the practical departments were so interrelated with the academic ones that it would be impossible to eliminate some courses without seriously damaging the school's effectiveness and purpose.[6] The trustees deferred to Charlie. The teaching program was never compressed even during the peak years of the depression.

The devaluation of the dollar in 1933 by President Franklin D. Roosevelt was a serious blow to the institution's financial resources. The dollar's fall on the European exchanges, accentuated in Greece by a rise on the part of the drachma, further reduced income.

To meet the deepening crisis, the board of trustees had been digging into its limited reserve since the beginning of the depression. Certainly the board's willingness to meet the problem in that way and its equal willingness to defer to Charlie's optimistic and sometimes purely intuitive judgments were vital forces in the school's ability to endure.

By 1935, at the peak of the crisis, expenses had been pared down to a minimum. In 1931 the Greek government, in the same bill that granted scholarships, had included a provision extending exemption from duty on farm machinery and equipment, purebred livestock, building materials, fuel, lubricants, and food supplies.[7] Since there was no way to decrease expenses further, the director and his staff studied ways to increase income. Expanding the sale of produce from the

---

[5] John Boudourouglou, interview with author, Athens, 26 February 1974. Lee Meyer came to Greece with the Near East Relief after the refugee exchange. He served as engineer at Anatolia College and later joined the staff of the farm school.

[6] Ann Kellogg House, "Memoirs," 134.

[7] "History of the Founding and Operation of the American Farm School," 30 August 1954, AFS Archives.

productive farm was one method. The dairy was another important source of income. The demand for the school's milk far exceeded production since it was the only pasteurized milk in Greece. As a tangent it is amusing to note that the method of milk delivery in town caught the attention of the *New York Times*, which described the procedure in a lively article. The milk had to be delivered by stealth "because of the unorthodox nature, to Greek minds, of the milk wagon's construction, constructed by Greek boys at the school." The milk wagon was the traditional American four-wheel type with a door in the middle to allow the milkman access to his supply. Seeing this strange vehicle on the streets of Thessaloniki, farm horses grew unmanageable. "Besides, the police disliked the way the wagon had to cut widely when turning corners through the narrow unpaved streets when it made its rounds."[8]

Another method of increasing income was to rent staff houses on the campus to outsiders. For instance, the American consul general, James Hugh Keeley, Jr., rented Hastings House while Harold (Doc) Allen rented Metcalf House. Hastings was built as a residence for the director but Charlie felt it was much too elegant for him at a time when an assured rental would be beneficial to the school. The Keeleys gladly bought a cow for the dairy when Charlie pointed out that they could thus assure themselves of a regular supply of fresh milk with some left over for the needy.[9] In order to make the two most attractive houses available for renting, Charlie and Ann, rather than occupy rent-yielding cottages, moved to a series of small apartments during the 1930s.

In 1935 the Greek government passed a bill granting a yearly subsidy to the school, a welcome addition to its scanty income. By 1937 the income from the various departments had reached the following levels:

| | |
|---|---|
| Dairy | $7,973.75 |
| Poultry | 859.00 |
| Pig | 1,918.00 |

---

[8] Both quotes are from the "American Milk Wagon Too Much for Saloniki," *New York Times*, 30 September 1934, 1.

[9] James Hugh Keeley, Jr., to the author, Winter Park FL, 14 March 1974, AFS Archives.

| | |
|---|---|
| Sheep | 6,355.20 |
| Horse & wagon | 1,380.00 |
| Total | $18,685.95[10] |

That same year, the treasurer reported a balanced budget for the first time since the onset of the depression. He wrote: "We are proud that we had the courage to draw upon our reserves to keep the School going during the lean years and we are happy that our current receipts are sufficient to meet our current expenses."[11]

While the construction of the physical plant had slowed during the 1930s, Charlie, the engineer, whose instinct was to construct, and his wife, Ann, turned their hands to another craft: that of building a loyal staff that would reflect the ideals of the founder. It would be composed of some educated men and others with particular skills, most of them Greek, to assure continuity and strengthen the institution's Hellenic nature. When we consider the survival of the school, we should focus on this hard core of devoted staff that Ann and Charlie began to groom in the 1920s and brought to culmination in the 1930s. Many were farm school graduates whose childhoods had been scarred by upheaval and parental loss; the school became their home and the House family their foster parents. Thus their devotion was filial. Others who gravitated to the school as adults were idealists and humanitarians who discovered there a worthwhile cause to which they could devote their lives. They merged with the House family to establish a tradition of behavior and philosophical outlook that continues to be transmitted to the younger staff and students today.

The anchor of the farm school team was Eleftherios Theoharides. An Asia Minor refugee, he joined the farm school staff as a teacher in 1923. Having graduated from Anatolia College in 1897 when it was in Merzifon in Asia Minor and from the University of Edinburgh's School of Theology in 1903, he was a man of liberal religious attitude with an unflagging devotion to the school. He taught English and Bible study. It was actually at his suggestion that the salaries were cut in half during the depression so that none of the staff

---

[10] Hollingsworth Wood, "Treasurer's Statement," 1937, AFS Archives. Unfortunately comparative figures for previous years do not exist.

[11] Ibid.

would have to be released. He died in 1958 and is buried in the farm school cemetery.[12]

Also of fundamental significance to farm school history is Theodoros Litsas. Upon terminating his work with Near East Relief, he joined the staff in 1930 to head the extension department as well as to supervise hygiene, recreation, and athletics. There was hardly an aspect of school life of which he was not an integral part: he recruited students, followed up graduates, entertained visitors, and directed school plays. After the untimely death in 1936 of Dimitris Georgiadis, who was office manager, he assumed Georgiadis's title. His dramatic role as caretaker when the Houses were removed to detention camp during the German occupation made him a central figure in the school's survival. His tragic death in a car accident in 1963 as he was returning from a visit to graduates cut him down in the fullness of life.[13]

Chief also among the loyal staff was Christo Starche, an orphan of Bulgarian parents who was brought to the school as a child. He studied at the farm school and after graduation was sent to Princeton through the generosity of Charlie's classmates. Entrance examinations had to be waived, for it was quite clear that he would be unable to pass them because of the culture gap and the language barrier. Once matriculated, he fulfilled Charlie's expectations and graduated as a civil engineer in 1929. He then returned to the farm school to head the industrial arts department. More than just a member of the staff, he was regarded by Ann and Charlie as their own child. When the prospect of war seemed positive in 1939, he moved to the United States with his American wife, Martha Parrott.[14] It was feared that foreign parents and lost birth records might cause him political difficulties in the event of war.

---

[12] Harry Theoharides, interview with author,Thessaloniki, 15 October 1971, written, AFS Archives. Eleftherios Theoharides'son, Harry, became head of the industrial department, and his two grandsons, Nikos and Eleftherios (Terry), although born in America, were brought up at the farm school.

[13] Chrysanthi Litsas, interview with author, Thessaloniki, 8 January 1973, written, AFS Archives.

[14] For an excellent character sketch of Martha Parrott and an explanation of her role in relief work in Greece and at the farm school, see Harold Allen, *Come Over into Macedonia* (New Brunswick: Rutgers University Press, 1943) 137–38.

A 1959 retiree, John Boudourouglou was also an Asia Minor refugee who came to Greece in 1922. After a seven-year period working as a teacher in orphanages under Near East Relief, he came to the farm school in 1931. He served as teacher, meteorologist, and finally as head of the boarding department. He was the only Greek on the farm school staff who became a Quaker. The story of his escape from Asia Minor and the fate of his family is material for an Odyssey.[15]

Another cornerstone of the staff was Dimitris Hadjis, who later became one of the best, if not *the* best, cattlemen in all of Greece. Brought to the school by an aunt in 1918, he was a refugee displaced by the upheaval of the First World War. After graduating in 1923, he became dairy supervisor. He felt that Ann had a special feeling for him and was responsible for recognizing his potential to become, with proper education, the farm manager. This all-important position, she felt, had to be held by a Greek, for it needed continuity as much as, if not more than, any other position.[16] As a result of the Houses' efforts, he received a scholarship to Cornell University's School of Agriculture, from which he graduated in 1932. He remained with the farm school until his retirement in 1970.

By retaining graduates as teachers, the Houses were in effect creating a dynasty of third generation heirs to Dr. John Henry House's philosophy. Exposed as they were to American education and mentality, the Greeks became bicultural in many ways, transmitting to their students the rich mix of the two cultures that Charlie consciously blended in the late 1920s and the 1930s.

With the assemblage of such a strong staff, the school was eager and able to contribute to the quickening community life of Greece. Charlie joined the Royal Agricultural Society (a semi-official national philanthropic society) and served on the managing committee of the Soup Kitchen of Thessaloniki, whose central kitchen distributed nourishment to 6,000 people. He was also a member of the governing board of the Touring Club of Greece, and the only foreigner to sit on

---

[15] John Boudourouglou, interview with author, Athens, 1 April 1975, written, AFS Archives.

[16] Dimitris Hadjis, interview with author, Thessaloniki, 8 September 1972, and Dimitris Hadjis, biographical note to the author, 1 November 1974, written, AFS Archives.

the city council of Thessaloniki. One of the staff volunteered to act as editor of an official church publication; another worked with the Society for the Prevention of Cruelty to Animals; a third acted as liaison between the School and the Aristotle University of Thessaloniki.

In 1932 the Academy of Athens, the country's scientific and scholarly body established in 1926 as a continuation of Plato's Academy, conferred on the school its Silver Medal for Virtue and Self-Sacrifice for providing agricultural education to rural youth. That same year, Charlie was decorated with the Gold Cross of the Order of the Savior for his participation, with the Patriotic Foundation for Social Welfare and Aid, known as PIKPA, in the construction of a fresh air home for children in the neighboring seaside village of Aghia Triada.

The economic problems of Greece in the 1930s were compounded by internal political tensions that finally led to the establishment of a dictatorship, a form of government widely known in Europe during the 1930s. On 4 August 1936 Prime Minister Ioannis Metaxas persuaded King George II to sign two royal decrees establishing a dictatorship. Metaxas called his dictatorship "The Regime of the 4th of August," a somewhat awkward and pretentious title that he concocted to give gravity to his government and the "civilization" he planned to create. The "4th of August" was actually a dyarchy of king and dictator in which the king played a comparatively passive role. It lasted until Metaxas's death in January 1941 during the war against the Italians.

The philosophy of Ioannis Metaxas's "new state"—or "civilization" as he called it—was designed to cultivate certain moral conceptions that would reinforce national consciousness. The details of his regime are important in terms of farm school development as they converge with the latter in the area of education. Since the youth of the country had become Metaxas's pet concern, he directed educational authorities on every level to mold the minds of young Greeks.

The dictator's view of the role of youth was colored in much the same hue as Mussolini's and Hitler's, at least with respect to ideology, for he wanted to ensure that youth grew up as staunch nationalists. The difference, however, was that Metaxas's brand of nationalism was more restrained and humane than Hitler's or Mussolini's. It did not encompass grandiose schemes for conquest, nor was it designed to create a generation of brutal, swaggering automatons. Since Greek family life was so tightly knit, any attempt to loosen it would have caused a revolution. Even in its conceptual stage, the idea of political inculcation was considered by many Greeks to be a real threat to the family unit, and the inclusion of girls in the Metaxas youth program was met with stiff resistance.

The dictator expressed the opinion that the "state has the unquestionable right to raise youth for its own end.... Children must be ready to offer themselves with pride to the social collectivity that is the country. The children of Greek parents belong to Greece."[17] Organizations such as the Boy Scouts and the YMCA that were international, and implied resistance to Metaxas's exclusively national thrust, were finally dissolved while some foreign schools became so infiltrated with Metaxas supporters and so manipulated by the government that they were little more than centers for the regime's propaganda. The National Organization of Youth (EON) was the instrument through which the government worked to cause the eventual downfall of the Scouts and the YMCA. It became the mechanism that bound foreign schools to the indoctrination of youth. Built roughly along the same lines as the youth apparatus in other countries ruled by dictators, it was the culminating expression of Metaxas's plan for Greece's children. The EON was actually formed in 1937 when a band of "volunteers" assembled to greet Metaxas on a visit to the Peloponnesian city of Patras. A month later, a unit was formed in Athens and other cities. Immediately a special body of laws was created to define the functions, responsibilities, and privileges of its members.[18]

By 1938, over 200,000 young Greeks had joined the movement. They were given uniforms and military ranks; members participated

---

[17] Excerpt from a speech by Ioannis Metaxas as quoted in Lincoln MacVeagh (minister to Greece), Athens, 5 January 1939, to the Secretary of State (Hull), 868.00/1068, Department of State, National Archives. For a study of the relationship between King George II and Metaxas see Everett J. Marder, "The Second Reign of George II: His Role in Politics," *Southeastern Europe* 2/1 (1975): 53–69.

[18] Everett J. Marder, "The Regime of the 4th of August: The Dictatorship of John Metaxas" (master's thesis, University of Cincinnati, 1969) 76.

in parades and were allowed free entrance to movies and other places of entertainment. Holidays were declared in their honor; excursions were sponsored by the government, which had put large sums of money at EON's disposal: legislation was passed to free the organization from accounting for its funds. Lincoln MacVeagh, head of the United States' legation in Athens, gives a graphic account of the children's appearance: "On March 25th all Greece was swarming with youthful figures clad in attractive navy blue uniforms with white ties, spats and rakish 'overseas' caps bearing the insignia of the organization—the double axe, which strikingly resembles the Italian Fasces, encircled with laurel surmounted by a royal crown.... Effective looking truncheons dangled from their belts."[19]

Membership was drawn principally from school students. Squads made up of older youths given great liberty of action were sent into the schools in order to enlist recruits.

Through this organization foreign schools were badgered. For example, the director of the French School in Athens refused to allow the formation of a unit of the EON in his school and as a result was subjected to great difficulties by the government. Also, Athens College, led by an American president, represented an extreme case of an institution that bent to the will and whims of the dictator. Since Athens College had two full boards of directors, one in America and one in Greece, the government, by replacing members of the Greek board with Metaxas appointees, was in a position to exploit the school for its own purposes. On Tuesdays the students were marched off to church; one whole day was devoted to the youth movement. In addition, the department of education at the University of Athens sent two lecturers each week to talk to the students on nationalistic subjects. Since it had become an outright tool of the government, Athens College seriously considered closing.[20] World War II and the

[19] Lincoln MacVeagh, 4 April 1938, to Secretary of State (Hull) 868.00/1040, Department of State, National Archives.

[20] Homer Davis (president emeritus, Athens College), interview with author, Athens, 5 October 1971, written, AFS Archives. Also Homer Davis, "History of Athens College," chapter 7, pp. 1–32, unpublished manuscript, AFS Archives. Athens College was founded in 1925 by a generous group of public-minded Athenians who turned to Americans for leadership. At that time it educated boys from elementary through secondary school. Thousands of Greek children have

death of Metaxas in January 1941 solved that dilemma. Athens College did not close and even continued to function throughout the German occupation.

Metaxas resented any foreign operation that attracted gratitude from the Greek public. The YMCA was particularly vulnerable, since it had a board of directors in Athens who were, according to MacVeagh, "well known anti-Metaxists and the whole list of the organization is from the point of view of the present government tainted with Venizelism."[21] In an interview with the then national director of Greek YMCAs, Herbert P. Lansdale, Jr., Metaxas told him that he was opposed to all forms of international intercourse for youth except when sponsored by the Greek government.[22] Finally, in November 1939, the Greek government dissolved the YMCA and YWCA, the Boy Scouts and the Girl Scouts.

While the American Farm School could not avoid all the pressures exerted on the schools and other organizations during the Regime of the 4th of August, it did for a number of reasons escape the most extreme measures to which other institutions were subject. The fact that in 1918 the farm school had been placed under the ministry of agriculture rather than the ministry of education most assuredly separated it from the fate of other educational institutions. When Metaxas made himself the minister of education and religion in 1938, the schools felt the full brunt of his educational mania. The farm school was spared his personal supervision; it continued to grow in cooperation with the ministry of agriculture. In 1938 a new law gave authority to that ministry to assist with increased financial aid. In 1939 the ministry's contribution was augmented by one-third.[23]

---

attended and still attend this and other schools directed by foreigners but regulated by the Greek Ministry of Education.

[21] Minister Lincoln MacVeagh, Athens, 15 December 1938, to Secretary of State (Hull) 868.00/1065, Department of State, National Archives.

[22] Herbert P. Lansdale, Jr., "Summary of the Conversation between His Excellency Ioannis Metaxas and Herbert P. Lansdale, Jr. in the Office of the Prime Minister on Tuesday, 13 December 1938, 5–6:30," from unpublished document relating to the YMCA of Greece during the dictatorship of Ioannis Metaxas, AFS Archives.

[23] Charles Lucius House, "The Memorandum of the Director Dealing with the Statute of the American Farm School under the Ministry of Agriculture of the Kingdom of Greece," 18 March 1938. AFS Archives

The very structure of the board of trustees proved to be another vital asset in removing the farm school from the purview of the Greek dictator. In the case of Athens College, which had two full boards of directors, one in Greece and one in the United States, the government had found it expedient to replace some members to achieve a more compliant composition. The farm school, however, had only one board with all but three members residing in the United States. Policy making and fundraising—the real power base—were vested in the American board members, making it impossible for the Greek government to tamper with the direction of the board.

It should also be borne in mind that by the 1930s the farm school, because of its Hellenic orientation, was considered by many sectors of Greek society to be as much Greek as American. While some foreign schools appeared to be centers of foreign propaganda and were hostile to Greek efforts to regulate or supervise their programs, the farm school responded to Greek authority with a cooperative spirit.[24] This fact went far, no doubt, in defusing the hostile attitude that Metaxas exhibited toward other foreign institutions, and accounts for his toleration of the farm school.

Certainly another mitigating factor in relations between the farm school and the government was Charlie's personal friendship with the king. This formed a sort of protective ring around the school. Yet the monarchy's protective capacity was limited, as shown in the case of the Boy Scouts and the YMCA. Although Crown Prince Paul was head of the Boy Scouts and the monarchy in general supported both the Scouts and the YMCA, its influence had no leverage when the prime minister decided to do away with these organizations. In spite of the royal family's strong opposition to EON, it became the dominant organ of the regime. It is of some interest to note that royalty and key government officials have traditionally visited the school, but that Metaxas, although he was in Thessaloniki many times, never did so.[25]

---

[24] James Hugh Keeley, Jr., to the author, Winter Park FL, 12 March 1974, AFS Archives.

[25] Metaxas's daughter feels that the reason he did not visit the school was tied to his general principle of not attending receptions and to the probability that when he went to Thessaloniki his program was more than full. She also felt that there was merit to the conclusion that he did not go because he would not care to make a

Of equal if not greater importance to the farm school's security during the dictatorship was the cooperative relationship that it established with local authorities. For example, one official, the director general of Macedonia, appointed by Metaxas, was a retired army general and professed to be a "democrat." He accepted the appointment from Metaxas because he believed that in such a position he could make many changes both in administrative matters and in mentality. In the conduct of his own affairs he did indeed set an example that he hoped his subordinates would emulate. One of his first innovations was to dispense with his bodyguards and chauffeur-driven limousine, as well as with many of the other ostentatious trappings that surrounded his office. He also proclaimed that he would personally receive citizens who felt the need to see him, and he instructed his subordinates to set aside one day a week to do the same. He traveled about by public conveyance, stopping in here and there unannounced to observe various aspects of Macedonian life. He was attracted to the American Farm School and throughout his administration became its firm supporter.[26] In a small country such as Greece with a strong central government, connections with key officials that link the outlying provinces to the capital serve as lifelines.

Still, in order to adapt to the new conditions that the dictatorship created, the school had to make certain accommodations. The most important of these was the formation of a group of the EON on the school campus. Although the organization was supposed to be purely voluntary, in reality the boys were ordered to join. Farm school students, for the most part obedient, joined without offering any resistance. Teachers were assigned to aid in EON projects. It was considered more efficacious to cooperate with the government in the youth movement than to present objections that might provoke a drastic official reaction. Theodoros Litsas joined with the staff to aid the EON group, which had supplanted his Boy Scout troop, but he continued to teach the same principles that he had taught as Boy Scout leader. The boys marched in parades and went on outings but

personal appearance in a foreign-supported institution. Nana Foka-Metaxa, interview with author, Ekali, 3 April 1974, written, AFS Archives.

[26] Keeley, Jr., to the author, 12 March 1974, AFS Archives.

were spared the more forceful indoctrination program that ruined the Athens College curriculum.

One of the most formative elements of farm school life, its religious orientation, was modified by pressures from the Metaxas government. Dictator Metaxas emphasized the Greek Orthodox religion as a basic foundation of education. This emphasis caused the farm school to change the Protestant form of worship to Orthodox.

Metaxas regarded the Church as a unifying factor in the nation. He expressed his strong opinion to Herbert Lansdale: "The [Orthodox] Church determines the faith and dogma and says to the individual 'Believe'." Therefore there is no place for a religious organization outside of the Church."[27] In 1937, the government issued a circular meant to insure that all children throughout the country attended church every Sunday. To be sure, the farm school boys had been attending religious service every Sunday, but the liturgy was according to the Protestant format, although Greek religious instruction was concurrently part of the curriculum. Rather than send the children off campus to parish churches, a chapel was fitted out on the top floor of Princeton Hall in compliance with the government circular, and on 21 November, the bishop of Thessaloniki dedicated it to Aghios Ioannis Chrysostomos. The establishment of the Orthodox chapel and the celebration of the Greek religious holidays did in fact assert the Hellenic roots of the farm school boys. The need for an Orthodox church had been recognized long before by Charlie. As early as 1924, when he was present at a board meeting in New York, he suggested that "a Greek interested in religious life be appointed on the committee and that the government be invited to send a representative to meet unofficially with the committee."[28] In the early 1930s he had made some tentative plans for the actual construction of a church on the campus. Gennadios, the bishop of Thessaloniki, a liberal and tolerant man, had mentioned in a discussion with Charlie that he had heard that the farm school might build a church. The bishop advised Charlie that unless it was an Orthodox church there could be no construction at all. Charlie answered that he had no objection to the building of an Orthodox church on the school grounds. Unexpectedly, the bishop told Charlie

---

[27] Lansdale, "Summary of the Conversation."
[28] Minutes of the board of trustees, 23 May 1924, AFS Archives.

not to build the church until Father House had died, for the erection of a church dedicated to any religion but the Protestant on his own grounds would offend the founder.[29] Bishop Gennadios's response is indicative of the kind of relationship that the farm school had established. This bishop can also be reckoned as one of the forces that shielded the school from Metaxas's penetration.

Given the narrow nationalistic confines of the Regime of the 4th of August, it is amazing that the school and the government did not come into conflict. A close analysis of some government programs reveals that in many instances the American institution and the government actually shared common educational goals. For example, the school's annual report for 1936–1937, written by Charlie, underlined conspicuously and purposely one of the principal areas of concern that was common to both the school and the state. "One of the special efforts of the present government," he commented, "is to stimulate education and especially practical training such as we are giving."[30] In his speeches to teachers, Metaxas often insisted that he did not want children to be "dry carriers of knowledge." He instructed that they be taught "to use the material of knowledge toward constructive and productive work.[31] Metaxas was fighting the same battle in 1936 that John Henry House had waged decades before in the Balkans: that of educating the young, especially the rural classes, along practical lines. It was in fact one of Charlie's enduring gifts that he could readily identify government programs that could be implemented by the school. Potentially dangerous conflicts were often defused by this talent. Theodoros Litsas, too, had the knack of detecting areas of irritation between government or community in the most sensitive situations and to work at them until some compromise, adjustment, or redefinition was found to satisfy the interests of all parties involved. Both men were consummate diplomats in a world where ignoring opportunities and petty frustrations could result in the sweeping away of all that had been carefully built since 1904.

---

[29] Ann Kellogg House, tape to the author, 20 September 1972, AFS Archives.

[30] "Thessalonika Agricultural and Industrial Institute, Annual Report, 1936–1937," AFS Archives.

[31] "*Lógos ekphonithís pros tous ekpaideftikoús leitourgoús*" ("Speech to the Teaching Personnel") in Ioannis Metaxas, *Lógoi kai skepsis* (*Words and Thoughts*), 2 vols. (Athens: IKARON, 1969) 1:96–99.

Another special concern that the school shared with the regime was community responsibility. Metaxas hoped that by stimulating a sense of mutual obligation he could in turn boost national consciousness. The idea of interrelated loyalties would make the Greek nation like "a guitar with strings stretched to such a fine point that if the slightest breath were to strike it, it would emit feelings, ideals, enthusiasm."[32] Both Metaxas and the farm school understood the intense individualism of the Greek peasant. After 400 years of Turkish rule, the man in the provinces had developed a distrust of national government. For centuries, the moral obligations of the Greek people had been confined almost exclusively to the family; unrelated persons tended to view each other with distrust or outright hostility. Any form of economic or civic cooperation was constricted by these attitudes. Thus the farm school's emphasis on service to the community at large and its program for teaching boys to be community leaders was a unique feature appreciated by the Metaxas government.

Another professed goal of the Regime of the 4th of August was to improve the lot of the peasant. With a population of just over seven million in 1936, Greece registered 1.5 million souls suffering from malaria and 250,000 from tuberculosis, of which 11,000 died annually. Life expectancy was only thirty-four years compared, for instance, to sixty-five in the United States, sixty in Germany, and forty-six in Poland. Some 60 percent of the people lived on the land in 1936, with a ratio of about 210 persons per square mile of cultivatable land.[33] The only possible way of improving the standard of living in rural areas was to increase the productivity of the land. The agricultural policy of the Metaxas régime was based on a plan of intense cultivation by the introduction of mechanization, improved seed, and modern scientific methods of farming. The farm school, with thirty years of continuous experience in wrestling with the land, served as an important boost to the government's agricultural

---

[32] Nikolaos D. Koumaros, "*I perí krátous antilípsis tis metabolés tis 4^is ávgoustou*" ("The View of the Changes of the 4th of August Regarding the State") *To Néon Krátos*, 18 February 1939, as quoted in Marder, "The Regime of the 4th of August," 71.

[33] *British Survey Handbook*, Greece, comp. Kathleen Gibbard (London: Cambridge University Press, 1944) 72.

program. Together with its emphasis on agriculture, the government instituted a rural health and housing program in an effort to alleviate the misery of thousands of refugees who were still eking out a pitiful existence in hovels.

The school's healthy contribution to many sectors of Greek life was known to the government. Besides graduating between twenty-five and thirty boys each year, it spent substantial sums of money in the country. By the late 1930s it had built from foreign contributions a relatively model plant while thousands of dollars flowed through the Farm School into the Greek treasury each year, a boon to Greece, since a lack of foreign exchange was one of the country's most serious economic problems. By giving steady employment to about ninety Greek citizens, the American institution was a source of income in an a region where unemployment was a chronic problem.[34]

In 1939 a new Greek law called for private agricultural institutions to submit a detailed explanation of every phase of their organization to the ministry of agriculture. Until the Metaxas years the American Farm School had developed almost entirely independently of government supervision; this was the first instance when every detail was put under government scrutiny. The report was considered so important that Ann and Charlie returned to New York to discuss its draft with the board of trustees. The government reacted favorably to the report and the continued operation of the school. When a new law renewed the school's legal status, doubtlessly all of the factors mentioned above had been taken into consideration, convincing the government to give its categorical approval to the thirty-three year old Thessalonica Agricultural and Industrial Institute.[35]

In 1935 Dr. House celebrated his ninetieth birthday. The list of those who sent congratulatory letters from all over the world—social, religious, and welfare organizations as well as statesmen, diplomats, and government officials—attests to how far his fame as humanitarian and educator had spread. His isolated but green pasture in Macedonia, had come to the world's attention. *Time* magazine in its coverage of the event, wrote:

---

[34] Charles House, "Memorandum," 1938.

[35] "Government Gazette of the Kingdom of Greece," No. 140/A, 4 April 1939, Vivliothiki Voulis, Athens

By slow chuffing train from Athens U.S. Minister MacVeagh and a quorum of the Greek cabinet traveled up this week to the northern seaport of Salonika, base of Allied operations during the war. Salonika was shelled again during the abortive Venizelist revolt last March. This time, however, diplomats and statesmen were going north on a more peaceful mission—to honor one of the most permanent institutions in the Balkans, bearded little old John Henry House.[36]

The magazine recalled that, when Dr. House came to the Balkans in 1872, Bulgaria was still a Turkish province and the Cretans had just undertaken an unsuccessful three-year revolt to free themselves from the Ottoman Empire. In addition: "A curly-haired moppet named Eleftherios Venizelos was just 8 years old.... He lived to see the complete independence of Serbia, Rumania and Montenegro; saw a Hollenzollern prince proclaimed Rumania's first king; later saw the Duke of Edinburgh's daughter Marie started on her way to becoming that country's most famed Queen. The first Balkan War and second came and went followed by the World War."[37]

A photo study of Dr. House taken at this time shows him clothed in an academic gown, his skin transparent with age, his bones showing in his hands. The years of outdoor work had not hardened his features; his exhausting trips through the Balkan wilderness had not fatigued him; the length of his years had not made him dull. He seems in the picture to have a quiet strength, a gentle dignity, and a composure that the years had not diminished.

On 19 April 1936, Easter morning, a month short of his ninety-first birthday, Dr. John Henry House died from coronary thrombosis. The Greek government passed a resolution that he be given a state funeral and that the director-general of Macedonia act as official representative of the Greek government and lay a wreath on the coffin. Because he had been decorated with the Silver Cross of the Order of the Savior and the Silver Cross of the Order of the Phoenix, a Greek military detachment was sent to render honor at the funeral. The House family requested Consul General James Hugh Keeley, Jr.,

---

[36] "Greece Farm School," *Time* 25/23 (10 June 1935): 21–22.
[37] Ibid.

to inform the Greek authorities that since the funeral would take place at the school and Dr. House would be buried in the school cemetery, no expenses would be incurred. The family requested that friends, instead of sending wreaths, contribute to a scholarship fund in Dr. House's memory. The request found favor in the government, which donated to the fund.[38]

Through the years, his simple grave has become the orientation point around which scores of other graves have been placed in the cypress-filled cemetery. The plain marble slab bears the epitaph composed by Charlie: "John Henry House/ An adoring servant of God/ A faithful apostle of Christ/ A practical friend of man."

Mother House and Ruth left for the United States in 1937, not realizing that a world war would preclude their return. Mother House died in 1947 in New York at the age of ninety-seven, all her faculties still intact; Ruth died in 1968. Father House's death and Mother House's departure from Greece with Ruth marked for the school the end of an era.

From mid-decade the international situation had been menacing. The indications that the winds of war were blowing toward the Balkans had been strong for a long time. Italy's aggression against Greece when the former took possession of Corfu in 1923 was only the first step in many that Mussolini took before he invaded Albania in April 1939 and then Greece in October 1940.

The Italian invasion of Albania was viewed as a prelude to the invasion of Greece. Tensions in Greece grew as Mussolini embarked on a policy of increasing hostility and aggression. Large units of Italian troops massed on the Greek-Albanian border; Italian overflights penetrated Greek territory; Greek ships were bombed; the Greek diplomatic staff in Rome was harassed; Italian propaganda programs broadcast lies to stir up popular indignation against the Greeks. The Greek government had maintained a policy of strict neutrality as Axis aggressions multiplied. The signing of mutual defense pacts with Balkan neighbors, the strengthening of the military, and an extensive public works program were all designed to meet the threat.

---

[38] "Report of Acting Director Georgiadis to President of the Board of Trustees re: Last Illness and Death of John Henry House," Thessaloniki, 19 April 1936, AFS Archives.

Charlie and Ann were in the United States in 1939 when on 1 September the Germans marched into Poland. They scrambled to get passage back to Europe on an American export freighter. Finally on 14 September, two weeks after the outbreak of war, they sailed on the *SS Executive* through the mine-infested waters: "We were glued to the radio! Russia, Germany, France, Poland and Great Britain were all at war. Parts of the Mediterranean had been mined and we were ordered to have our bags packed for the lifeboats and for several nights I slept in my bathing suit! At Gibraltar we formed a convoy of 8 tankers and 4 gunboats."[39]

The Houses arrived at the farm school on 9 October. Although many Greeks had been mobilized, none of the school staff had been called. Life on the campus was still continuing at a normal pace. However, preparations for civil defense began immediately. Theodoros Litsas was in charge of air raid precautions. He had a thousand sandbags filled and stored in the basement of Princeton Hall. It turned out that the fear of air raids in Thessaloniki was fully justified. Since Princeton Hall was the most solidly constructed of all the buildings and the most centrally located, it was designated as an air raid shelter. The supervisors' room was fitted out as an infirmary with cots, stretchers, and first aid supplies.

Shortages, such as occurred during the previous wars, were beginning to appear. Greece, which always had to import raw materials as well as manufactured items, was hit by a fuel shortage. Coal was very expensive and almost unobtainable. Sugar and coffee were increasingly hard to find and such simple necessities as thread were disappearing from the market. Charlie built an oven on the school grounds so that in an emergency the school would be able to bake its own bread.

By 1939 the departure of the foreign community from Greece seemed ominous. Consul General Keeley, who had been renting Hastings House, left on home leave with his family. At the outbreak of war, the US Department of State reassigned him so that his household effects had to be stored at the farm school for the duration of the war. Christo Starche, whose papers had always been a problem since he was orphaned in the First World War, felt it would be safer to

---

[39] Ann Kellogg House, "Memoirs," 160, AFS Archives.

take his American wife and two children to the United States. Other foreigners were leaving one by one to return to their homelands.

By April 1940, when the first blackout practice was held in the Thessaloniki area, nerves were beginning to wear thin. Electricity was cut and inhabitants were obliged to hang blankets in their windows. News from Denmark and Norway intensified the foreboding. "We are trying to live as normally as possible under the circumstances, but everyone is extremely anxious," wrote Ann House to her family.[40]

In June 1940, four months before the Italians invaded Greece, an American ship, the *Exmore*, put into the port of Thessaloniki. Before it left, Ann and Charlie loaded it with Mother House's dining room furniture and other treasured pieces owned by the family. The *Exmore* was the last American vessel to leave the port until after the war. The Houses watched it depart. Ann wrote in her memoirs, "We had the sudden realization that we were saying farewell to our last visible link with our America for perhaps many years."[41]

But there was no thought given to their return to the United States. The farm school was their home, their love, their responsibility. Only arrest and incarceration by the Germans would separate them from the school.

---

[40] Ann Kellogg House, "Memoirs," 164.
[41] Ann Kellogg House, "Memoirs," 165.

# CHAPTER 11

# WORLD WAR II:
# INVASION AND OCCUPATION

One of Winston Churchill's most trenchant observations in his volumes on World War II is the passage in which he describes the vivacious and irrepressible Greek spirit.

> The Greeks rival the Jews in being the most politically minded race in the world... [Their] stormy and endless struggle for life stretches back to the fountain springs of human thought. No other two races have set such a mark upon the world. Both have shown a capacity for survival in spite of unending perils and sufferings from external oppressors, matched only by their own ceaseless quarrels and convulsions. They have survived in spite of all that the world could do against them and all they could do against themselves.[1]

The Greeks' "capacity for survival" at the farm school during World War II hinged on a keen perception of reality, enormous flexibility, nimbleness of mind, and the simple desire to endure. Boundless loyalty to the farm school transcended personal issues. Outside forces that were intolerably hostile drove the staff as a group to their keenest sense of community.

The school's World War II experience divides itself into two periods: first, the invasion by the Italians in October 1940, followed

---

[1] Winston Churchill, *Closing the Ring*, vol. 5 of *The Second World War* (Cambridge MA: Houghton Mifflin Company: 1951) 532.

by the German invasion in April 1941; and second, the German occupation (1941–1944). The war presented a totally new set of circumstances to which the school's staff responded with an ingrained pattern of behavior that was the result of forty years of conditioning. The staff's reaction stands not only as a triumph of Greek endurance but also as a monument to the House family ethos.

Directing an educational institution is very often like leading a nation—each is an exercise in politics and statecraft. Ann and Charlie knew how to attain leadership and hold it. Like all good leaders they understood how to extract loyalty from their subordinates. With keen perception, they directed this loyalty not to themselves but to the institution. When they were deported by the Germans, the staff was thus able to continue functioning as a cohesive community.

The wartime conditions abruptly altered the school's entire purpose and structure. With the invasion of the Italians on 28 October 1940, almost the entire staff was mobilized immediately, along with the older students. Thus, within a day, ceasing to be an educational institution, the school shifted its operation to production of farm produce and items that could be made in its workshops. Unexpectedly but logically, it became a hostel and shelter for hundreds of refugees fleeing from the Italian bombs that rained down upon Thessaloniki from the first days of the war. Later, with the German invasion of Greece in April 1941, the declaration of war between Germany and the United States in December 1941, and the deportation of Ann and Charlie on 14 November 1942, the communication flow between America and the school was stemmed. When the dark night of German occupation descended on Greece in the spring of 1941, it descended alike on the school. The darkness into which Greece was plunged was a long night of horror. The great *pína* (starvation) of the winter of 1941–1942 can hold its own on the grotesque scale of human suffering engendered anywhere by German occupation. In addition, throughout the whole country entire villages were burned to cinders, often with inhabitants locked inside the flaming buildings, hostages were shot, and the conditions of life were made unbearable through the systematic destruction of all human necessities.

World War II began for the Greeks on 28 October 1940 at 3:00 A.M. That morning Italian Minister Grazzi appeared at the door of

Metaxas's home in Kifissia, a fashionable Athenian suburb. He brought with him an ultimatum demanding that Italy be allowed to occupy "a number of strategic points in Greek territory."[2] "Mr. Minister," the seventy-year-old dictator is said to have replied, "*Ohi*" (no). With this defiant refusal to allow the Italians gratuitous entrance into Greece, Metaxas, the dictator of a small country with limited status, became at once a national and international hero. His people united behind him and thus Greece became the first European country besides Britain to oppose the Axis with effective results.

The invasion undertaken by Mussolini with great optimism and launched from Italian-occupied Albania ended in the Italians' humiliating defeat. Fighting alone on the ground with air cover provided by the British, the Greeks stemmed the offensive thrust of the Italian armies on 9–11 November 1940 in the last phase of the battle of Pindos. By the end of that month, the Greeks had pushed the Italian line back into Albania, capturing the principal southern Albanian towns of Koritsa, Aghia Saranda, and Argyrokastro. By the middle of January 1941, the Greek army had taken Klissoura and was pushing northward. The Italians at this point were able to stabilize their line. In spite of their vigorous counteroffensive, personally observed by Mussolini, they could not dislodge the Greeks from the line they had established in Albania. The Albanian campaign had reached a stalemate. After four months of fighting, the Greeks had occupied the southern third of Albania and had taken thousands of Italian prisoners. Both Greek and Italian soldiers settled down for the winter of 1941 on the frozen wastes of the Albanian mountains, where young men on both sides suffered more from the agonies of hunger and frost than from the duress of combat.

The Italians had been seriously wrong in their judgment of the Greek will to resist. At a meeting of the Grand Council on 15 October, 1940 when the Duce asked his advisers about the morale of the Greek people, Count Ciano answered: "There is a wide rift between the people and their plutocratic leaders who support the British cause and are back of whatever spirit of resistance there is in the country. The

---

[2] *Diplomatic Documents: Italy's Aggression Against Greece* (Athens: Royal Ministry for Foreign Affairs, 1940) 133. Document 178 gives the full text of the ultimatum.

leaders form a small and wealthy minority while the people in general are indifferent to everything including our invasion."[3]

Just the opposite was true. Intensely patriotic, the Greeks coalesced against this external force. The then crown princess of Greece, Frederica, alluding to both the internal squabbling and the splendid quality of her people's resistance, stated: "I saw them at their best, and if their second best is worse than anybody else's their best is quite unsurpassable."[4] Thus the Greek will to resist and the Italian lack of military finesse cost the Italian Duce his anticipated victory.

On the morning of the Italian invasion, the farm school was in turmoil. Workers and staff who were hurrying to be off to the front clamored for their pay. They hastily packed small bags of food and other necessities and started out to the mobilization centers accompanied by some of the older students who were of draft age. Dimitris Hadjis was among the first to leave, along with about thirty others who were in the first call. The boys who could get home were sent off immediately while others who lived at a distance were kept at school until transportation could be arranged. Since all vehicles and beasts of burden throughout the country had been requisitioned by the government, the problem of getting the youngsters home was nearly insoluble. One boy had to walk for a week to reach his village. Meanwhile the depleted staff, consisting of older men, had to manage the dairy, machine shop, poultry, pigs, livestock, and farm fields. Yorgos Pseftakis, one of the four orphan boys who stayed at the school, took charge of the electric light plant and did a man's job even though he was only fifteen years old.[5] Years later he joined the staff.

On 29 October 1940, the day after the declaration of war, all schools were closed by government directive. For the first time since the initial boys arrived in 1904, the doors of the American Farm School swung shut to the rural youth of Greece, not to open again

---

[3] Mario Cervi, *Storia della guerra di Grecia* (Milano: Sugar Editore, 1965) appendix 41 ("Verbale della riumone tenuta nella salla di lavoro del Duce a Palazzo Venezo, 15 Ottobre 1940, ora 11"), 419.

[4] Queen Frederica of the Hellenes, *A Measure of Understanding* (London: Macmillan, 1971) 27. Frederica became queen in 1947.

[5] Ann Kellogg House, "Memorandum: Experience in Europe During the War, 1940–1944," Thessaloniki, 28 September 1944.

until 1 November 1945. Most other schools were converted into hospitals or storage facilities for military supplies, or were set aside for anticipated emergencies. The farm school's infirmary was requisitioned by the Greek army as a hospital; the doctor and his family were given quarters on campus.

At the beginning of the war, Charlie approached the Greek government about the status of the school, asking the authorities not to requisition it. He proposed that the school's mission be redirected to help in the war effort. In addition, he reported to the government officials in December 1940 that there were twenty-five families sheltered at the school; the heads of six were serving in the army, and four of the wives were pregnant with no place to go. Among these families were fifty children plus some students who had not been able to return home or were orphans. Aside from these compassionate individual circumstances, Charlie emphasized that the school was a flexible, varied operation that could remain productive even under the new situation. Three hundred and fifty acres of land were there to cultivate, 56 head of cattle could continue to breed and give milk, 40 pigs, and 100 sheep, and some hens would serve as an important food source when other stocks diminished. The Greek government was in total agreement; it allowed the institution to proceed with its farm and industrial enterprises and even to retain its draft animals, farm machinery, and truck so that food production and industrial output should not be curtailed.[6]

Almost immediately the Italians sent their bombers over Thessaloniki. The city was struck for the first time on 31 October 1940. In this raid direct hits were scored on the post office, the Mediterranean Hotel, and a clinic, frightening the people, who fled the city. It was with this initial raid that the farm school was turned into a shelter for frightened, dazed refugees, most of them relatives or friends of the staff, as they streamed out of the city. On 1 November there were five more raids, causing even more refugees to crowd into the school. The bomb shelter that had been constructed to hold only farm school personnel was packed beyond hygienic capacity. Another shelter, actually a dugout under an old Roman aqueduct that had been discovered on the school's fields years before, was provided for the

---

[6] Ann Kellogg House, "Memoirs," 169, AFS Archives.

farm workers who could not reach Princeton Hall in the event of a raid.

Ann House and Chrysanthi Litsas, calling the children of the staff together, assisted them in tearing strips of newspaper and pasting them crisscross on the windows to prevent shattering. Flour sacks, rugs, and blankets were secured in place at the windows to insure that no light leaked into the night.

If anyone doubted the necessity of securing the school against a direct hit, all doubts were erased when, on 2 November a small bomb exploded on the campus close to the water reservoir. Fortunately no one was hurt and no damage was done. It was left to the Germans to earn the distinction of being the first to destroy the school's property, which had survived four decades of almost constant warfare without being damaged.

The next day it was decided that the consul general of the United States, John Johnson, should move to the farm school with his staff since the consulate was located in an especially vulnerable spot on the waterfront. Ann told of the hectic task of clearing Hastings House of former consul James Hugh Keeley, Jr.'s, belongings to make way for the consular staff: "Every few minutes the siren would blow and we'd have to lock up, run, and spend an hour or so over at Princeton Hall in the shelter, then open the house, do some more packing and moving, then go to the shelter again for a number of hours."[7]

Ann's own reaction of the Italian bombing was:

> If they had started right in on Monday when thousands of soldiers were crowding all the streets and thoroughfares coming into the city to be mobilized, there would have been some excuse, but they waited until there are practically no soldiers left, all having gone to the front miles away from Thessaloniki, and then bombed homes; a large number of the casualties are women and children and a large number of the houses ruined are in the poorest part of town.[8]

---

[7] Ibid., 167.

[8] Ann Kellogg House to her family, Thessaloniki, 4 November 1940, AFS Archives. This letter was of great interest for its eyewitness contents and was sent to the Department of State where it was initialed by officials in the Division of Near Eastern Affairs and Foreign Service Administration.

In spite of the havoc caused by raids, the remaining staff wives organized relief work on campus. The women met frequently to sew for the Red Cross. Fourteen sewing machines were gathered and twenty-six women set themselves to turning out hospital gowns and triangular bandages.

As the war gained momentum with the Greek advances, casualty reports indicated a huge death toll. Grim remarks such as "our casualties were so heavy that the doctors asked us to pray for snow to cover the corpses"[9] make one realize the horror of the times.

Snow did come early; unfortunately, the winter was uncommonly severe. By the middle of December the cold had settled in to stay. Ann recorded, "We are having a bitter winter and this morning it is snowing hard. My thoughts are constantly with our boys at the front. More and more of the soldiers are coming to the hospital with frozen feet."[10] Other sights in Thessaloniki that she witnessed demonstrated the toll that the victory was exacting from the youth of Greece: "So much of the fighting had been up in the mountains where the snow is deep and the soldiers' feet froze. They swelled up so that they had to take their shoes off and couldn't get them on again. So many have had to walk many miles with no protection and the feet of many had to be amputated."[11] There were civilian casualties from the bombing: "So many children and old people. It almost breaks one's heart," Ann wrote after visiting a hospital in Kalamaria.[12] The pacificist in Ann House cried out against the killing and maiming she was observing in Macedonia. In one letter she blurted out, "War really is Hell."[13]

Since such a large proportion of the population was trekking in the direction of the school, the Greek police set up a station on the grounds. Their office and sleeping quarters were installed in Chamberlain Hall. Their presence—along with the American consulate, Greek medical personnel, and the groups of refugees—filled the campus to overflowing.

---

[9] Ann Kellogg House, "Memoirs," 165.

[10] Ibid., 168.

[11] Ann Kellogg House to her family, Thessaloniki, 15 December 1940, AFS Archives.

[12] Ibid.

[13] Ibid.

On 30 November, the American Red Cross nominated Charlie to act as its representative in Greece in the position of executive officer. He felt that since the school had ceased functioning as an educational institution its responsibility now was to fulfill emergency services and to continue its farming activities as much as possible. Since Theodoros Litsas, who was past the draft age, had remained, Charlie concluded that the institution's now abbreviated mission could be handled by Litsas and other capable, devoted individuals. On 1 January 1941 Charlie appointed Litsas acting director with full legal power to conduct the official and financial business under a power of attorney that had been issued as early as 1938. Litsas was the man whom Charlie had been grooming since 1930 to take his place should the need arise. Thus until the American Red Cross office in Athens was shut in June 1941, Litsas managed the school's affairs during Charlie's frequent absences in Athens, at the front, and at various Red Cross distribution points throughout Greece.

The school continued its farming and industrial activities notwithstanding the lack of manpower and replacement parts for machinery. In spite of its depleted staff, it also rendered aid to neighboring farms by ploughing land belonging to fifty-two families who were unable to do the work themselves owing to wartime conditions. By January 1941, the industrial department had made 250 wire cages for the Red Cross to be placed over the legs of soldiers to keep the weight of the blankets off their frozen feet. It also produced sundry hospital items such as bandages, canes, crutches, and wheelchairs.

Most notable was the fact that the school's building program could continue under these dire circumstances. Chamberlain Hall, for which Charlie had obtained supplies and cement before the war, was extended; refugees who had been crammed into Princeton Hall were removed to Chamberlain. Princeton Hall was then turned into a Red Cross storage center for relief supplies distributed in Thrace and Macedonia. A huge red cross made of colored brick with a white lime background was laid in the center of the playground so that it could be visible to aircraft.

At the end of January, John Metaxas died of natural causes. From the beginning of the conflict with Italy, he had refused to have British troops support him on Greek soil, one of the reasons being that

Britain had relatively few troops to offer, another being his fear that its entry might provoke the Germans to invade Greece.

With Metaxas's death, the new prime minister, Alexandros Koryzis, and King George II reevaluated the more critical international scene and invited the British to send troops to fight in Greece. By the end of March, 31,000 British Commonwealth troops had arrived, the number finally rising to 57,000. The German invasion commenced on 6 April.[14]

Ann House remembered the first sight of these troops as she and Charlie made their grueling trip back to Thessaloniki from Red Cross headquarters in Athens: "I remember the 23rd of March, 1941.... The day before, a British plane had been brought down near Athens by the Italian airforce and quickly Greek girls completely covered it with flowers. On our drive we passed 326 British trucks and 19 cannons. The soldiers waved at us and shouted 'thumbs up'...young girls threw flowers...boys wove branches into nets covering their trucks to camouflage them."[15]

Most Greeks felt that if they defeated the Italians then war with Germany would be inevitable. Both the civilian and the military population dreaded this prospect. When the attack came it was indeed a nightmare, but "one that had to be faced for the cause of Greek honor and for the sake of liberty."[16] No one expected a victory over the Germans; yet at no time did the Greek people themselves want to give up the struggle, although defeatism and quisling activity were detected in the highest echelons of the Greek government and among some of the highest ranking officers in the military. Even the German high command complimented the Hellenic troops' tenacity and bravery in defending Macedonia at Rupel Pass, where they fought against overwhelming odds.

The German invasion started when Prince Erbach-Schönberg called on Prime Minister Koryzis on 6 April 1941 at 5:30 A.M. to declare that German troops were going to cross the frontier immediately. The provocation for this action, the German official

---

[14] Edgar O'Ballance, *The Greek Civil War, 1944–1949* (London: Farber and Farber, 1966) 40.

[15] Ann Kellogg House, "During the War," AFS Archives.

[16] The words in quotes are those of Floyd Spencer, ed., *War and Post War Greece* (Washington, DC: US Government Printing Office, 1952) 13–14.

explained, was the landing of British troops in Greece. Actually, most of the evidence supports the thesis that the real motive for the Nazi assault was Mussolini's failure to subdue the Greeks. But the arrival of British troops in Greece endangered the German right flank (i.e., the Balkans) for the projected "Operation Barbarossa," the code name for the German invasion of Russia.

The combined British and Greek defensive campaign was a withdrawal action from the start. The Nazis had planned to invade through Bulgaria and Yugoslavia. The Greek High Command falsely assumed (hope against hope) that the Yugoslavs would be able to defend the Macedonian passes leading from Yugoslavia into Greece and had therefore not established strong defense points in this most vulnerable area. It is also true that a large proportion of the Greek army was tied down in Albania, leaving few Greek troops to serve on the Yugoslav border. But the Yugoslavs could not hold against the elite, fully motorized German troops. The Germans smashed through the Yugoslav line, meeting a British force between Monastir in Yugoslavia and Florina in Greece (the Monastir Gap). The Germans broke through the British resistance at this point. Once they were through the Monastir Gap, their way from the north was clear to Thessaloniki. On 9 April, despite resistance from three Greek divisions in Eastern Macedonia and Thrace under the command of General Bakopoulos, the Germans entered Thessaloniki.[17]

George Weller, the last American correspondent to leave Thessaloniki, gave the following eyewitness account of the final day before the city fell into enemy hands: "At 6:00 p.m. the last train had departed from a tiny boxlike little station in Salonika carrying what it could from the huge packages of army stores which had been piled up on the platform. In the meantime the heaviest Greek motorized equipment went pouring through the main boulevards…. Departing soldiers tried to cheer the crowds by shaking their fists toward the north but the pedestrians merely stood and watched."[18] All the while, oil tanks were being blown up by Greek authorities to prevent them from falling into German hands. The burning fuel filled the sky with

[17] I. S. O. Playfair et al., *The Mediterranean and the Middle East* (London: Her Majesty's Stationary Office, 1954) 1:85–86.

[18] George Weller, "Strategic Retreat from Salonika—First Story of a Modern Anavasis," *Boston Evening Transcript*, 14 April 1941, 1.

columns of black smoke and flame. "By 6:00 p.m.," Weller observed, "it would have been natural to expect some such mass rush upon the few vessels at the waterfront as occurred at Smyrna and upon the occasion...of the burning of Salonika but none occurred."[19]

The city was being evacuated not only by the military but by the gendamerie and civil employees such as those of the telephone and railroad services. They all crowded aboard two big freighters in the bay of Thessaloniki, while the overflow boarded *caiques* and other diesel-powered ships. At 7:00 P.M., a new and "deeper note suddenly came...from the western marshes: bridges going heavenward."[20] These were the bridges spanning the Axios River, destroyed by the Greeks to impede the German advance. Thessalonians cognizant of the location of these last explosions understood them to be the final notes in a brilliant but tragic symphony. The war in Northern Greece was over. Another line of defense was being established by British Commonwealth and Greek troops toward the south—unfortunately one that would not hold either. By 8:00 A.M. on 8 April all except military communication with Athens had been cut. The next day Weller described his last moments in Thessaloniki before he headed toward Athens: "Occasional shots were audible and as the city fell completely silent, illuminated by brilliant flames, the distant reverberations of artillery began to be heard, volley after volley. Already German motorcyclists who were seen on Salonika streets Wednesday morning were nearing the city keeping compact formation and encountering no resistance."[21]

Weller said his goodbyes to Consul General Johnson, Vice Consul Edmund Gullion, and Charles House. When he asked the latter when he and Ann intended to leave Greece, Charlie answered, "I guess we'll stay. We've lots of people on our hands."[22]

While Weller was relaying his description of the last moments, Ann House was recording her thoughts as she sat at the farm school, composed but apprehensive: "The German Army marched into Macedonia this morning! We do not know where they are approaching Salonika! Our windows keep rattling with the explosion

---

[19] Ibid.
[20] Ibid.
[21] Ibid.
[22] Ibid.

of the Benzine at the Airdrome, which the Greek forces set on fire before they left. Except for the thuds, everything seems strangely quiet without the humming of the planes at the moment."[23] It is remarkable that Ann's correspondence during the entire span of the war seldom betrays any fear or anxiety. The lines that follow, although written in low key, are a rare expression of her forebodings: "Do you remember, Mother House, when you were the only American woman in Salonika? I guess I'm that way now and in all Macedonia, if the Riggs, Gardners and Meverette Smith succeeded in getting across the Vardar bridge before it is blown up."[24]

The departure of these Americans was the final opportunity for the Houses to leave ahead of the approaching Germans. But their decision to remain at the school had been made long before and all evidence leads to the conclusion that neither of them ever seriously entertained the idea of leaving. Charlie's official declaration on this matter to the board of trustees in America reads:

> The question naturally comes up whether Ann and I should follow the advice of the legation and return to the United States.... We have not given this suggestion serious consideration because first of all, while the Red Cross activities continued in Greece I was not free to leave; second, when they cease, opportunities for service and my responsibility for the school remain; and third, as long as we have responsibilities and opportunities we do not feel that we should leave.[25]

His concluding lines in this declaration strike closer to the heart of the matter: "The appreciation shown by all our Greek friends is

---

[23] Ann Kellogg House to her family, 8 April 1941, as quoted in Ann Kellogg House, "Memoirs," 175, AFS Archives.

[24] Ibid. By "American," Ann House meant Americans not of Greek descent, for there were many Americans of Greek descent who remained in Thessaloniki. Some of them were arrested and deported by the Germans when the United States and Germany went to war.

[25] Charles Lucius House, "Report," as cited in *Newsletter*, October 1941, AFS Archives.

very touching. We have cast our lot with these people and these difficult times may prove our greatest opportunity to serve."[26]

An additional point—that the years of domesticity and responsibility had not dulled his capacity for adventure—no doubt played a role in his decision. By electing to stand with her husband, Ann, devoted to him and the school, made the first of a series of courageous decisions that strengthened the spine of the school not only during World War II but also during the civil war that was to follow. Of the many American educators and those involved in philanthropic services in Greece, the Houses were among the few who remained into the period of German occupation.

The occupation of the farm school by the Germans began on 19 April 1941 with the commandeering of Princeton Hall, Hastings House, and Bayard Dodge House only. The official requisitioning of the entire school was not made until 17 January 1942, after Germany had declared war on the United States and had arrested Ann and Charlie.

From 19 April 1941 until 17 January 1942, therefore, the farm enterprises and the entire plant, with the exception of the principal housing quarters, remained under Charlie's direction. He was supported by his operating staff, most of whom had returned on 27 April when the Germans entered Athens. After the director's arrest, the control of the buildings was taken over by the German army, while the livestock and agricultural sectors came under the jurisdiction of the German air force. The various units billeted on campus (Austrian troops being mixed with German) were as follows: "19 April 1941 anti-aircraft battalion of 250[;] 22 April 1941 Replaced by air force[;] May 1941 Headquarters of army intelligence unit remained until evacuation (included Signal Corps elements)."[27]

Immediately after the Germans' arrival in April, Consul General Johnson had addressed himself to the German authorities over the problem of their commandeering of American property. He demanded of the Germans, to little avail, that the requisitioning of American property be accomplished in accordance with an official requisition

---

[26] Ibid.

[27] "American Farm School during World War II: Chronological Statement of Principal Events Affecting the School," AFS Archives. This document was probably written by Charles House after the war.

procedure that required a receipt and gave promise of compensation.[28] Even though the United States was not at war with Germany at that point, Johnson reported that his efforts to halt German occupation of Anatolia College and the farm school failed. Although the German consul general was cooperative, giving repeated promises that the schools' premises would not be occupied, it was a futile gesture on his part since the occupation policies of the Germans were fixed not by consular but by military authorities. The American consul general's job of protecting American property was made all but impossible by the fact that the different military units occupying Thessaloniki, at least in the first days, were acting independently of each other and that central headquarters had not gained control over roving detachments who demanded billets at the school.[29]

Almost immediately after their arrival, the German authorities ordered the American consulate to remove itself from the school and reestablish itself in town. The consul general complied; the German military did not wish the American official to remain situated in the midst of their operations.

By the middle of May, although the school buildings had been occupied by three different military units, no formal receipt or requisition slip had been presented by the Germans. The earliest requisition form held by the American Farm School Archives bears the date of 16–30 November 1941. The forms were issued thereafter every two weeks. They detail the time period of requisition and list the ranks and numbers of men in each rank who were billeted at the school, but do not designate or number the buildings occupied. Besides this information, an instructional paragraph directed the owner of the requisitioned property to apply to the Greek tax office for compensation. Charlie, following the American consul general's advice, instructed Litsas as of 25 September 1941 to charge the occupation forces for rent, lights, water, and requisitioned space.

---

[28] Consul General John Johnson to secretary of state, 26 May 1941. 740.0011 European War 1939/1941. National Archives.
[29] Ibid.

There is no evidence to suggest that either the German authorities or the Greek tax office ever made any payments to the school.[30]

The morning the German army appeared in Thessaloniki, Charlie called on a German general and another of the commanding officers. They assured him of their cooperation, first with a general statement, then with particular reference to the Red Cross relief work, which they both agreed must be continued. But whatever cooperation they offered apparently did not include putting the farm school off limits to commandeering forces.

The first troops to demand billets came on 17 April. They were a small group composed of a sergeant and several soldiers who entered Ann and Charlie's home bearing rifles and demanding quarters. Charlie's temper flared at the sight of the weapons in the sanctity of his own home. He glowered at them, "You can shoot me, but you can't order me around like this."[31] The soldiers apparently lacked the conviction to press the issue. Backing down, they left.

On 19 April the first occupiers, whom Consul General Johnson had mentioned in his report, arrived at the school and demanded quarters for 250 officers and men to be granted within the hour. When the Germans arrived in such force with officers in command, it seemed the better part of valor to make some attempt to enter into dialogue with them. Theodoros Litsas, with his wife, Chrysanthi, acting as interpreter, met the ranking officer at the gate. As the farm school staff gathered around, the officer became more arrogant and asked for all the beds at the school. Litsas asked the crowd to disperse and invited the officer to his home for a quiet glass of ouzo. The German had lived in Athens for some time and spoke Greek. In these more quiet conditions, Litsas offered the German only the beds that were free, resisting giving over beds that were needed for farm school inhabitants. The officer agreed and the matter was settled in an amicable way. Thus the occupation of the school was initiated, but the means by which it occurred illustrates the modus vivendi that was to characterize the relationship between the Germans and the school staff for the duration of the occupation. This modus vivendi was based

---

[30] For example, see *Quartier Anweisung* ("Requisition Form"), 16 November 1941, AFS Archives. The farm school archives house an interesting collection of German documents concerning the occupation of the school.

[31] Ann Kellogg House, notes to the author, 18 July 1974, AFS Archives.

on mutual need. The Germans required the farm school both as a base for various operations and as a source of food supply. It was well within their interest to conduct themselves with a modicum of flexibility and a semblance of reasonableness. The farm school staff, by the same token, was determined to sustain itself as well as possible in the face of possible starvation and to protect the farm school from harm. Charlie stated that the school's policy was to "circumvent the German authority wherever it seemed contrary to the interests of the School or the Greek people, while remaining true to our Christian principles and the purposes of the school. We planned to make use of the authority of the Germans, where it could contribute to our objectives without relying on their force."[32] In sum, tolerable relations with the Germans became a fixed goal in farm school policy.

We would do well to remember Charlie and Ann House and their values. They were deeply religious pacifists; they considered war and violence an evil and tended to perceive all those who participated in the war on either side as victims of the evil more than as perpetrators. The idea that a whole nation, such as the Germans, might share a "collective guilt" was a concept untenable for the Houses. Neither was capable of a deep-seated hatred against anyone including the "enemy"; individual Germans were treated by them as individual human beings. It is true that on particular issues Charlie, who loved a good argument, would often get into long, heated discussions with the Germans. He was particularly vocal about their treatment of the Jews, which enraged him. But it must be admitted that his attitude toward the Germans at that time was, to say the least, unique. Although the United States was not at war with Germany, American sentiment had hardened against the Germans before the invasion of Greece. Some Americans who had escaped from Greece to America with the advance of the Germans were suspicious of the Houses' attitude toward the occupying forces. A case in point is the following incident related by Petros Stathatos, aide de camp to King George II. When the king went to the United States in 1942 to generate support for Greek war relief, he and Stathatos were invited to the White House. At dinner an American official turned to the king's aide and asked if what he had heard were true: that Charles House was

---

[32] *Newsletter*, July 1944, AFS Archives.

collaborating with the Germans. "Certainly not," roared Stathatos, "the Germans are collaborating with House."[33]

The following case serves to illustrate the "who is collaborating with whom" dilemma. During the extreme famine conditions of the winter of 1941–1942, every effort was made to expand the farm's productivity in order to supply farm school families. Vegetable gardens were planted in every available space, but such expansion demanded water in substantially increased amounts. Fuel to propel the irrigation pump was nearly unobtainable. It came to the staff's attention that, some 60 miles away, a large group of hungry woodcutters was willing to trade its wood for wheat, but not with the Germans. The Germans at the same time lacked firewood for house fuel and were willing to pay the farm school for it if the school could obtain this scarce commodity. The school proposed that the Germans provide the transportation (the rarest commodity of all) to haul the 300 tons of firewood from the woodcutters. A deal was struck. The Germans provided the transportation, the farm school staff gave food supplies to the woodcutters in exchange for wood, the Germans received 60 tons of firewood in exchange for transportation, while the farm school received the remainder of the wood allotment to run its pump to irrigate the vegetable gardens.[34] The most important personality in working out a modus vivendi with the occupiers was Theodoros Litsas. He was also a pacifist, having accepted most of the Quaker philosophy even though he remained a member of the Greek Orthodox Church all his life. In spite of the horrors he had witnessed during the Anatolia disaster, he was an incurable optimist and believed that human beings can create a better world. He was infinitely patient with others, for he was convinced that an approach could be devised to sway any difficult person. Given all these strengths, he became a superb diplomat and there is no doubt that the school and its staff owe much to the indefatigable and courageous efforts of Theodoros Litsas during this difficult period.

The staff had returned with the collapse of Greece. Dimitris Hadjis, the first to be mobilized, walked steadily back day and night

---

[33] Petros Stathatos, interview with author, Athens, 15 February 1972, written, AFS Archives.

[34] Chrysanthi Litsas, interview with author, Thessaloniki, 8 January 1973, written, AFS Archives.

with no food or water, covering 160 miles. John Boudourouglou, who had left for Athens ahead of the invading Germans, returned and took on as one of his principal chores the recordings of the meteorological station. Pericles Papadopoulos, a graduate, worked the farm fields while Argerios Dirmertzis worked the dairy. Iraklis Iasonidis, a Pontian Greek who had come to the school as a refugee student, acted as messenger, driver, and "man Friday." Eleftherios Theoharides, who had approached retirement age, also remained, for a short time running a school for the small children living on campus.

As the Germans settled into Macedonia, the effects of their occupation began to take hold. All over Greece industries were suppressed or exploited by the occupiers for their own needs. An intolerable charge was levied on the Greeks to pay the occupational costs. This taxation—combined with an issue of worthless paper money, German looting, and confiscation—wreaked havoc in every section of the country.

Since the drachma had lost its value and shortages were acute, real property became the medium of exchange. City dwellers seeking to exchange their possessions for food trudged past the farm school toward the villages laden with lamps, dishes, and pieces of furniture, some of it wheeled in baby carriages for lack of other conveyance. As the villages close to the city exhausted their food supplies and had no more to barter in exchange for the real property of the city people, the latter had to extend their range further into the countryside, their own stock of real property rapidly diminishing. On one occasion some of the farm school staff driving a horse and wagon into the city gave a ride to a woman and a little boy who were staggering back to the city after one such foray into the countryside. The child died in the cart before reaching the city. In Athens, people were dropping in the streets by the thousands from hunger and exhaustion as the process of bartering itself ceased to be viable. Although Thessalonians never suffered to the same extent, the situation was desperate. To leave the campus, where food was short but life was sustainable, was to enter a world of horror.

The Germans devised almost satanic methods to insure mass starvation of the Greek population. One of their most destructive acts was to divide the country into three occupation zones: German, Italian, and Bulgarian. In this system the Germans themselves held key points such as Thessaloniki, Athens, and Crete as well as retaining control of all communication networks. Thessaly, Epiros, Central Greece, the Peloponnesos, and many Aegean islands were given to the Italians. Bulgaria was awarded the historically coveted areas of Eastern Macedonia and Western Thrace. The inhabitants of one zone were refused passage into other zones. As no zone was self-sufficient, the small country's compartmentalization became a catastrophe for the Greeks.[35]

The farm school continued to produce food although under the most inhibiting conditions. Machinery broke down and was repaired by makeshift methods. No worn out parts were replaceable except by scavenging, or salvaging something from one machine to put into another. An indication of how elemental life had become is afforded by the following extract from Ann's diary. All through the war Ann jotted down a line or two almost every day so that she could be reminded in years to come of the most important happening of any given day. Her entry for 26 July 1942 recorded that on a walk through an old ammunition dump in the direction of neighboring Sedes, Charlie found forty-eight screws that he hoarded for future use. This was recorded as *the* event of the day.[36]

From the first, the farm school staff was put on a severely limited food ration by the Germans. They received a half-liter of milk a day for each child and a quarter-kilogram of bread per person. The staff was also paid a small salary for their labor but it was a token amount. John Boudourouglou recalls that on his daughter's birthday in 1943 he gave his month's wages for seventeen pieces of candy for the child's party.[37]

On 22 April 1941, Ann and Charlie's home was occupied by Germans but the Houses were permitted to retain their own bedroom and bath. At the end of that month, Charlie informed the farm school staff that he could no longer meet the payroll since money from the United States was not entering occupied Greece. Families were invited

---

[35] Konstandinos Doxiadis, *Thisíes tis Elládos, aitímata kai epanorthósis ston b' pangósmio pólemo* (*Sacrifices of Greece, Requests and Reconstruction in the Second World War*) (Athens: n.p., 1947) 13.

[36] Ann Kellogg House, "Memoirs," 190, AFS Archives.

[37] John Boudourouglou, interview with author, Athens, 4 September 1974, written, AFS Archives.

to stay in their homes but Charlie reminded them of the likelihood of requisition.

On 11 December 1941, Germany declared war on the United States. Within the hour the Germans arrested the Houses, taking them to Dulag 183 prison in Thessaloniki, a camp where British wounded were detained. Edward Howell, who later served (1970–1973) as the school's vice president for community relations, wrote about the camp: "For those of us Allied prisoners of war shivering in the snow and ice of Dulag 183 in Thessaloniki it was a hard Christmas. We were the last of some twenty thousand who had passed through here after the short-lived battles of Greece and Crete. We had all been wounded and had not been fit enough to be moved earlier. An emaciated collection of cripples, we found it hard to keep up our spirits."[38] He never forgot the Houses' entrance into the camp: "The man was tough, wiry, leather-faced with light blue eyes which always contained a twinkle. His wife must have been an exceptionally lovely girl. Always a radiant, smiling figure, she shone in our gloom like an angel of light."[39]

If ever the board of trustees had contributed to the survival of the school, certainly its sustained interest during the war years was part of that service. When Germany declared war on the United States, the board had questioned the need for a New York office. It appeared on the surface that the expense of such an office could no longer be justified. But the board chose to see beyond the bleakness of the moment and "felt it would be a great pity to lose contact with the fine group of friends which has been built up with such care and patience during the past ten years."[40] The decision to keep the New York office open was a wise one. Later, as the war clouds began to lift, this operation in New York became a rallying point and nerve-

[38] *Athens News*, 29 December 1953, AFS Archives.

[39] Ibid. Wing Commander Howell of the RAF had been in the Battle of Crete. Badly wounded, he was captured by the Germans and placed in Dulag 183. Shortly after meeting the Houses he escaped and made his way eastward through the mountains toward Turkey. As he passed through Macedonian and Thracian villages he was often aided and protected by graduates and friends of the American Farm School. This experience is described in Edward Howell, *Escape to Live* (London: Longmans Green & Co., 1958).

[40] Minutes, Board of Trustees, Finance Committee, New York, 21 December 1942, AFS Archives.

center for postwar reconstruction activities and plans. The farm school's success in picking up its former momentum and entering its postwar reconstruction phase was due in large measure to the determination of its board. The arrest of Ann and Charlie and their removal from the institution, with all that this implied, made the integrity of the board even more indispensable.

When the Houses were arrested, Theodoros Litsas as acting director assumed full responsibility of the farm school. Early in January 1942, following the Houses' removal, the Germans informed Litsas that the school would be taken over by the army. They forbade school personnel to leave and ordered all food products to be distributed by the military authorities. Until that time, while the farm was still being run by Charlie and the staff, only certain buildings had been commandeered by the German army. On 16 January, a group of officers made an inventory of all school land, machinery, supplies, and livestock. The authorities ordered that at the end of each month the staff furnish the Germans with the following:

1. A statement of receipts and expenditures
2. a list of salaries and wages
3. a list of milk produced and handed over to military authorities
4. a list of poultry produced and handed over to military authorities
5. a livestock inventory showing increase and decrease of stock, with explanation.[41]

It was plain that the Germans were serious about having tight control over the school. On 17 January the American Farm School was occupied by them in its entirety.

During this critical time, Ann and Charlie remained imprisoned. They were kept at Dulag 183 until 15 January 1942, when they were moved to a flat at 17 Velisariou Street used by the Germans as a holding center for American citizens before they were sent off to prison camps. Here they were under guard but allowed visitors. During this period Ann was permitted to go on frequent visits to the school

---

[41] Theodoros Litsas, "First Report," quoted in *Newsletter*, 1945, AFS Archives.

lasting a few hours each. Finally on 7 March 1942 both were released on restricted liberty to the jurisdiction of the German army commander at the school. With their return to the premises under these new conditions—as aliens in an occupied country, their own country being in a declared state of war with the occupiers—their position was no longer simply tenuous but outright dangerous.

The German air force had taken over the entire farming enterprise, while the army commandeered the housing units. The farm was placed under the control of the commander of the airport of Sedes. The school buildings, still under the army's jurisdiction, were used as the headquarters of a signal corps regiment. Also, since May 1941 the headquarters of the army intelligence service had been lodged on the campus. The table built by the Germans on which to spread out their maps to follow the battle of Crete still stands in the living room of Metcalf House. The school staff found considerable latitude for productive activity within the overlapping and blurred jurisdictions of these sometimes competitive military authorities.

A major setback for the proper cultivation of the farmland occurred when the air force appointed as field supervisor a corporal who caused considerable strain on relations between the Germans and the farm school staff. He had been a farm boy in Germany. Naturally, the Macedonian climate and soil were resistant to the German methods that the corporal applied; his applications caused a major disruption in farm productivity. The corporal, personally ambitious and a fanatical Nazi, was uncomfortable with his agricultural failures. He tried to shift the blame from himself to the farm school staff, accusing them of sabotage. Tensions were exacerbated when a soldier who was not even a farmer but a tradesman was assigned as the corporal's assistant. These two soldiers made every effort through distortion and accusation to fire farm school personnel in a bid to assume complete control of the agricultural operation. They did succeed in having two or three workers dismissed before the higher authorities finally relieved the corporal and his assistant of their duties, assigning in their stead competent, more educated men who correctly gauged the ability of the school staff. Thereafter the school personnel were given free rein to cultivate the fields.[42]

---

[42] Ibid.

Charlie's position rendered him helpless during these times. After internment in the city and before his removal to Germany he was relegated to the status of ordinary worker. He did, however, find the means to repair, clean, and extend the whole water distribution system, adding a 1500-foot extension to connect the central water system with a well located on a piece of property purchased just before the war. Major repair was also effected on the electric transmission lines. Foremost in Charlie's mind as he took note of the enormous destruction of every sector of the country was the fact that the "school plant is still intact and may be counted upon to do its share in the reconstruction of Greece. We must continue to prepare for that responsibility."[43]

In the summer of 1942, tension in Thessaloniki, a city of over 48,000 Jews, grew unbearable. On 11 July all Jews were ordered to report to the German authorities and forced to put signs in their store windows. Later they were rounded up and deported. In a country where anti-semitism had never been part of the cultural or official tradition, the deportation of such a large population of the city of Thessaloniki came as a severe trauma to the already despairing people.[44] Many are the stories of Greek Christians who saved and hid Jews throughout the occupation. Yet this old center of Jewish life was erased when the Germans deported the Jews to concentration camps for extermination. Few returned to Thessaloniki after the war. In 1943 the Germans confiscated Jewish money and blew up 300,000 gravestones in the cemetery where today's Aristotle University of Thessaloniki stands.[45]

Inevitably, deportation finally came for the Houses, too. On 4 November 1942 they left for internment in Germany. Charlie was sent to Ilag VII at Laufen in Upper Bavaria and Ann was sent to Liebenau in Wurtenberg. They were finally reunited at a family camp in Vittel, France; shortly thereafter they were exchanged for German prisoners of war on 6 March 1944 and sent back to the United States.

---

[43]Charles House, *Newsletter*, July 1944, AFS Archives.

[44] Steven Bowman, "Messianism in Greece," 1972 (unpublished paper, author's possession) gives an excellent brief account of the historical and cultural milieu of Jews in Greece. Also see S. Victor Papacosma, "The Sephardic Jews of Salonika," *Midstream* (December 1978): 10–14.

[45] Ibid.

The deportation of Ann and Charlie was a large discouragement to the farm school staff. The next big blow to their sense of security came when the Germans requisitioned their homes in the summer of 1943. A destructive bombing raid carried out by the RAF on the airport of Sedes had caused considerable damage to the billets of the German air force who were stationed there. In order to provide alternative quarters for their personnel, the air force requisitioned all of the houses on the campus. The staff was given a choice of moving into town to homes vacated by the departed Jews or moving into wooden barracks built by the Germans on farm school property where the girls' school later stood. With the exception of Eleftherios Theoharides, who, having reached retirement age, preferred to live in the city, all the staff elected to move to the barracks.

To suit its needs, the German air force made many alterations to Princeton Hall and James Hall as well as minor ones to some other buildings. Near the main entrance to the campus in the pine grove they erected three brick barracks. The occupying forces used Princeton Hall and James Hall for offices and Kinnaird House as a supply depot. The basement of James Hall was converted into an installation center for the German signal corps, an arrangement that was later to cause that building's partial destruction. Metcalf House was, of course, designated as the general's quarters and guest accommodations; Hastings House was refurbished as an officers' mess. Into Dodge House moved officers, while Farnham and Mosely Cottages were reserved for a general and his chief of staff. The other buildings and the three barracks were given over to soldiers for billeting.[46] Ironically, the farm school building complex proved a tailor-made installation for military purposes. Beyond this, it was a pleasant place for the occupiers to live, isolated from the hostile glare of the population and protected from the horrific scenes of human suffering that their occupation created.

The airforce's temporary role as comptroller of the farm was short-lived. In fall 1943 competition and antagonism developed between the airforce and the *Befehlshaber* (commander) of the Occupation Forces of Thessaloniki-Aegean for control of the farming enterprises. The situation became so inflamed that finally, according

---

[46] Litsas, "First Report."

to Litsas, the high command in Germany had to intervene. It was decided at some higher headquarters that the *Befehlshaber* was responsible for the farm, but that the products were to be split evenly with the air force. The high command fixed the prices under which the products could be sold.[47] The margin of profit was too narrow to allow a decent wage for the staff, and even if it had been otherwise it is hardly conceivable that Greek employees would have received anything other than starvation wages under such a destructive occupation policy.

On 28 October 1944 the Germans evacuated Thessaloniki. Those who had occupied the farm school had conducted themselves tolerably, in a relative sense, during their three and a half years of residence. All farm school personnel interviewed agreed that the Austrians, among the occupiers, conducted themselves as humane, civilized human beings. Another noteworthy point mentioned by the interviewees was that after the first year of occupation there was a clear belief among the Germans and Austrians stationed at the farm school that the Axis was bound to lose the war. A few fanatics believed otherwise, but the general consensus was overtly defeatist. One officer, for example, asked Charlie if he could send a large amount of money he had saved to the United States for safekeeping; he feared that since Germany was losing the war his money, too, would be lost.[48]

Although the actual departure of the Germans from the farm school was not until October 1944, evidence of an imminent departure became apparent months earlier. During the previous August, Allied pressure in other areas on both land and sea was forcing the Germans to displace their occupation troops. Rumors in Thessaloniki circulated that the air force at the farm school was to be replaced by a demolition cadre. A change of routine became apparent when various units began to requisition foodstuffs, draft animals, carts, and donkeys, and officers were departing in large numbers. At that time, the school staff discovered powerful bombs placed in James Hall and Princeton Hall. Litsas called on both the air force commandant and the *Befehlshaber*, begging them not to destroy these two principal

---

[47] Ibid.

[48] Chrysanthi Litsas, interview with author, Thessaloniki, 8 January 1973, written, AFS Archives.

school buildings since the institution served the needs of a now desperate rural population.[49] The German authorities first answered with a transparent lie: that they had no such intention. When Litsas pressed the fact that there were huge bombs planted in the buildings, they promised to investigate. As the days dragged on, other potentially destructive actions were taken by the Germans. For example, the departing air force was allotted by regulation 50 percent of the year's hatched chickens, which it intended to transport out of Greece. Litsas pleaded with them to consider that these were the only purebred chickens in the country. At last the Germans agreed to leave as many chickens as the farm school had possessed on 17 January 1942 when the military took over the farming projects. The air force personnel also demanded some heifers and horses, but finally at the last moment they ran off with only two draft horses and two wagons. It is certainly to Theodoros Litsas's credit that the livestock was not more seriously depleted.[50] His argument was that the heifers were necessary for the rural population of Greece. It is hard to believe that he could have thus persuaded the representatives of a country whose deliberately wanton behavior had destroyed the food supply of a helpless people. Yet his pleading must have had some effect, since the actual destruction and thievery were infinitely less than what the Germans had originally planned.

Finally in the middle of October, almost two months after the bombs had been discovered, an officer of the *Befehlshaber* visited the school at Litsas's insistence and asked the officer in charge of the demolition squad not to destroy the buildings. The demolition officer claimed that the bigger buildings had to be destroyed since they were self-dependent installations and valuable to the Allied Forces. Litsas countered with the historic example that during World War I the Allies had never used the school's premises.[51] Again Litsas repeated the argument that damage to the buildings would only be to the detriment of the Greek people. In the end the demolition officer insisted on the destruction only of the eastern portion of James Hall, where the troops had installed telephones and other communication equipment. Actually, two 500-pound bombs were set to explode in

---

[49] Litsas, "First Report."
[50] Ibid.
[51] Ibid.

James Hall and three similar bombs were placed in Princeton Hall, powerful enough if detonated to render the buildings useless.

By now the German evacuation of Northern Greece was an immediate reality. On 12 October, Athens and Piraeus had been evacuated and on 14 October, British troops under the command of Major General Scobie landed in Greece. On the same day, George Papandreou, the Greek prime minister, returned to the capital. The German departure from Northern Greece could hardly be retarded much longer.

As the Germans prepared for their withdrawal, they blew up the telephone posts they had earlier erected all over the city. The agony of the farm school staff heightened as the occupiers seemed to delay their departure from day to day. On 27 October they posted guards around the school grounds. The next morning, exactly four years after the Italians had crossed the Greek-Albanian border, the Germans withdrew from the school. In a last minute frenzy they committed acts of destruction against the school that they had never committed during their three and a half years of occupation. They smashed doors, broke lightbulbs, and shot wildly into the air.[52] The worst destruction was caused by one of the 500-pound bombs that had been placed in James Hall. The explosion blew up the southern portion of the building, inflicting damage also to the northern section. If the other 500-pound bomb in the other part of the building had been detonated, James Hall would have been destroyed. Princeton Hall was left undamaged; the bombs there were never detonated.

During the entire evacuation procedure, the farm school staff lived in a state of anxiety. From their barracks they could hear the explosions of communication points. The knowledge that huge bombs lay planted within the walls of James Hall and Princeton Hall heightened their sense of doom and terror. But a worse and closer danger to their lives occurred when a group of *andártes* (resistance fighters) took up positions around the school in order to fire upon the departing Germans. The *andártes* selected as their target a German outpost on the southern outskirts of the town. The Germans returned fire. In the center of the crossfire were the wooden barracks in which the staff were living. With the help of the Swiss consul, Theodoros

---

[52] Ibid.

Litsas communicated with both groups, urging them to cease their fire lest they murder the inhabitants of the barracks. Finally the firing ceased when the Germans departed and the *andártes* came out from their firing positions and entered the farm school premises. The occupation had come to an end.

The Germans' departure created a terrifying void of civic administration. As oppressive as Germans rule had been, it functioned at least as some sort of government. The new situation was fluid; other elements drifted into the school from both north and south. First came the *andártes*, who bivouacked on the campus, helping themselves to the scanty supplies. The farm school staff, psychologically and physically depleted, stood by helplessly as they watched them take their fill. No sooner did the *andártes* leave than, at the beginning of November, British commando troops moved into the brick barracks that the Germans had constructed. Actually, the evening when the departing Germans were exchanging fire with the Greek *andártes,* a British commando had arrived at the school but had left immediately for the city. But at least this new development meant that the staff cottages were vacated for the school's tired and ragged families.

But the agony was far from finished. As abysmal as the war and occupation had been, the Greeks had yet to reach their nadir: the civil war was gathering momentum. The first group of *andártes* to visit the campus on the day of the German evacuation would not be the only such group to violate the school. The second group, making their appearance five years later, added a chapter in farm school history that is both brutal and heroic.

# CHAPTER 12

## RECONSTRUCTION AND
## CIVIL OR GUERRILLA WAR

The *andártes,* who descended on the farm school after its liberation from the Germans were participants in a civil or guerilla war,[1] the seeds of which had been planted during the Axis occupation. By 1946, it had grown into a fratricidal conflict lasting until 1949.

The vicious contest for political ascendancy in Greece between communist and government forces called to the fore once again the sensitive issue of Macedonia. In fact it was upon this issue, among others, that both the cohesion within the *Kommunistikón Kómma Elládos* (Communist Party of Greece), known by its acronym KKE, and its relations with the international movement faltered. Thus the Farm School once again found itself in the eye of the storm, by virtue of its location, when Macedonia became not only an ideological issue, but one of the battlefields upon which the issue was contested. Yet against the tragic and barbarous spectacle of brother set against brother the reconstruction of the farm school gained surprisingly strong momentum.

Charlie and Ann had been repatriated; they landed in New York on 15 March 1944. Charlie was immediately tapped by the United

---

[1] I use the subjective term "civil war" in this discussion, but recognize that the term "guerilla war" is preferred by some. Those who prefer the term civil war usually see the conflict from the insurgents' viewpoint: civil war has implicit in its definition a broad numerical involvement and a full-fledged ideological struggle. The opposing view, held by rightists, implies that a rebellious minority spurred by outside forces caused the war. I have not found an objective term acceptable to all.

Nations Relief and Rehabilitation Administration (UNRRA) and appointed chief of the Greek-Armenian branch. Because the war was still raging in Europe, he had decided to turn his attention to relief organizations forming in the United States until he could return to the school. Since House's departure from Greece in 1942, Theodoros Litsas had been acting director. The board of trustees requested that Sydney Loch, a relief worker, journalist, and former teacher at the school—relieve Litsas of the crushing responsibility he had borne throughout the war until the Houses could return to Europe. Loch, who had spent the war years, doing relief work with the Quakers, was able to return to Greece on a British ship with an early party of relief workers. He came officially in his capacity as a representative of the Friends Relief Services, arriving in Thessaloniki on the heels of the departing Germans. Joice Nankivell Loch, his wife, followed him to Greece some three months later. "The Farm School," she wrote, "looked as if an evil hurricane had blown through it. Most of the staff were morose, but at that time nearly everyone was unsmiling and after the first excited welcomes had been said they broke down again into extreme unhappiness. There were none who had not lost relations, and lost most horribly."[2]

The staff begged to be paid but the economic situation in Greece was catastrophic; even if money from America could be obtained, there was nothing to buy. Joice Loch wrote that Thessaloniki was "knee-deep in German marks which people flung out of the window as the Germans left."[3] Banks were not functioning. Her husband had been allowed to bring 500£ into Greece sent him by the board of trustees. He asked permission of the Bank of Greece to lodge it there. According to Loch, when he deposited the sum it was the only money in the entire bank.[4] Barter still prevailed: gold, sugar, cigarettes, rice, and flour were the standards. Thessaloniki was a ghost town with "useless German marks fluttering everywhere and…thick underfoot, while rubbish bins flowed over. There were no trams or buses; people plodded on their ill-shod feet huddled in ragged outer garments against the bitter cold."[5] Means were quickly found to transmit funds to

---

[2] Joice Loch, *A Fringe of Blue* (London: John Murray, 1968) 211.
[3] Ibid.
[4] Ibid.
[5] Ibid.

Greece to pay the staff, but the wages set by the Greek government in 1945 were only three times the prewar level whereas the rise in the cost of living had exceeded the 300 percent level. In fact, Charlie reported that although the farm school had begun to pay wages triple the prewar amount by the close of 1944, essential commodities were from ten to fifteen times greater.[6] In a strongly worded report to the finance committee, he suggested that financial resources be strained to the utmost in order to rehabilitate the staff, which had suffered "five years of deprivations, uncertainties and tragedy. It is little wonder that there is unrest, misunderstanding and frayed nerves. If on a bare subsistence basis we find that our staff is being 40 percent underpaid, it would not be surprising if they were 50 percent inefficient and ineffective."[7] He also urged that the Greek government, the secretary of state, UNRRA, and the Society of Friends be contacted with a view to setting fall 1945 as the school's opening date. At a meeting of the board of trustees in the spring, members questioned the wisdom of opening the school in the midst of Greece's political turmoil. Charlie's opinion was that the school should be opened as soon as possible since "the Greeks had suffered so much and were in such a state of unrest that a little kindness would go a long way in helping them, and...reopening the School would have a stabilizing effect on the people."[8] Two years in German detention had not weakened the resolve of Charlie and Ann House.

In view of the internal problems, the prospect for Greece's rapid rehabilitation was apparently all too discouraging. Under the direction of Loch, supported and advised by the slowly resuscitated staff, the school focused its attention on improvements that would involve the least expenditure of human energy and financial resources yet at the same time start the institution on the road to rehabilitation. The livestock sector of the farm was intact thanks to the care of Dimitris Hadjis; thus full attention could be given to it with the promise of substantial reward. By March 1945, 700 chickens had been hatched in the school's incubators. Pig breeding, always a staple activity, was accelerated and many young pigs were sold to the public for eating or

---

[6] Charles Lucius House, "Report to Members of the Finance Committee of the Board of Trustees of the American Farm School," 25 March 1945, AFS Archives.

[7] Ibid.

[8] Minutes of the Board of Trustees, 8 May 1945, AFS Archives.

further breeding. By March, the dairy, which had been supervised by Argerios Dimertzis, was producing between 100 and 150 quarts of milk per day.[9] Considering the devastation caused by the occupation and the added destruction created by the civil war, the farm school's productivity was unique in Greece. Even UNRRA had hardly begun to implement its program. Stopgap relief measures carried out by the British Military Liaison immediately after the Axis withdrawal were the only help that the Greek people had received. In the late winter of 1945 the American Farm School was one of the few organizations able to make even a token contribution to relief and rehabilitation. It was all too clear to any observer that help would have to come from outside if the country were to move forward at all, for the Greek people themselves were being devoured by another war, this time of their own making, but propelled and exploited by foreign interests.

The various issues that gave rise to the Greek civil war are too numerous and complicated to include in this history. It is important, however, to know some of the political factors that underlay the social conditions prevailing in Greece in the 1940s.

By spring 1943, divisive political issues had become apparent among the various resistance bands. ELAS, by far the strongest, was directed by a parent political group, EAM, the *Ethnikón Apeleftherotikón Métopon* (National Liberation Front), whose leadership was communist dominated. This group was at political odds with EDES, the *Ethnikós Dimokratikós Ellinikós Syndesmós* (National Democratic Greek League), whose political affinity was initially republican, or at least non-communist. By 1943, the friction between ELAS and non-communist guerrilla bands (there were other groups besides EDES that were non-communist although EDES was the most significant) had escalated into bloodshed. EAM was bent on neutralizing rival powers before the German evacuation so that it could be in an uncontested position to impose its political will on the newly liberated country.[10] ELAS's method was simply to attack rival

---

[9] *Newsletter*, March 1945.

[10] Edward S. Forster, *A Short History of Modern Greece* (London: Methuen & Co., 1960) 217. There is a variety of interpretations in English on the civil war period. For examples of varying opinions consult: John O. Iatrides, *Revolt in Athens: The Greek Communist Second Round 1944–1945* (Princeton: Princeton University Press, 1972); Constantine Tsoucalas, *The Greek Tragedy* (Baltimore: Penguin Books,

bands in order to eliminate them. Meanwhile, to add to the bloodshed, the Germans, clashing with both ELAS and EDES, moved into population centers carrying out harsh reprisals that included destruction of villages and shooting of hostages. In Macedonia they were particularly severe, mounting vigorous operations against the resistance fighters in the area of Thessaloniki. A vicious step taken by the Germans had been to turn over a large portion of Macedonia to Bulgaria, which, in turn, engaged in the wholesale murder and deportation of Greeks. In the last 18 months of the occupation, 50,000 Greeks from Macedonia and Thrace were deported to Bulgaria for forced labor.[11]

The position of Yugoslavia made life in Macedonia even more unbearable. From 1944 to 1948, Tito had committed his country to helping the Greek communists. Using Yugoslavia as a favorite sanctuary (Bulgaria and later Albania were also open to the communists), the Greek insurgents were able to strike into the heart of Macedonia, then retreat over the border into Yugoslavia for rest and resupply. This freedom of movement over communist borders enabled the communists to devise a sinister program, one that was to harm the American Farm School: the removal of Greek children. On 3 March 1948 during a conference of Balkan youth in Belgrade, the communists resolved to take children between the ages of three and fourteen and bring them over into Cominform countries for reasons of "safety." By the end of the civil war, the Greek government estimated that over 28,000 children had fallen victim to guerrilla bands. The government claimed that the communists' objective was

---

1969); Evangelos Averoff, *By Fire and Axe: The Communist Party and the Civil War in Greece* (New Rochelle: Aristide Caratzas Publisher, 1978); C. M. Woodhouse, *The Struggle for Greece, 1941–1949* (London: Beekman/Esanu, 1976); D. M. Conduit, *Case Study in Guerrilla War: Greece during World War II* (Washington, DC: US Government Printing Office, 1961); US Command and General Staff College, *Internal Defense Operations, A Case History, Greece, 1946–1949* (Fort Leavenworth KS: US Army Command and General Staff College Reference Book, 1967); Edgar O'Ballance, *The Greek Civil War, 1944–1949* (London: Farber and Farber, 1966); D. George Kousoulas, *Revolution and Defeat: The Story of the Greek Communist Party* (London: Oxford University Press, 1965).
[11] O'Ballance, *Greek Civil War*, 169.

to brainwash these children.[12] According to a 1948 UN Balkan Commission report, 10,000 had been detained in Yugoslavia by that year, as well as 3,000 in Hungary, 2,600 in Bulgaria, 2,235 in Czechoslovakia, and 2,000 in Albania. At this time, Frederica, then queen of the Hellenes, corresponded with the United States secretary of state, George Marshall, regarding what pressures could be placed on these countries to resolve the problem. On 30 May 1949, the American delegate to the United Nations, Warren R. Austin, transmitted in writing to the secretary general of the United Nations the United States' formal regret that by 1949 no Greek child had been repatriated.[13]

Concurrrently the Greek government opened camps away from the danger zones for border area children. Whole families in vulnerable areas moved to common shelters for protection from marauding bands. Moreover, government troops forcibly removed the population from many areas in order to give the army more maneuverability. The disruption to Greek youth was total.

This aspect of the civil war, although perhaps the most fiendish, was only one of the socially demoralizing factors. Forced recruiting of youth into communist bands, carrying off of young girls to mountain hideouts in a country where the role of woman was fixed and passive, and the ravaging of villages for food and supplies were both symptoms of and cause for further degeneration in the Greek countryside. In terms of human upheaval and the quest for territory, this new war in the Balkans was reminiscent in some respects of the old Macedonian Struggle. Broadly speaking, the political and social conditions were creating the same kind of havoc as in former times, but with the added twist that youngsters were now direct victims.

Before the world war, the communists had envisioned a free Macedonian state, but by 1947 any façade of unity that the Cominform countries had presented on this issue began to crack. Tito suggested that Greek Macedonia become part of the Yugoslav federation, while Bulgaria entertained her age-old dream of seeing it

---

[12] Ibid. The Communist Information Bureau, founded in 1947, consisted of representatives from the communist parties of the Soviet Union, Yugoslavia, France, Italy, Poland, Bulgaria, Czechoslovakia Hungary, and Rumania.

[13] Queen Frederika of the Hellenes, *A Measure of Understanding* (London: Macmillan, 1971) 127.

absorbed into a greater Bulgaria. The Greek communists, too, were split on this issue; nationalist-minded ones were plainly against detaching Greek Macedonia from the motherland while diehard international communists were prepared to accept the Bulgarian "solution," which became the Cominform's formula after Tito was abandoned by the Cominform in 1948.[14]

In late 1944, in the midst of this chaos, Theodoros Litsas had enlisted the cooperation of the newly functioning UNRRA, the Greek army, and others to open a scouting program for youngsters. The youth were housed in the German barracks on the campus and trained to be Boy Scout leaders. Girl Guides were also added. The 200 youngsters who were accommodated in 2-week periods enjoyed a respite from the hunger and terror filling the countryside. The program was tailored to the needs of the time. It emphasized sanitation methods, which had always been appalling in Macedonian and Thracian villages and which, under the more stressful circumstances of the war years, had deteriorated even further. The scouts were also taught how to lead recreational activities and, most importantly, how to form patrols to protect against the guerrillas.[15]

A second project conducted simultaneously with scouting was called the "preventorium." Joice and Sydney Loch—with the aid of the Society of Friends, the Swiss and Greek Red Cross, the International Red Cross, UNRRA, and the British army—installed a number of children for three-week periods in the wooden barracks that had been built by the Germans in 1943 to house the farm school staff. These children were undernourished. It was hoped that pre-tubercular conditions might be eliminated if they were put on a proper diet for this period of time. The Boy Scouts patrolled the preventorium keeping an eye out for guerrilla encroachments.

Other meaningful and cooperative efforts were under way. The Society of Friends aided the farm school in its poultry breeding by contributing to the cost of feeding more than 2000 chicks, which were immediately distributed to villages, where virtually all poultry stocks had been decimated by the war. Both pig production and garden produce were increasing despite the most discouraging conditions. There was no proper feed for the cattle until an army veterinarian

---

[14] O'Ballance, *Greek Civil War*, 189.
[15] *Newsletter*, July 1945.

suggested that the bumper crop of spinach raised in the gardens might do for the cows, which because of poor diet were bearing stillborn calves. The spinach proved to be such a successful cattle feed substitute that it was also given to the pigs.[16]

Under Sydney Loch's guidance, the school took up its mission of relief immediately after the Germans' departure. The voices of youngsters rang out on the campus; the pinched faces of children and staff took on new life as the winter passed; summer blossomed with full fields ready for harvest.

The board decided to open the school's doors in October 1945. On 22 September 1945, Ann and Charlie House returned to the farm school after an absence of two years and ten months. Ann was shattered by her first view of Greece at the port of Piraeus: "The miserable shacks, the ruined buildings, the piers and warehouses roofless, the broken concrete pillars, great piles of twisted iron, formerly cranes, ships everywhere about standing on end, tipped over on their sides, their smoke stacks half sticking out of the water, the ragged sailors in their shabby little boats, children salvaging bits of board and boxes thrown overboard with the garbage—a desolate scene."[17]

An almost superhuman effort was made to get Princeton Hall back into condition so that it could house the boys for the school's reopening. Although the building had not been blown up by the Germans, its rooms had been partitioned; furthermore, its amenities had been ruined by the destructive spree of the departing occupational forces. It was hardly habitable. Many organizations made the refurbishing possible. Members of the British army donated wood, building materials, beds, blankets, mattresses, pots, pans, stoves, and other practical objects. They also fenced in some of the fields. The Greek Red Cross and PIKPA, *Patriotikón Idrima Kinonikís Prónias kai Andylípseos* (Patriotic Foundation for Social Welfare and Aid) gave food and flour while UNRRA donated medical equipment.

On 31 October 1945 the first two boys arrived at the school. On 5 November 1945 Charlie cabled to a concerned board of trustees, "Classes started today. Boys'enrollment 52, girls' 40. All well."[18]

---

[16] *Newsletter*, October 1945.
[17] Ibid.
[18] Ibid.

That year the entrance of village girls into an educational program was a milestone in the history of the American Farm School. The innovation was activated by Joice and Sydney Loch working with the British Friends Relief Service in cooperation with several international organizations. Named the Quaker Domestic Training School, it was known more commonly by its nickname, the *Scholí Quakerón* (the Quakers' School) or simply the Girls School. They brought to reality a life-long dream of Mother House, a dream shared almost from the beginning by the board of trustees and many American, English, and Greek friends of the school. It had taken over forty years to bring the seed to fruit. The sad irony connected to the opening of the Girls School was that it took a global war and colossal devastation to generate the necessary financial and human energies to initiate such an undertaking. The Quakers supplied the funds and leadership while the farm school lent the land and the barracks that had been constructed by the Germans for the staff during the war.

The first class of girls were traumatized by the war; several were orphans. They came mainly from Halkidiki, where Quakers had made previous contacts owing to their relief activities. Many of the children arrived suspicious or nervous; all were in poor health. The girls were taught to live and work together even though political division had set their families one against the other. As family feuds were a tradition in some parts of Greece, the guerrillas readily exploited this feature of village life. One girl, for example, had attacked the daughter of her father's murderer. Removed from their violent environment, the youngsters learned to function in a more relaxed manner, responding to the spirit of the school, which was teaching them to rise above particularism and to set the common good as their standard.

The attempt to reverse behavior patterns was difficult. During the first year the civil war was particularly active throughout Northern Greece. Postal services were irregular, so at times the students did not hear from home for months. When letters did arrive they often told of murder and horrors. The girls could not be removed totally from the realities of Greek life, but their reactions were conditioned and tempered by the spirit prevailing at this new school.

The spirit that animated the Girls School under the direction of the Quakers ran parallel to that of the farm school. The directors

exhibited a distinct sensitivity to and appreciation of the best qualities
of the Greek people and attempted "to nurture the finer feelings that
must otherwise be battered out of the boys and girls of this generation
by the cruelties of civil war."[19] Many years' experience among these
people had taught Joice Loch that "one only has to give away half an
aspirin to a Greek to receive it back a thousandfold. Their generosity
fills me with shame to receive anything when I know how little they
have to give. I remember that in '41 they died in hundreds from
starvation and that nowhere is there a record to be found of a single
British or American soldier escaped and hiding among them, who died
during that time of starvation."[20]

The Quakers tailored the academic program to fit the immediate
needs of village girls. Aside from academic subjects, the students were
taught the practical elements of homemaking, particularly sanitation.
One of the most innovative programs concerned childcare. An
orphan baby, often in delicate health, was taken into the school to be
cared for by the girls and staff. A baby specialist attached to the
health division of UNRRA supervised the project, so the infant was
under professional surveillance. After this period of nurture, the baby,
robust and healthy, was given up for adoption. For the girls, this
practical instruction by a specialist broke down ignorant, superstitious,
and often injurious practices associated with childcare in the villages.

As for the boys, third-and fourth-year students whose education
had been interrupted by the war were brought back to finish. At the
same time, a first-year class was started for new boys and each
consecutive year another first-year class was begun until the school
had reestablished a regular four-year program. Charlie stated: "In their
academic studies the boys are below the standard of their ages, which
is natural for boys whose schools have been closed for four or five
years during the war, but they were eager to learn and full of zeal in
their practical lessons with the livestock and in the fields and
shops."[21]

Besides these programs, the farm school also quartered, at the
expense of a philanthropic agency, a group of intractable Jewish

<hr>

[19] *Newsletter*, April 1948.
[20] Ibid.
[21] Charles House to H. A. Gardner, Thessaloniki, 21 November 1945, AFS
Archives.

youth, their childhood marred by years in the mountains, hounded by the Germans. They were put to work in various departments to train for weeks as farmers preparatory to their emigration to Israel.

The school's first academic year after World War II (1945–1946) was filled with alarming events. In fact, these events were crucial even on the international scene. In Churchill's judgment: "When three million men were fighting on either side of the Western front and vast American forces were displayed against Japan in the Pacific, the spasms of Greece may seem petty, but, nevertheless, they stood at the nerve center of power, law and freedom in the Western world."[22] Churchill was speaking of a moment when Britain was bent on destroying the communist factions in Greece. By December 1944 the communists had gained control of important sectors throughout the country. Strong British military intervention at that time had caused ELAS severe loss of both military and political support. Finally, in the middle of February 1945, after intensive fighting between British and ELAS forces around Athens, ELAS was forced to negotiate. The result of that negotiation, which proved later to be merely a hiatus and not the end of the hostilities, was the Varkiza Agreement worked out between ELAS and government forces. Both parties agreed to the following:

1. Democratic liberties of the people were to be restored.
2. Martial law was to be partially relaxed.
3. There was to be an amnesty for political crimes committed since 3 December, but those who did not lay down their arms by 10 March and those guilty of crimes against common law were excluded.
4. All hostages were to be released by ELAS.
5. A national army was to be formed.
6. All ELAS forces were to be disbanded within fourteen days.
7. A civil administration was to be reestablished.

---

[22] Winston Churchill, *Triumph and Tragedy*, vol. 6 of *The Second World War* (Boston: Houghton Mifflin, 1953) 325. For an account of the British forces that landed in Greece in 1944 see Henry Maule, *Scobie, Hero of Greece: The British Campaign 1944–1945* (London: Arthur Baker, 1975).

8. A probe was to be conducted into three civil servant categories: those who held office under Metaxas; those who functioned under the occupation; those who took part in the revolt on the side of EAM.

9. A plebiscite was to be held to decide on the future of the monarchy and a general election was to be scheduled with Allied observers present during both of these occasions.[23]

One of the vital points of the Varkiza Agreement was the election plan. This election did take place on 31 March 1946, followed by the plebiscite in September to learn if the Greeks wanted the return of the monarchy that had established itself outside of Greece when the country fell to the Germans.

At this historic moment the farm school was able to serve as one of the centers in which Allied observers could be lodged. An advance force of American officials designated specifically for the purpose of preparing details prior to the arrival of the main body of observers was dispatched to Greece. Among those assigned to Thessaloniki was Professor Peter Topping from the University of Cincinnati. Fluent in the Greek language, he was among the first of the preparation group to arrive. His mission was twofold: to find interpreters to assist the Allied observers and to select suitable quarters where officials and their drivers could lodge.

The farm school was chosen as one of the principal centers and a "heroic effort was made by the Houses," Professor Topping recalled, "to house these people."[24] In the end, sixty to eighty mission members were quartered on the campus. The barracks built on the main campus by the Germans were repaired with the installation of windows, doors, and bath facilities, the cost borne by American government aid.

The school was not only one of the few comfortable oases in war-stricken Greece; it also supplied luxuries such as butter, eggs, pasteurized milk, and an occasional hot bath. Topping believed that

---

[23] Edward S. Forster, *A Short History of Modern Greece* (London: Methuen & Co., 1960) 224.

[24] Peter Topping, interview with author, Ekali, 14 May 1972, written, AFS Archives.

the presence of Americans on the campus contributed much to the institution's moral and material regeneration. The staff's morale was buoyed up by the sight of well-dressed, well-fed, well-organized American officials engaged in an enterprise that promised to end the chaos that was engulfing Greece. Although the school was Greek in many respects, decades of American involvement had been its source of stability. The material benefits were substantial. The school was reimbursed for the room and board of these officials by AMFOGE (the Allied Mission for the Observation of Greek Elections, composed of French, British, American, and South African representatives. The Russian ally had been invited but had replied "*nyet.*"). In addition, the Allied Mission became a source of needed machinery. The American observers arrived with much of their self-sustaining equipment. Automotive spare parts and repair parts, communication paraphernalia, and electrical apparatus were all part of their baggage. Much of this equipment was left at the school when the observation team withdrew.[25] A portion was given to the institution outright while other parts were purchased from the American government at a substantially reduced price.

The elections were held on the last day of March 1946, affirming the democratic, legal base of the government. The plebiscite held in September of that year called overwhelmingly for the return of the monarchy. The boys who had entered the school in October 1945 and the new ones in 1946 had some cause to believe that Greece was at least moving toward normalcy even though the war-torn countryside belied this. Many *andartes,* although badly mauled by the British military intervention in late 1944 and early 1945, had not turned in their arms as promised, but had scattered to the mountains to resume their struggle. The renewed warfare lasting from 1946 to 1949—much of the violence centered in Macedonia and in North Central Greece—was one of the most tragic periods in modern Greek history.

The contrast between the serenity within the boundaries of the campus and the savagery in the countryside was stark. Yet the students could not be sheltered from the tragedy outside. One boy, whose father was president of his village, was killed. Bad news filtered in daily. Ann wrote about the boys' experiences: "There was a battle

---

[25] Ibid.

last week near the village of one of the boys; many homes were burned. Theirs was not, but they lost through looting all the animals, farm equipment, furniture and clothing, the house left an empty shell."[26]

She commented on the boys' buoyancy: "It is remarkable how calm this boy and many others are in taking things. They really have a tremendous amount of grit."[27] She added, "Another boy had to spend three nights of his holiday hiding in a hole in the ground."[28]

The violence took its toll on the youngsters in many other ways as well. For example, one boy who was being tested for entrance into the school was asked to read for Ann and other admissions staff. The youngster stood tongue-tied, unable to articulate. When the staff learned that the traumas he had suffered had caused him such severe difficulties that he could not talk face to face with strangers, he was allowed to turn his back to the listeners and then to read his selection. He read well and was admitted.[29]

The British had thought that the Varkiza Agreement, the elections, and the plebiscite would settle the country's internal problems, but it had become apparent by early 1947 that the communists respected none of these. Great Britain, which for centuries had protected its sphere of interest in the eastern Mediterranean, was deeply involved in Greece's internal affairs, both political and military, during the early days of the civil war when, by means of a full-scale intervention, it had deflected the initial ELAS thrust. By early 1947, however, when over 25,000 ELAS troops rallied again in the mountains, gravely endangering the nation's security, Britain's energies began to flag. By February of that year it was forced to reckon with domestic realities. Britain's prewar sphere of interest could no longer be supported by its shrinking power base. In December 1946 the overburdened London government reported to Washington that it was no longer able to sustain the load of resisting communist incursions in the Balkans. Turkey, too, was feeling communist pressure both within her own borders and from Russia, her neighbor to the north.

---

[26] *Newsletter*, April 1948.
[27] Ibid.
[28] Ibid.
[29] Ibid.

British influence in and responsibility for Greece and Turkey was being gradually dropped into the laps of the Americans. Britain was convinced and the United States government agreed that the Greek communists in collusion with communist countries to the north would in time seize control of all of Greece unless the government forces received immediate and substantial reinforcement. Turkey, if squeezed on the one side by Russia and on the other by a communist-controlled Greece, would collapse owing to the force majeure.

After conferring with some members of Congress, President Truman, in a surprise appearance at a joint session of Congress, presented on 12 March 1947 a precedent-shattering plan later named the Truman Doctrine. He proposed that the United States should "support free people who are resisting attempted subjugation of armed minorities or by outside pressures."[30] To implement his plan to aid Greece and Turkey, the president requested a $400,000,000 congressional appropriation for economic and military aid. Of this sum $100,000,000 was for strengthening Turkey's military capacity and the rest was earmarked for the establishment of stable political conditions and a sound economy in Greece. Half of the $300,000,000 designated as aid for Greece was for military purposes and the other half for economic programs to restore transportation and electrical utilities as well as to enhance agricultural rehabilitation and industrial reconstruction.

The Truman Doctrine with its accompanying appropriation was, in reality, only a stopgap measure for helping independent countries resist communism and restore war-shattered economies. However, the Marshall Plan, authored by Secretary of State George C. Marshall, was designed to insure the viability of the Truman Doctrine on a vast scale. The secretary of state proposed that the countries of Europe confer in order to draw up a long-range plan for economic recovery, including ideas for mutual assistance, since the United States, under the Marshall Plan, was ready to support European nations with financial help. Along with fifteen other nations, Greece immediately

---

[30] As cited in Thomas A. Bailey, *A Diplomatic History of the American People* (New York: Appleton-Century-Crofts, 1969) 797. For an account of American aid see William Hardy McNeill, *American Aid in Action, 1947–1956* (New York: The Twentieth Century Fund, 1957).

seized the opportunity to be integrated into the new plan, officially known as the European Recovery Program.

When the Truman Doctrine and the Marshall Plan were applied to Greece, their success was spectacular. Their value as a military factor in turning the tide against the communists cannot be overestimated; as C. M. Woodhouse has written, "Greece is one of the few countries in the world and the only one in Europe where Communist attempts to seize power by armed force have been successfully confronted and defeated."[31] In addition, the effectiveness of these measures in the humanitarian sphere may be counted as one of the most constructive in the history of philanthropic ventures. As a consequence of American involvement, an economic and political environment was created in which democracy had a strong chance for survival.

What American interest in Greece did for the American Farm School in terms of survival, regeneration, and future development was so meaningful that it deserves detailed attention. Since the new plans included financial support for agricultural and educational enterprises, the farm school, one of the more widely recognized, solid, and enduring institutions in Greece, was singled out as a recipient institution. AMAG (the American Mission for Aid to Greece), an agency of the European Recovery Program in Greece, worked through the Greek ministry of agriculture to rehabilitate institutions considered crucial in the agricultural sector. The farm school's participation with AMAG and the ministry of agriculture in Project Agreement 14, signed in 1948, was one of the first efforts that not only returned to the school its prewar capability, but changed its status from a small institution with limited connections known to a small group of friends in the United States, Britain, and Greece to that of a well-publicized center to which agriculturalists, sociologists, government officials, social workers, veterinarians, and even poets and writers were attracted. The school received more unsolicited, effective publicity during the years of the Marshall Plan than in its whole previous history.

Since funds from AMAG were channeled through the Greek ministry of agriculture to the farm school, the three participants were

---

[31] C. M. Woodhouse, foreword, in O'Ballance, *Civil War*, 13.

drawn together into an operation that demanded tight coordination, constant consultation, and smooth cooperation. Warm relationships forged during this time, particularly between the Greek ministry and the farm school, produced a treasury of good will that endured long after the Marshall Plan expired. The farm school's opportunity to show what it could do to American agriculturists connected to the project enlarged its reputation, bringing it professional recognition in the United States.

The sub-projects of Project Agreement 14 provided for the following:

1. rebuilding James Hall;
2. building, adding to, and equipping a training center;
3. procuring equipment for training and demonstration;
4. disposing of the old dairy herd, much of which had contracted tuberculosis, and procuring a new herd.
5. providing short course training in tractor use and machinery for refugee boys.[32]

The ten-month project began in March and terminated in December 1948. According to the terms of sub-project 1, James Hall, which had been bombed, was redesigned on the same ground space, retaining the major lines of its style, but was fireproofed, adapted to classroom and laboratory needs, and restyled to afford larger space (specifically an extra study hall and a lecture hall) for increased enrollment.

Under sub-project 2, a training center was constructed using the German barracks on the main campus as a shell to provide facilities for courses in agriculture and related fields. According to the agreement, the farm school itself installed furnishings, fixtures, and equipment to feed and house fifty trainees. Also, the school was responsible for constructing additions to the barracks to make the training complex feasible. The Greek ministry of agriculture and the

---

[32] Report of the Director of the American Farm School on Operations and Financial Transactions Involved in the Execution of Project Agreement No. 14 entitled "Agricultural Training, American Farm School, Under the Cooperative Agricultural Program of the Greek Government and the American Mission," 30 December 1948, 2, AFS Archives.

American Mission agreed to pay for these improvements. Brick was selected as the building material and the center was extended to provide a total of approximately 1000 square meters of floor space. Although the barracks had been temporarily repaired in 1946 for AMFOGE, two years later they were considered dangerously substandard: the tar-papered portion of the roof leaked, the tiled section sagged. Yet, despite all the deficiencies of these structures, rehabilitating the barracks saved the school the enormous cost of replacing them. This saving, plus the dimensions and location of the complex, justified the project. By 31 December 1948, two varieties of short courses for adults, conducted by the farm school in conjunction with the ministry of agriculture, had been tested in the newly refurbished buildings to determine if these structures were suitable. The teachers involved were satisfied. Not only did the enhanced barracks become an asset to the farm school, they also served as a model for similar training centers throughout Greece.[33]

Sub-project 3 was one of the most far-reaching, since it provided equipment for training and demonstration. One can easily surmise the atmosphere at the school as it sped forward to regenerate itself from the stagnation and dilapidation into which it had fallen during the war years. The institution's dynamic quality contrasted with the destructive activities occurring in the countryside. The civil war was not only retarding the nation's economic recovery, it was also causing increased destruction. While other European countries could direct all their energies to exploiting the possibilities of the Marshall Plan, Greece was diverting its energies to the execution of a ruinous civil war. To understand the impact of sub-project 3 on the country, one needs to review some facts and examples of rural life during the last phase of the civil war. Hundreds were displaced as a result of guerrilla raids or of clashes between the Democratic Army of Greece (successor to ELAS in early 1946) and government forces. Whole towns and villages were destroyed or damaged, their inhabitants streaming to refugees centers by their own volition to find some protection from the savagery, or moved there by the Greek government.[34]

Ann House recorded her view of the pattern of refugee life that had developed in Macedonia: "We passed the school in the village of

---

[33] Ibid., 4.
[34] Forster, *A Short History*, 233.

Liti, which is full of refugees from villages that have been evacuated by the army. Families had brought with them some bedding, pots and pans and some pigs, chickens and one or two calves all tied in the schoolyard. Families were crowded together, and although the villagers said they gave food to the children, they all looked pretty peaked and very ragged and dirty."[35]

Descriptions by farm school parents reveal the tragedy in its full human dimensions. A mother of an entering boy wrote:

> We are deported from our village and are living in some barracks, which means no pasture for our cow, the pig and the horse; no garden for vegetables and fruit; no weed cutting or any other means to make our living. To make the situation still worse, Pavlos, my son, is sick and my son-in-law [has been] mobilized and sent to Corinth. Although I am used to hardship, having to work hard as a widow to raise my three children, who were orphaned at a very early age, this time I feel myself hopeless and desolate, having nowhere to look for any help except to God.[36]

The influence of the sub-projects on the regeneration and modernization of an almost totally ruined agricultural environment was broad and deep. It was geared to reshaping rural life directly by introducing into village homes methods and equipment that would raise the standard of living to a more tolerable level. Sub-project 3 provided equipment, instructors, demonstrators, and specialists for training farmers and their families in the following areas: "farm machinery, processing equipment for agricultural projects, equipment for village homemaking, pumping equipment, pipe laying, and well casing, a variety of equipment necessary for rural education."[37]

Sub-project 4 was confined to one goal: strengthening Greece's dairy industry. It should be repeated that the American Farm School was a pioneer in that field, producing the first pasteurized milk in the entire country. Greece's topography is not suited to easy cattle breeding; therefore, farmers had been historically resistant to the

---

[35] *Newsletter*, 1947.
[36] Ibid.
[37] "Project 14," 10.

development of cattle breeding on a scientific or large-scale basis. Milk and cheese requirements were traditionally fulfilled by the breeding of goats and sheep, also on an unscientific basis. The unrestricted grazing of goats and sheep throughout the countryside had been destructive to forestation and groundcover and a direct contributor to erosion. The laws passed to restrict grazing were both unpopular and difficult to enforce. War, corruption, and civil strife had decimated stocks of sheep and goats as well as the always meager herds of cattle. The sub-project's objective was to provide training on the care and breeding of dairy herds through the following: "1. demonstrating procedures to eliminate tuberculosis in dairy herds; 2. establishing a service within the Ministry of Agriculture to inspect and accredit herds, including a program to eradicate bovine tuberculosis; 3. introducing Guernsey and Jersey cattle into Greece."[38]

The farm school benefited tremendously thanks to the implementation of sub-project 4, as it was able to improve its own herd to a degree that would never have been possible without this type of funding. The school's herd had been started in 1908 with a purebred Jersey bull imported from England and two native Greek cows. Other good stock had been introduced through the years, but the herd was never ascertained to be 100 percent free of tuberculosis since the school never possessed the means to test for the disease. Some progress had been made toward testing before the war, but during the occupation the Germans introduced infected cattle, causing a sharp increase in bovine tuberculosis. Before sub-project 4 was implemented, tests showed that of the whole herd, 53 percent were reactors, and of the milking cows 87 percent were reactors. In accord with the terms of sub-project 4, the school received the funds to destroy all infected cattle, to segregate exposed non-reactors, to perform the necessary disinfecting of stables and surrounding areas, to furnish new quarantine facilities for incoming animals and, of supreme importance, to provide from the herd offspring at least six months old to distribute free of charge to institutions or organizations in accordance with the ministry of agriculture's authorization.[39]

In addition to Project Agreement 14, the school was supported in two important ways in 1948 by the United States Educational

---

[38] Ibid., 17.
[39] Ibid.

Foundation in Greece (the Fulbright Program). First, this program granted food maintenance scholarships for fourteen new students each year. The youngsters, coming as they did from war-devastated farms, could not afford to pay their own board. The food maintenance fund was most generous in that it provided for up to fifty-six students at any one time. Secondly, the foundation granted three fellowships, covering full salary plus travel expenses, for teachers to come from the United States. Among the first to arrive under the Fulbright plan were two old hands in Greece. Edmund Keeley, son of former Consul General James Hugh Keeley, Jr., taught English and athletics. He went on to become a chaired professor at Princeton, a founding member of the Modern Greek Studies Association (which he served as its first president) novelist, translator of Greek poets, and a farm school trustee. The second Fulbrighter, Lee Meyer, had come to Greece originally from Turkey in 1922 with the Near East Relief and had subsequently been resident engineer at Anatolia College until 1935, when he returned to the States. He assumed duties at the farm school as engineer, teacher, and head of the industrial arts department. His kindness, generosity, and closeness to the boys did much to foster the school's basic values among them. After his Fulbright grant expired, he remained at the farm school until his retirement. He is buried in the school cemetery.

The most fateful feature of this first Fulbright grant to the school was the fact that it also brought Bruce M. Lansdale, a third old Greek hand, to act as administrative executive. Young Lansdale had lived in Greece as a boy while his father, Herb, was national director of the Greek YMCA. Bruce had returned in 1946 as an official on the Allied team for the observation of Greek elections and again, later, under a Ford Foundation grant. A character evaluation report (an obligation owed to the Ford Foundation and written by Charlie), honestly and shrewdly appraised young Lansdale's performance, describing both his strengths and weaknesses. Beyond that, Charlie had analyzed Bruce's performance thoroughly from every angle for a subjective reason. Having realized that Bruce Lansdale possessed the requisite training, education, and traits of character to direct the farm school, Charlie favored his candidacy to succeed him as director.

Although AMAG terminated at the end of 1948, it was followed by the European Cooperation Administration (ECA). Through this subsequent agency, funds for repair and expansion continued to flow into the school. In 1949 the ECA gave farm machinery and equipment worth $20,000, a new dairy barn, two silos, three staff cottages, and some industrial buildings. UNRRA had also contributed about $20,000 worth of equipment for agricultural and vocational training. The Greek government, continuing its endorsement of the school, passed Omnibus Law 1286 granting the School duty exemptions for farm machinery, equipment, and supplies.[40]

Thanks to its rebuilding and expansion, the school gave all who came in contact with it a sense of security, setting a tone of assurance and permanency that was unique in an otherwise fear-ridden and desperate climate. Since 1945, the school had expanded its curriculum to include a variety of courses that attracted farmers of all ages and interests, thereby touching almost every surrounding village and reaching across the full breadth of Northern Greece.

By 1946, the year after the school had reopened, and immediately after the elections and the king's return, Greece had complained to the United Nations Security Council that the communist guerrillas were being allowed to cross back and forth from Albania, Bulgaria, and Yugoslavia. A UN committee set up to investigate this charge confirmed the Greek accusations. With the aid of the Marshall Plan, the Greek government was able to send five divisions into the field by the winter of 1948–1949 to counter the rebels, and from that time onward the communist thrust was thwarted. But much to the guerrillas' advantage was the 600-mile frontier over which they could slip into friendly communist countries. Owing to the Marshall Plan, the Truman Doctrine, Tito's split with Stalin, and his closing of Yugoslavia's borders to the Greek insurgents in July 1949, the communists' effort was flagging. Between May 1947 and February 1949 they had failed at Florina, Konitsa, Grevena and Karpenisi, but their attempts had convulsed the social order of the countryside.

The fact that communist guerrillas had neither violated the school nor threatened to do so had lulled government authorities and the school administration into a false sense of security. No extra guards had ever been posted; there was only the single, usual watchmen for the children's protection. In November 1948, the *New*

[40] "Government Gazette of the Kingdom of Greece," No. 2961A, Vivliothiki Voulis, Athens.

*York Times* mirrored this complacency: "The raiding guerrilla bands of General Markos Vafiades, which can hardly be regarded as pro-American, have given this corporation wide berth in their operations. Nor have the Communist rebel leaders uttered one word of public criticism against it."[41] The journalist went on to explain that, in his view, the farm school was so dearly loved by the peasants of northern Greece that any group that attacked the institution would lose the sympathy of many of its supporters.[42]

However, the illusion of privileged sanctuary was shattered two months later when the communists kidnapped forty-one farm school boys, thereby erasing the hope that "we could have peace and quiet to do our task."[43]

On the night of 29 January 1949, one of the coldest and longest of the Macedonian winter, a force of communist guerrillas lay hidden in the school cemetery. Toward 10:00 P.M. they crept under a moonless sky toward a thicket of trees that surrounded the irrigation pool across from Princeton dormitory. They watched John Boudourouglou, the teacher on duty that night, make the last check on the boys; they then remained in hiding until the lights were turned off.[44]

About twenty members of the band ran into Princeton Hall, awakened thirty-seven boys and four young assistant instructors. At first the boys thought it was a prank but, as they came more fully awake and saw the armed men clearly, they realized it was not a joke. The insurgents ordered the boys to take one blanket and one loaf of bread from the kitchen before marching them off into the freezing night.[45]

Meanwhile, the leader of the band, dressed in a military overcoat and carrying a Tommy gun, entered the home of Director House. Holding Ann and Charlie at bay and ripping out the phones, he explained that he was taking the boys. Charlie urged the guerrilla chief

---

[41] "American School in Macedonia Busy," *New York Times*, 21 November 1948, 16.

[42] Ibid.

[43] Ann Kellogg House to her family, Thessaloniki, 29 January 1949, AFS Archives.

[44] John Boudourouglou, interview with author, Athens, 5 September 1974, written, AFS Archives.

[45] Ibid.

to take him instead of the youths or, barring that, at least to allow him to accompany the boys wherever the chief was taking them. The leader retorted to Charlie that the struggle was purely Greek and that he could not let a foreigner join the group. Furthermore, he elaborated that he was not operating independently but on orders from a higher command. Charlie asked the leader pointedly if he did not understand that there existed a yet higher command that had to be obeyed. The guerrilla leader answered that he did indeed realize what Charlie meant since his own father had been a village priest. But he could not be dissuaded. Upon departing, he warned Charlie that if he set foot from his house to inform the authorities it would cost him his life.

No sooner had the insurgents departed than Charlie ran to the dormitory and was met by the sight of forty-one empty beds. He made a hurried dash to the Girls School, where he found the girls all safely asleep. He raced home, had Ann and Ruth pack him some clothes and a sleeping bag, climbed into a jeep and started out to inform the authorities and pursue the guerrillas.[46]

It is Boudourouglou's analysis that the government forces that started in immediate pursuit of the guerrillas assumed that the band would take the youngsters immediately toward the mountains. Instead, they had herded the boys along the flats and marshes toward the airport. In Boudourouglou's opinion, the error was actually a piece of good fortune. Had the army and the guerrillas confronted each other, some of the boys might have been shot either by accident or by their captors.[47]

The boys marched across the black marshes accompanied by some eighteen to twenty communists, a factor that was likely to impede any effort to escape. The communists, however, were exhausted. They had trekked three days steadily to reach the school. Now, without respite, they were continuing their fatiguing march through cloying muck and were soon to turn into the mountains. Later, some of the escaping boys reported that many of the guerrillas had dropped from exhaustion along the way.

---

[46] Ann Kellogg House to her family, Thessaloniki, 29 January 1949, AFS Archives.

[47] John Boudourouglou, interview with author, Athens, 5 September 1974, written, AFS Archives. Archimedes Koulaouzides, interview with author, Thessaloniki, 5 June 1975.

The youngsters had been kidnapped for the purpose of swelling the communist ranks. Some of them would be sent over the borders to be indoctrinated. The exploit's propaganda value was considered high by the communists. Their underground press proclaimed in bold type: "American Farm School boys join ELAS in the fight for freedom."[48] The feat was also considered a direct challenge to the Greek army, which was responsible for the control of guerrilla movements in its area. To have a force of twenty insurgents kidnap a group of forty-one boys a few kilometers outside Greece's second largest city was to shake the confidence of the population and to question the capability of national forces. Along with this motive, the kidnapping was intended to intimidate villagers. Each boy was instructed that should he escape from his captors and be recaptured he would be classified as a deserter from the rebel army and executed. To reinforce this point, the captors forced some of the boys to observe the execution of an eighteen-year-old village youth who had fled and been retaken.[49]

The first two youngsters to escape did so the first night on the marshy flat. In the black of night one boy and a companion sank to the ground and remained motionless until the band had passed on. One of the two lived near the spot where they were hidden. As the hostages and their captors disappeared into the dark, the two youngsters ran toward the house where the boy's father stood, rifle cocked, having heard the muffled steps. The son called to his father, who, recognizing the voice, lowered his weapon.

During the next week, the boys continued to escape one by one and in small groups, greatly aided by the exhausted condition of their captors. Also, frequent skirmishes that occurred between the guerrillas and national groups along the way caused enough distraction to allow some of the boys to abscond in the confusion. Another factor that worked to the boys' advantage, according to Boudourouglou, is that the guerrillas hesitated to shoot escaping boys since the firing would

---

[48] As quoted in Miriam Vaughn Dubose, "American Farm School in Greece Carries on Program Despite Guerillas: Youths Captured by Andartes Return to Study Agriculture," *Winston-Salem Journal and Sentinel*, 26 June 1949, 6c.

[49] Ibid.

reveal their presence to villagers who had been armed by the army to act as guards.[50]

Ann and Charlie House both felt that the fact that the leader was a priest's son made him sympathetic. He confessed to the boys, "The words of your director are tearing my heart open, but I have to do it."[51] The leader was referring, of course, to Charlie's remark about a still higher command. The escapees hid in caves during the day and at night made their way back to school, foraging for food along the way, their clothes torn to rags, their shoes worn and in many cases lost.

Meanwhile the staff made every effort to find the boys. Dimitris Hadjis and Theodoros Litsas explored the countryside in jeeps, in many cases actually picking up some of the escaping boys, most of them looking more like scarecrows than children. From the robust and healthy creatures they had been when they left, they returned starved, tattered, and haunted. Twenty-four hours after the kidnapping took place, twenty-five of the forty-one had managed to escape, walking from sixteen to eighteen hours. On the campus, meanwhile, the rest of the boys gathered to greet the escapees as they arrived. The youngsters reported their hardships. For instance, some said it was so cold in the mountains that they could chop up bits of ice to slake their thirst. They told stories of death: during one of the skirmishes with the national forces one of the guerrilla leaders was killed.[52]

On 3 February, all but five of the abducted students were back at school. Now with only five captives to guard, the guerrillas were better able to prevent escape. Also, the deeper the group advanced into the mountains the harder it was for the boys to find their way back through the bewildering and hostile terrain. To add to their plight, they noticed that the mountains were infested with other bands that might capture them as they escaped.

The last boy to escape returned to the school on 6 April after over two months as a captive. His father had been on the rebels' side and some feared that he might join the insurgents. Yet members of the staff such as Theodoros Litsas who knew the boy well were

---

[50] John Boudourouglou, interview with author, Athens, 5 September 1974, written, AFS Archives.

[51] Ann Kellogg House reported this information related by an escaped student in *Newsletter*, 9 February 1949, AFS Archives.

[52] John Boudourouglou, interview.

convinced that he would return. He did. His odyssey was one of severe physical suffering. February and March had been bitterly cold in the mountains and the plucky youngster had stood in water up to his neck for more than two hours during his escape.

Spyros, the next to last to return, suffered badly from flat feet. The bandits had put him on a donkey because he could not keep up the pace. This, it might be added, was the only consideration he received at the hands of his captors. Later he was tossed into a prison and kept there for five days with no food or clothing. His survival and return exemplified the determination and extraordinary courage of these boys.

However, there was no doubt that the school had received a psychological jolt. The blatant violation on the part of the communists had unnerved the institution in a way that no other single event had done. Ann House wrote, "We here have been passing through what a very large number of village people in Greece have been living through during the past few years."[53] "We're praying," she continued, "that through the experience we may be more understanding and that we may be given the strength and wisdom to carry the spirit that will bring about peace and brotherhood among these suffering people. Violence that breeds hatred has touched this peaceful spot."[54]

Ann's comment indicates that the more violent realities of Greece's calamity had cut to the very quick of farm school life. Because of this violent intrusion, the school had lost its Garden of Eden aura. Yet, at the same time, Ann's comment reaffirms the spirit of the school. In no way were the boys allowed to become embittered. Their rehabilitation, she concluded, would be based on "the spirit of Christ's Kingdom."[55]

Nowhere is the spirit of the school more clearly articulated, the source of its survival more graphically illustrated, than through the story of the kidnapping. The courage of the boys was largely innate, but many of them reported that the strong religious underpinning supplied by the school had served as the cornerstone of their strength. One boy, for instance, kept repeating a hymn he had been taught.

---

[53] *Newsletter*, 3 February 1949.
[54] Ibid.
[55] Ibid.

Their attraction to the school was magnetic, as if some physical pull was drawing them homeward.

Characteristically, the school remained true to its mission, training the rural youth in things of the heart, hand, and mind. Classes for the younger boys had continued with only slight interruption during the kidnapping incident. Those who were kidnapped were sent home for a few days if their parents wished, but resumed their studies after a short visit.

The graduation held in spring 1949 was the most emotional in the farm school's history. Every member of the senior class that had been abducted was present to receive his diploma, and not a few of them as well as many in the audience felt that they were present through the grace of God. The graduates were ready to take their place on the soil, not unscathed by their experience, but with a heightened sense of the precious quality of human life.

The communists lost the war in 1949. That year marks the end of the violence in Greece that had been an ever-present reality since the school's founding in 1904. The country's recovery and the school's development in the postwar decades were accomplished in a peaceful way but by no means in a static environment. Greeks, kinetic and individual, when they are not the victims of history, are making it.

# SELECTED BIBLIOGRAPHY FOR BOOK ONE

*I. Primary Sources*

A. The history of the American Farm School has been drawn chiefly from materials in the American Farm School Archives. Some of these items lack dates, titles, page numbers, or authorship, including the newspaper clippings and magazine articles. Items that lack complete citations are identified in the notes and here as AFS Archives to indicate that they are filed in the American Farm School Archives.

Selected materials from the archives used in research for this book include:
1. American Farm School annual reports since 1907.
2. Missionary correspondence from the late nineteenth century.
3. Much of the House family correspondence and records such as Ann Kellogg House's diary from World War II, Ann Kellogg House's letters to her family beginning in 1923, Charles House's letters to his wife before their marriage in 1923, and Susan Adeline Beers House's correspondence to her children stretching over three decades.
4. Letters written by the Reverend Edward B. Haskell, the co-founder.
5. Unpublished manuscripts such as Joice Loch, "Dear Bruce Letters," Catherine Owen Pearce, "Lighthouse on a Grecian Shore," and Everett Marder, "The Regime of the Fourth of August: The Dictatorship of John Metaxas. Master's thesis, University of Cincinatti, 1969.
6. Official farm school documents and correspondence with government and philanthropic agencies such as the Greek Ministry of Agriculture, Head of American Legation in Athens, the American Consul General in Thessaloniki, the Bible Lands Mission Aid Society,

the American Board of Commissioners for Foreign Missions, UNRRA, the Ford Foundation, the Fulbright Agency and German Occupation forces.

7. Minutes of the American Farm School Board of Trustees.

8. Articles from Greek, English, and American newspapers and magazines beginning with 1901, on specific farm school topics or significant events in Greece.

9. Farm School public relations brochures and house-organ media such as *Newsletter* and the *Sower*.

10. Chronologies and summaries of school history by Joice Loch and anonymous writers.

B. Interviews conducted through correspondence or in person gave a valuable dimension to the understanding of the school.

C. The Macedonian Archives, Thessaloniki, furnished the original deed for the American Farm School land purchase and the first tax records.

D. The National Archives in Washington, D.C. were consulted for information concerning the Metaxas years and the early World War II period.

E. Other Primary Sources

Barry, G. C. "Relief Work Among the Villages of Mount Pangaeon: Report of the American Red Cross Commission to Greece" 1919. AFS Archives.

Davis, Homer. "History of Athens College." AFS Archives.

House, Ann Kellogg. "Memoirs," unpublished. AFS Archives.

Lansdale, Jr., Herbert P. "Summary of the Conversation Between His Excellency John Metaxas and Herbert P. Lansdale, Jr., in the Office of the Prime Minister on Tuesday, Dec. 13, 1938, 5–6:30." Unpublished. AFS Archives.

Metaxas, Ioannis. *Lógoi kai sképseis, 1936–1941* (*The Speeches and Thoughts of John Metaxas*). Volume 1. Athens: IKARON, 1969).

Morgenthau, Henry. *I Was Sent to Athens*. New York: Doubleday Doran & Co., 1929.

Royal Ministry for Foreign Affairs. "Diplomatic Documents: Italy's Aggression Against Greece." Athens: Royal Ministry of Foreign Affairs, 1940.

## II. Secondary Sources
## A. Monographs, General Histories

Allen, Harold. *Come Over Into Macedonia*. New Brunswick NJ: Rutgers University Press, 1943.

Averoff, Evangelos. *By Fire and Axe: The Communist Party and the Civil War in Greece*. New Rochelle NY: Aristide Caratzas, Publisher, 1978.

Bailey, Thomas. *A Diplomatic History of the American People*. New York: Appleton Century Crofts, 1969.

Barker, Elisabeth. *Macedonia: Its Place in Balkan Power Politics*. London: Oxford University Press, 1950.

Barros, James. *The Corfu Incident of 1923*. Princeton: Princeton University Press, 1965.

Bowman, Steven. "Messianism in Greece," 1972, unpublished paper. Author's possession.

Campbell, John, and Philip Sherrard. *Modern Greece*. London: Ernst Benn, 1969.

Cervi, Mario. *Storia della guerra di Grecia*. Milano: Sugar Editore, 1965.

Dakin, Douglas, *The Unification of Greece, 1770-1923*. London: Ernest Benn Ltd, 1972.

Dakin, Douglas. *The Greek Struggle in Macedonia, 1897–1913*. Thessaloniki: Institute For Balkan Studies, 1966.

Doxiades, Constantine. *Thisíes tis Elládos: Aitímata kai epanorthósis ston b' pangósmio pólemo* (*Sacrifices of Greece: Applications and Restoration in the Second World War*). Athens: n.p., 1947. AFS Archives.

Ford, Ford Maddox. *A Mirror to France*. New York: A. & C. Boni, 1926.

Forster, Edward. *A Short History of Modern Greece, 1821–1956*. Revised edition. London: Methuen & Co., 1960.

Grabill, Joseph. *Protestant Diplomacy in the Near East: Missionary Influence on American Policy, 1810–1927*. Minneapolis: University of Minnesota Press, 1971.

Hall, W. W. *Puritans in the Balkans, the American Board Mission in Bulgaria: A Study in Purpose and Procedure.* Sofia: Cultura Printing House, 1938.

Hofstadter, Richard. *Anti-intellectualism in American Life.* New York: Alfred A. Knopf, 1970.

———. *The Progressive Historians.* New York: Alfred A. Knopf, 1968.

Horton, George. *Recollections Grave and Gay.* Indianapolis: Bobbs-Merrill, 1927.

Housepian, Marjorie. *The Smyrna Affair.* New York: Farber and Farber, 1970.

Howell, Edward. *Escape to Live.* London: Longmans, Green & Co, 1958.

Iatrides, John O. *Revolt in Athens: The Greek Communist Second Round, 1944–1945.* Princeton: Princeton University Press, 1972.

Kousoulas, D. George. *Revolution and Defeat: The Story of the Greek Communist Party.* London: Oxford University Press, 1965.

Lash, Joseph. *Eleanor and Franklin.* New York: W. W. Norton & Co., 1971.

Lewis, Bernard. *The Emergence of Modern Turkey.* London: Oxford University Press, 1968.

Loch, Joice Nankivell. *A Fringe of Blue: An Autobiography.* London: John Murray, 1968.

Maule, Henry. *Scobie, Hero of Greece: The British Campaign, 1944–1945.* London: Arthur Baker, 1975.

Moore, Frederick. *The Balkan Trail.* New York: Macmillan Co., 1906.

Morgenthau, Henry. *I Was Sent to Athens.* Garden City: Doubleday, Doran & Co., 1929.

McNeil, William Hardy. *Greece: American Aid in Action, 1947–1956.* New York: The Twentieth Century Fund, 1957.

Nankivell, J. M. *A Life for the Balkans: The Story of John Henry House of the American Farm School, Thessaloniki, Greece.* New York: Fleming H. Revell Co., 1939.

O'Ballance, Edgar. *The Greek Civil War, 1944–1949.* London: Faber and Faber, 1966.

Palmer, Alan. *The Gardeners of Salonika*. London: Andre Deutsch, 1965.

Papacosma, S. Victor. "Minority Questions and Problems in East European Diplomacy Between the Wars: The Case of Greece." Paper read at American Association for the Advancement of Slavic Studies, October 1978. Author's possession.

Parker, Charles.*Return to Salonika*. London: Cassell, 1964.

Pentzopoulos, Demitrios. *The Balkan Exchange of Minorities and Its Impact Upon Greece*. Paris: Mouton, 1962.

Playfair, I. S. O. et al. *The Mediterranean and the Middle East*.Volume 1 of *The History of the Second World War*. London: Her Majesty's Stationary Office, 1954.

Psomiades, Harry. *The Eastern Question: The Last Phase*. Thessaloniki: Institute for Balkan Studies, 1968.

Queen Frederika of the Hellenes. *A Measure of Understanding*. London: Macmillan, 1971.

Sanders, Irwin. *Rainbow In the Rock*. Cambridge: Harvard University Press, 1962.

Spencer, Floyd, editor. *War and Post War Greece*. Washington, DC: European Affairs Division. Library of Congress, 1952.

Stavrianos, L. S. *The Balkans Since 1453*. Hinesdale IL: Dryden Press, 1958.

Stevenson, Adlai F. *Call to Greatness*. New York: Harper, 1954.

Stoianovich, Traian. *A Study in Balkan Civilization*. New York: Alfred A. Knopf, 1967.

St. Clair, William. *That Greece Might Still Be Free*. London: Oxford University Press, 1972.

Tsoucalas, Constantine. *The Greek Tragedy*. Baltimore: Penguin Books, 1969.

US Command and General Staff College. *Internal Defense Operations: A Case Study of Greece, 1946–1949*. Fort Leavenworth KS: US Army Command and General Staff College Reference Book, 1967.

Vacapoulos, Apostolos. *A History of Thessaloniki*. Thessaloniki: Institute for Balkan Studies, 1963.

Vouras, Paul. *The Changing Economy of Northern Greece Since World War II*. Thessaloniki: Institute for Balkan Studies, 1962.

Wolff, Robert Lee. *The Balkans in Our Time.* New York: W. W. Norton and Co., 1967.

Woodhouse, C. M. *The Struggle for Greece, 1941–1949.* London: BEEKMAN/ESANU Publishers, 1976.

Zotiades, George. *The Macedonian Controversy.* Thessaloniki: Institute For Balkan Studies, 1961.

B. Articles

Chatfield, Charles. "World War I and the Liberal Pacifist in the United States." *The American Historical Review* 75/7 (1970): 1920–37.

Dakin, Douglas. "British Sources Concerning the Greek Struggle in Macedonia, 1901–1909." *Balkan Studies* 2/1–2 (1961): 71–84.

Damaskenides, A. N. "Problems in Greek Rural Economy." *Balkan Studies* 6/1 (1965): 21–34.

Marder, Everett J. "The Second Reign of George: His Role in Politics." *Southeastern Europe* 1/2 (1975): 53–69.

Papadopoulos, Stephanos. "Ecoles et associations grecques dans la Macedoine du nord durant le denier siécle de la domination turque." *Balkan Studies* 3/1–2 (1962): 397–442.

1069. Σκηναί καί τύποι της Μακεδονίας — Οἰκογένεια χωρικῶν
Scènes et types de Macédoine. — Famille de paysans.

Susan Adeline Beers House and Dr. John Henry House
(Mother House and Father House).

Dr. John Henry House with graduating class of 1914.

Mother and Father House and their six children, 1929. Top row: Ruth, Gladys, Father, Mother, Charlie. Bottom row: Jack, Ethel, Grace, Florence.

On our way to the
Aegean Sea. For a Swim.
The Camionette is in
a ditch. Mother is waiting!
Our Last day. 16 Aug. 1921.

Donka Ilieva    Marecka Stamenova    Miss Stone    Mrs. Kerefinka Oosheva
                Mrs. Tsilka            Athena Dimeva

MISS STONE'S BIBLE CLASS IN BANSKO, MACEDONIA
*Five of whom were with her when she was abducted*

McClure's Magazine published a five-part autobiographical account of the kidnapping and ransom of Ellen M. Stone and Katerina Tsilka entitled "Six Months Among the Brigands," May-September 1902.

By Claude Shepperson

"IF IT IS NOT PAID. . . ."

THE MOTHER AND HER BABY

*The first picture of them ever taken*

MRS. TSILKA, HER BABY BORN IN CAPTIVITY, AND MISS
STONE

*From a photograph taken specially for " McClure's "*

THE BABY AND GLADYS HOUSE, THE DAUGHTER OF
MR. J. H. HOUSE, SENIOR MISSIONARY AT SALONICA

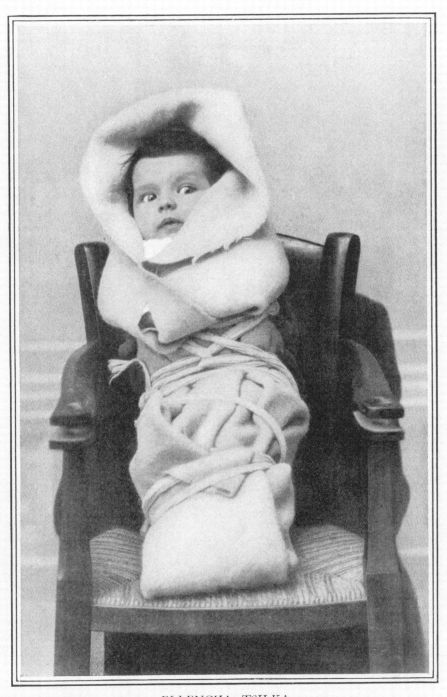

ELLENCHA TSILKA

*From a photograph taken when she was about three months old*

ELLEN M. STONE
FROM HER LATEST PORTRAIT

Some of the first boys at the Farm School, orphans from Monastir.

Haskell Cottage, the first Farm School building, was constructed in 1903.

The Albanian Guard, Hassan, in front of James Hall, completed in 1906.

The boys in their summer uniforms, 1911.

Graduating class studying
Bible with Mrs. House

A corner of the green house given by Dr. Cowles

Monastir Macédonienne en costume national

Essery Memorial Well. The inscription from the Gospel of St. John "Whosoever thirsts, let him come unto me and drink" was carved in Greek, Turkish and Bulgarian languages.

The caption in Ann House's photo album reads, "Some members of the Friendship Association. There is no place to tell of the unfailing help the School received from Officers, Nurses and men, not only in times of special need, but also in numberless friendly deeds from day to day. Those who went through those trying years together will never forget."

# Thessalonica Agricultural and Industrial Institute

When Peace comes we will be able to buy supplies in Salonica at **bargain** prices.

**Five Dollars** may provide us with FIVE or TEN hospital BEDS or **50** bags of CEMENT, etc.

Where to send the $5.00 ?   WM. B. ISHAM, Treas.

27 William Street

N. Y. City

For Information write

MISS ELLEN M. STONE   or   L. HOLLINGSWORTH WOOD

7 Gramercy Park. N. Y.   20 Nassau St., N. Y. City

From the beginning, the School solicited the help of supporters in the United States to keep its educational programs running and to make possible the School's relief efforts for the population in the countryside surrounding Thessaloniki.

# To Put Heart in the Balkans

Drawn to the Dardanelles, the *Queen Elizabeth,* greatest engine of destruction produced by Western civilization, batters the Turkish forts.

Drawn by the love of these ancient, backward people, Dr. J. H. House has brought the constructive seed of agricultural and industrial education to break down the walls of prejudice and ignorance.

The Thessalonica Agricultural and Industrial Institute has for 12 years been teaching new, Hampton methods in that old, ill-governed land.

The inhabitants love the school. It came through the Balkan wars without the loss of a dollar's worth of property, though neighboring farms were looted.

Famine prices rule in Salonica now. Perhaps the Allies' guns may bring down Russian corn. Will we have money with which to buy?

Will you send $5.00 (via William B. Isham, Treasurer, 27 William Street) to Dr. House and his wife and daughter and the native teachers to keep going this plant which will make real growth possible when the destruction ceases?

Ask any Balkan Relief Agency about the School.

LUCIUS H. BEERS
Vice-President
L. HOLLINGSWORTH WOOD
Chairman Finance Committee

Dated: New York
August 1st, 1915

# A Balkan Rescue

**The Allies**      The Allies abandoned their attempt to force the Dardanelles.

**Dr. House**      Dr. House has kept right on at Salonica in his work of bringing the forces of farm and home to the Balkan peoples.

**American Aids**      American bacteriologists showed the Balkan people how to stamp out the frightful typhus plague, and American experts in agriculture and homebuilding are showing the survivors how to live.

**Aeroplanes and Eggs**      The Allies' aeroplanes frighten our chickens, but the officers of the allied armies at Salonica pay 4 cents apiece for our fresh eggs.

**How you get there ?**      You can send $5.00 (via William B. Isham, Treasurer, 27 William Street) to Dr. House and his wife and daughter and the native teachers to keep going this plant which will make real growth possible when the destruction ceases. This will help our hens support a real work.

Ask any Balkan Relief Agency about our School.

LUCIUS H. BEERS
Vice-President
MARGARET ARMSTRONG
L. HOLLINGSWORTH WOOD
Finance Committee

Dated: New York
March 28th, 1916

Charles Lucius House at Princeton, probably 1909.

Charlie posing with "Teddy", left, and Father, Mother and sister Ruth, right, in front of James Hall, 1917.

This photo was called "Peace and War" by Charlie. His brother, John Henry House, Jr., poses proudly in uniform, while Charlie, conscientious objector, smiles playfully at the camera.

7

Nichola and Joanis, Greek boys-
refugees from Asia Minor to the
Caucasus, from there with Wrangel's
Army to Salonica. Just admitted
to the School. August 1921.

47

Russian Officers of
Wrangel's Army at
work on the new Laundry
as day Laborers.

Princeton Hall's cornerstone was laid in 1924. Refugees from Asia Minor worked to complete the building several years later.

The campus and fields of the American Farm School circa 1930.

Charles Lucius House and Ann Kellogg House on their wedding day, 2
December 1923.

Charlie and Ann House welcomed a constant stream of students and visitors to their Metcalf House living room.

The student body on the steps of Princeton Hall, late 1920s.

# In the Land of Aristotle

## *"Learning by Doing"*

THESSALONICA AGRICULTURAL
AND INDUSTRIAL INSTITUTE

SALONICA, GREECE

An informational and fundraising brochure for friends and supporters of the
School in the United States, 1927.

Student farming activities during the 1920s.

# ALL QUIET
## on the Balkan Front

ONE hears little nowadays of unrest in the Balkans—
that quondam hotbed of European troubles. All is
calm for the present. Even in the cat's paw region of
Macedonia, last stronghold of Turkey in Europe, now
divided among Serbia, Bulgaria and Greece, quiet
reigns. Vanished from Salonica, Near East base of the
Allied Armies after the Gallipoli fiasco, are the whir-
ring wings of planes in flight, the roaring machine
shops, the million marching feet of that armed camp.
Peacetime shipping crowds the harbor below Olympus.

And what of the million and a half hungry, ragged
refugees whose forced exodus from the flames of
Smyrna made harrowing headlines less than ten years
ago?

"The American Farm School is the most effective type of educational and social
service now being rendered in that troublous part of the world. It is related to the
daily needs of the people – their health, food, home life, recreation and religious
faith. These are the influences which on the long view will eliminate wars and
establish the peace for which we all yearn." From brochures explaining the mission of the American Farm School and its service to Greece and the Balkans,
1931.

# A GREECE
## Unknown to Baedeker

TOURISTS seldom wander far from the Acropolis. From a moonlit ruined temple looking eastward to the Isles of Greece, a spell of ancient romance seems to hang over the land. Mudholes in the roads discourage visits to the back country where grim reality keeps the nose of the peasant to the grindstone. Trudging over barren fields behind his gaunt cows and primitive stick plough, he is about as unromantic a figure as can be seen anywhere on the continent. His is the common lot in a land nine-tenths rural.

Here and there are signs of new leaven working. Modern steel ploughs are gradually replacing old Testament wooden ones. Water is sometimes piped instead of carried in jugs on the heads of prematurely old

---

# A FAREWELL
## TO ARMS

DIGGING in Greece is by no means confined to archeologists. Nine out of ten Greeks dig for their living, depend for bread and butter on the fruits of the earth. Time was when promising sons of peasants migrated to the city to seek their fortune in trade or shipping or turned to brigandage, whilst duller brethren stayed on the land in the clutch of primitive custom and superstition.

But Greece is undergoing a change of heart. Tillage is displacing pillage. The age old scorn of the townsman for his rustic cousin is gradually giving way to a healthier sentiment of pride in fields and flocks, as youthful rural leaders trained in the American Farm School demonstrate what can be done by rotating crops and modern dry farming methods. Word of the doings at this School near Salonica has gone the rounds of the villages despite wretched roads and illiteracy.

---

# TOMORROW
## and TOMORROW

LITTLE by little the need for emergency relief in the Near East is passing. Colossal sums have been poured out by American philanthropy for the alleviation of devastating misery. From now on our endeavor should be less to do things for these retarded people than to train them to do things for themselves. On the basis of his tours and survey of the Near East, Dr. Thomas Jesse Jones, Educational Director of the Phelps-Stokes Fund, feels that

"the American Farm School is the most effective type of educational and social service now being rendered in that troubled part of the world. It is related to the daily needs of the people—their health, food, home life, recreation and religious faith. These are the influences which on the long view will eliminate wars and establish the peace for which we all yearn."

Nearly thirty years ago the American Farm School at Salonica was first opened to give young Greeks

Graduation ceremonies and other celebrations were conducted on a grand scale during the 1930s, and demonstrated the myriad practical agricultural skills mastered by American Farm School students.

The School's milk wagon leads the parade. *The New York Times*, 30 September 1934, recounts: "The American Farm School prepares the only pasteurized milk in Greece, but is obliged to deliver its product to Saloniki by stealth because of the unorthodox nature of the milkwagon constructed by Greek boys at the School. The police, asserting that the wagon could not turn corners, long ago banned it from the streets."

The School was known for its fight against malaria in the Greek countryside.

# PAVING THE WAY
## *for* PEACE

> The peace of our part of the world depends very much on whether we can bring to a happy solution the agrarian question and I see in your institution a fundamental contribution to the ends we all so deeply desire.
>
> *E. K. Venizelo*
>
> ### Premier of Greece

### THE AMERICAN FARM SCHOOL
### SALONICA, GREECE

Premier of Greece Eleftherios Venizelos and his wife Helena were strong supporters of the School. Mrs. Venizelos made a gift in 1931 of $1,000 from her personal funds for the founding of the future Girls School.

# Quartier-Anweisung!

~~~~~~~~~~~~~ Die Amerikanische Farm-Schule

Ort **S a l o n i k i** Strasse ...........................................

hat Quartier zu leisten ohne Verpflegung für 30 Tage

ab **16.4.42** , bis **15.5.42**

| | | | |
|---|---|---|---|
| ./. Generale | | 5 | Unteroffiziere |
| ./. Stabsoffz. | | 21 | Mannschaften |
| 1 Hauptleute | | 5 | Geschäftszimmer |
| 3 Oberleutnant u. Lt. | | 1 ~~Mannschaftsheimerum~~ | |
| 2 Oberfeldwebel | | 8 | Garagen |
| 1 Feldwebel | | 1 **Raum als Offz.Casino** | |

~~...LICH RICHTIG~~ Kloppstnummer : 0.1690

Zahlung an das ... u. den 15.5.1942

~~...eueramt~~

~~...kommandantur~~ Standortkommandantur

~~...uartieramt~~ Quartieramt

## Merkblatt zur Quartieranweisung !

Quartierscheine sind am 1. und 15. jeden Monats beim
Quartieramt zu erneuern. Die vorhergehenden Scheine sind
dem Quartiergeber auszuhändigen, damit dieser vom Steueramt
das Quartiergeld ausgezahlt erhält. Beim Abrücken der Einheit
aus dem Quartier muss dieses beim Quartieramt abgemeldet
werden.
   Die Quartierscheine werden den Quartiergebern übergeben,
wobei die Tage des Aufenthaltes auf dem Quartierschein vom
Führer der Einheit zu bescheinigen sind.

## Παρατηρήσεις επί τῶν ἀποδείξεων ἐνοικίου

Αἱ ἀποδείξεις ἐνοικίου ἀνανεοῦνται τὴν 1ην καὶ 15ην ἑκάστου μηνὸς εἰς τὸ
Γραφεῖον Καταλυμάτων (Quartieramt). Αἱ ἀγόμεναι ἀποδείξεις δίδονται πρὸς τοὺς
χορηγοῦντας καταλύματα, ἵνα οὗτοι εἰσπράξωσι ἀπὸ τὴν Οἰκονομικήν Ὑπηρεσίαν τὸ
ὀφειλόμενον ἐνοίκιον. Ὅταν μία μονὰς ἀναχωρεῖ ἀπὸ τὸ κατάλυμα πρέπει νὰ γνω-
στοποιεῖται τοῦτο πρὸς τὸ Γραφεῖον Καταλυμάτων.
   Αἱ ἀποδείξεις ἐνοικίου παραχωροῦνται πρὸς τοὺς χορηγοῦντας καταλύματα
ἵνα σημειῶνται ἐπ' αὐτῶν ὑπὸ τοῦ ἀρχηγοῦ τῆς μονάδος αἱ ἡμέραι τῆς παρα-
μονῆς των ἐκεῖ.

Official requisition form for billeting German officers and men in Farm School
campus housing.

Ann House's note reads: "17 March, 1942. Captain Severin, prison commander, Thessaloniki, Mrs. Kaier, Ann and Charlie House in front of their home on the campus. Severin allowed food to be brought from the Farm School to starving prisoners who would have died were it not for his aid."

The swastika flying over Eleftherias Square, Thessaloniki (reproduced by kind permission from Hagen Fleischer's "Stemma kai Swastika," Vol. II, Athens 1995).

James Hall prior to World War; its destruction in 1944; its rebuilding in 1948.

ΒΑΣΙΛΕΙΟΝ ΤΗΣ ΕΛΛΑΔΟΣ

ΥΠΟΥΡΓΕΙΟΝ _____ ΓΕΩΡΓΙΑΣ _____

ΓΕΝΙΚΗ Δ/ΣΙΣ _____

Διεύθυνσις Γ. Μ. & Ἐρευνῶν

Τμῆμα Γ. Ἐκπαιδεύσεως

Γραφεῖον _____

Ἀριθ. πρωτ. ___23015___

'Εν 'Αθήναις τῇ *18*-5- 194 5

Πρὸς

15874 /H2/44

25-5-45

Τὸ Ὑπουργεῖον τῶν Ἐξωτερικῶν
Δ/σιν Πολιτικῶν Ὑποθέσεων
Τμῆμα Βαλκανοκῆς
Ἐ ν τ α ῦ θ α

Ἀπαντῶντες εἰς τὸ ὑπ'ἀριθ.4274 ἐ.ἔ. ὑμέτερον
φον, ἔχομεν τήν τιμήν νά σᾶς γνωρίσωμεν ὅτι ἡ ἐπαναλει-
τουργία τοῦ Ἀμερικανικοῦ Γεωργικοῦ Σχολείου Θεσ/νίκης
εἶναι ἐνδεδειγμένη ἵνα οὕτω συνεχισθῇ τό τόσον καρποφό-
ρον ἔργον αὐτῆς. Τό Ὑπουργεῖον τοῦτο λίαν εὐχαρίστως
θά ἔβλεπε καί πάλιν ἐπικεφαλῆς τοῦ ἀνωτέρω Ἱδρύματος
τόν κ. Κ. Χάουζ ὅστις μέ τόσον ζῆλον, ἀφοσίωσιν καί ἀπό-
δοσιν διηύθυνε τοῦτο μέχρι τῆς ἐνάρξεως τοῦ λήξαντος
πολέμου, ἐπιδεικνύων συνάμα κατά τήν ἐξάσκησιν τῶν καθη-
κόντων του αὐτῶν ὡς καί διά τῆς ἐν γένει συμπεριφορᾶς
αὐτοῦ τό θερμόν διά τήν Ἑλλάδα ἐνδιαφέρον του, τῶν συν-
θηκῶν τῆς ὁποίας εἶναι καλός γνώστης.

Ὁ Ὑπουργός
Π.ΚΟΥΤΣΟΜΗΤΟΠΟΥΛΟΣ

Ministry of Agriculture document recognizing the School's contribution to
Greece and requesting its reopening after the war, 15 May 1945. (reprinted by
kind permission of the Ministry of Foreign Affairs Diplomatic and Historical
Archives. Research courtesy of Hagen Fleischer.)

# BUILDING FOR PEACE
## *in the Balkans*

## AMERICAN FARM SCHOOL
### SALONICA, GREECE
1902-1947

The senior class was reunited in June 1949 after their abduction by a band of
*andártes* the previous January. Dimitris Hadjis instructs the class.

# Book 2

## 1950–2003

# FOREWORD TO BOOK 2

Thankfully, the second half of the twentieth century was not so destructive as the first half (at least not in Europe), yet the American Farm School experienced sufficient vicissitudes from 1950 to 2002 to keep it on its toes. Happily, however, the school has flourished spiritually and even sometimes economically owing to a creative energy that produced significant expansion and change.

The most important developments that stand out in this period are: (1) the advent of coeducation thanks to the incorporation of the Quaker Girls' School, (2) the lyceum option for secondary school students, (3) the founding of the Dimitris Perrotis College of Agricultural Studies, which led as well to the splendid upgrading of the Zannas Farm and to a new library on the second floor of Princeton Hall, (4) the enhanced program of rural development involving lifelong learning for adults, (5) the application of research to agricultural production, creating Greece's most advanced dairy herd, poultry unit, and in sum a model farm that is widely copied by others, (6) the servicing of an international student body at the post-secondary level, first with Africans and Middle Easterners enrolled in short courses starting in the 1950s, then with Balkan neighbors attending Perrotis College after 1996, (7) the European Union's financial and programmatic support to and trusting reliance on the school to carry out EU goals.

Of course one could add, despite fear that the list might become too long, the conducting of student exchange programs, the enhancement of English language instruction, the placement of young American interns in the school and college dormitories, and..., and...

Yet the vicissitudes were sometimes severe and many were overcome thanks to patience, political savvy, compassion, and luck. Interestingly, political problems derived from both the right and the left—from the former chiefly after the civil war, from the latter during Andreas Papandreou's socialist premiership when private schools were denigrated and sabotaged. Fortunately, the farm school's leadership was able to cope with government officials as well as student unrest during this radicalized period.

New issues have risen in recent decades, a major one being the encroachment of Thessaloniki. Father House's barren acres in the middle of nowhere are now an agricultural oasis in the middle of housing developments, supermarkets, ring roads, and automobile dealerships—the detritus of an ever-expanding metropolis whose population, some predict, will eventually reach 4 million. Neighbors complain of cow-smell; wells no longer suffice or are in danger of pollution, salination, or both. Can the farm—especially its cattle, pigs, chickens, and turkeys—survive in the present location?

And what about the school's exciting new college, populated in part now by Albanians, Bulgarians, and other neighbors? Although the institution is reviving Father House's vision of a pan-Balkan student body, it needs to be exceedingly proactive in order to attract Greek students, who have been accorded increased opportunity to enter the nation's state-supported universities.

These are some of the problems that remain to be solved in the twenty-first century. Book 2 of this twentieth-century history chronicles their origin and growth. It also traces the degree of success with which previous challenges were confronted. Most importantly, book 2, like book 1, memorializes extraordinary people, both Greek and American, showing how their belief in the school gave them fortitude, forbearance, useful cunning, and a general hardiness, all of which equipped them to fertilize and cultivate the institution.

Peter Bien
Riparius, New York
23 July–26 August 2002

# ACKNOWLEDGMENTS

Source materials for the writing of book 2 were based nearly as much on interviews as on written materials. For all those who granted me face-to-face interviews, as well as those who contributed by e-mail and telephone, I thank them for their thoughtful commentary. I interviewed many other people, such as Mimi Cobb and Zina Pantazis, whom I have not cited, but who gave me invaluable insights.

The Wayland Massachusetts Library through its connection to the Minuteman Network and access to other extensive systems offered me the advantage of borrowing books nationwide. I wish to thank all the research librarians there who helped me work through the multitude of electronic pathways. I was aided in my study by Brandeis University's extensive resources and by guidance from its research librarians to whom I owe a note of thanks. The staff of the Dimitris & Aliki Perrotis Library at the American Farm School researched certain important items and assisted with the photo research. Eleanor Brown extended herself to insure that computer resources were always at my disposal.

The book could not have been written without the reestablishment of the American Farm School Archives by Charlotte Draper. Her professional attention to that project made my research an orderly task.

I am especially grateful to the following people who read my manuscript at certain key moments in its development and offered incisive comments: Peter Bien, George Daoutopoulos, George and Charlotte Draper, John O. Iatrides, George A. Koulaouzides, for both the English and Greek editions, Edward Mendler, Bruce and Tad Lansdale, and Everett Marder. Ioannis D. Stephanidis deserves credit not only for his translation to Greek of book 1 and book 2 but also for his critical comments. And although I heeded much of their advice, I, alone, am responsible for all the opinions expressed herein. Randall Warner, who heads the publications office at the farm school, has aided me in almost every phase since I first undertook the project. She has acted as archivist, fact checker, editor, and friend from the very first. I owe her immense gratitude.

To the board of trustees of American Farm School, supportive of my project from the beginning, I wish to express my deep appreciation.

Accepting Charlotte and George Draper's invitation in 1998, I embarked on writing book 2 of *Stewards of the Land* to mark the school's centennial celebration. I also promised to participate in the translation to Greek of book 1 and book 2 for the Greek publication. It is now 2003 and the work is finally done.

It was Peter Bien who was instrumental in the publication of *Stewards of the Land: The American Farm School and Modern Greece* in 1979. In addition, he understood the value of bringing the 100–year story of the American Farm School together in this single volume, and assisted me in the total undertaking. Once again I express to him my deepest appreciation for his initiative, hard work, and encouragement.

For the research for book 2, I made many trips to Greece. In its legendary spirit of hospitality, the entire American Farm School community made my frequent stays on campus enjoyable and productive. If the book is at all successful in bringing alive the history of the American Farm School, it is due to the readiness of the school's presidents and their wives to share their experiences and special knowledge: Bruce and Tad Lansdale, George and Charlotte Draper, and David and Patti Buck.

For helping to underwrite the publication of this volume, I am indebted to the Bodosakis Foundation, the J. F.Costopoulos Foundation, the Costas and Mary Maliotis Charitable Trust, the Alexander S. Onassis Public Benefit Foundation, Peter Bien, and Euterpe Dukakis.

Brenda L. Marder
Hanover, New Hampshire
5 June 2003

# INTRODUCTION

To decipher Greece's historical progress from antiquity to modern times would be the work of a lifetime. But by isolating a century and studying it from a designated perspective, we can at least glimpse one aspect of the country's life. Perhaps this story may serve that smaller purpose, for the history of the American Farm School is a portal leading to a significant segment of twentieth-century Greece.

*Stewards of the Land: The American Farm School and Greece in the Twentieth Century 1950–2002* traces selected aspects of postwar Greece—and often refers to even earlier periods—that best lend context to the history of the American Farm School as it strove to improve the quality of education it offered to rural youth during this transforming period of modern Greek history. Strewn with barriers, this course was as rocky as the Greek terrain itself. Political events such as the imposition of a military junta from 1967 to 1974, with its dulling nationalism, and the first two governments of Andreas Papandreou (from 1981 to 1985 and 1985 to 1989), who advocated an anti-American policy, placed stumbling blocks along the farm school's path since as a guest in Greece, the American-chartered institution, determined as always to remain apolitical, nonetheless depended on the good graces of and smooth cooperation with the host government to conduct its affairs.

Glancing back at the widespread poverty and social strife that scarred Greek life from the time of the school's founding in 1904 until mid-century, and the upheavals that stained the postwar period until 1974, we note at the start of the twenty-first century that the misery of the past has been relegated mainly to history books; the glitter of Greek life today has largely blotted out the dark side of history.

As peace and prosperity spread across the land, the challenge for the farm school in the second half of the twentieth century became altogether more abstruse. Greece's accession to the European Union in 1981 forced the institution to think globally and the country to relinquish a measure of its sovereignty as it fell into step with the most advanced democracies of Europe.

In this altered atmosphere the school's purpose became larger than simply transforming hungry village boys into skilled tillers of the soil and impressing upon them the dignity of manual labor, or even expanding their spiritual sensibilities, projects that may seem to readers rather quaint from the vantage point of the twenty-first century. Instead the problem at hand became more nuanced, leading the director, the faculty, and the board of trustees to pinpoint the shifting challenges. After all, the institution's survival depended on the utility of its mission. Would American donors support a school in far-flung Thessaloniki that served no meaningful purpose? What Greek students would choose to attend the farm school unless it offered them a real benefit? And would Greek benefactors continue to contribute to an institution that was merely a touching anachronism? What Greek ministry of education or ministry of agriculture would cooperate with an irrelevant institution? When the traumas of war had healed, and the political situation had stabilized, the school's self-searching led invariably to the urgent question, "Why an American Farm School in Greece?" It is this question that must stand as a pivot point for book 2. In contrast, the question never surfaced in book 1, covering 1904–1949, despite the turmoil in the Balkans: the earlier times, for a multitude of reasons, were ripe for the existence of such an institution.

Adding to the complexities facing the school during the last decades of the twentieth century, the student body had undergone a profound sociological swing. In former times, the youngsters had completed at best six years of elementary school—mere farm boys—when they entered the farm school. After 1978, when girls were integrated, the students, now adolescent coeds, had completed nine years of school, were well-fed, more sophisticated, better educated, possessed by an urban lifestyle they viewed in the media that often conflicted with the more homey values inculcated by the school.

In its educational evolution, the school had metamorphosed over the century from a vocational junior high (1904–1965), to a four-year trade school (1965–1971), then to a two-year technical school operating simultaneously with a three-year vocational high school (1977–1998), and finally to a three-year technical school operating simultaneously with a three-year comprehensive high school, established in 1998. The inauguration of the Dimitris Perrotis College of Agricultural Studies in 1996 with its international student body not only brought post-secondary students to the campus, but also Albanians and Bulgarians entered the mix. Just after the harrowing Bosnia Herzogovina crisis and during the Kosovo calamity, students from Macedonia (called FYROM: Former Yugoslavian Republic of Macedonia or Skopje by Greece) also enrolled. And in 1998 adult education was reorganized into a formal department of Lifelong Learning.

The school's task from 1950 on was essentially to keep pace with the social, cultural, political, and economic changes that shook the country, to adapt creatively by putting its own stamp on the changing curricula mandated by the Greek government, and to maintain its role as a pioneer in the domain of agricultural education for youth and adults as well. Converting the latest agricultural technology to suit Greek conditions in a society thirsty for modern systems was also of primary importance. From its inception the school was known as a pacesetter, introducing various types of agricultural machinery, new crops and applications, and improved livestock; any lag in presenting the latest inventions, it was thought, could jeopardize this reputation. Equally pertinent, the school's farm, to maintain its function as a teaching tool and demonstration center, and to remain financially viable as an operation, had to keep current with the latest advances. Yet, the huge leaps in technology and the expense involved in staying apace created a quandary.

Throughout, the institution did strive to embrace the founder's original ethos: instilling a respect for the individual; cultivating a solid work ethic; developing a sense of cooperation and community; focusing on practical education; and underscoring religious values, but the last with decreasing emphasis as the leadership became more secular. Now that the older members of the staff who had personally known Charlie House or Bruce Lansdale have each retired, the transmission of these values may no doubt become problematic.

Basically, what came out of the periodic studies and frequent evaluations conducted on campus, usually by committees composed of board members joined occasionally by outside observers, was this conclusion: the imperative for Greek farmers to learn entrepreneurial skills that pertain to agricultural management. These include planning, quality control, packaging, accounting, and the uses of technology to manage both agricultural production and agricultural business and to market end products successfully. As Europe entered the information age, it was crucial for growers to participate in the new economy. As early as the 1960s the rural Greek was no longer a peasant snared in the limited world of subsistence farming, but a person "whom one approaches differently from a peasant," as sociologist and farm school trustee Irwin Sanders comments, "for his attitudes and values are different."[1] With the modern application of farm machinery, proper seed, fertilizer, irrigation, pest control, and marketing, in addition to land consolidation, price supports, and loans from the Agricultural Bank of Greece, improvement in living standards had become generally available to the average farmer. Clearly the time had come for Greece to educate future managers of agricultural enterprises, not peasants bowed by profitless drudgery like those doomed creatures depicted in Millet's evocative paintings.

The man who led the farm school from 1955 to 1990 was Bruce M. Lansdale, its third director, who first came to Greece in 1926 in a bassinet at the age of six months. A gregarious person and a natural leader who loved Greece and its rural folk, he knew the country more thoroughly than most foreigners ever do when they adopt a second land as their spiritual home. Bardic by nature, he shared his knowledge through storytelling, weaving together didactic myths concerning the farm school and the droves of people it touched. No characterization of Bruce would be accurate without reference to his use of Hodja parables. "For many years Hodja coveted a priceless gold coin owned by a simple uneducated monk who lived in a hermitage high above the village. Finally one day he decided to climb the hill..."[2] Through the medium of Hodja, a Near-Eastern wise man and folk figure—poor, simple, downtrodden—who lives by his nimble wit, Bruce drew an affecting

---

[1] Irwin T. Sanders, *Rainbow in the Rock: The People of Rural Greece* (Cambridge: Harvard University Press, 1962) 60. Professor Sanders's book is still considered a valuable resource for understanding rural Greece in the postwar period. As a rural sociologist and a member of the board of trustees of the farm school, he conducted a series of illuminating studies setting the school in the context of Greece's rapid postwar development.

[2] For the rest of the anecdote, see Bruce M. Lansdale, *Cultivating Inspired Leaders: Making Participatory Management Work* (West Hartford CT: Kumarian Press, 2000) 2. For a standard collection of Hodja stories, see Alice Greer Kelsey, illustrations by Frank Dobias, *Once the Hodja* (New York: Longmans, Green, 1963).

profile of the Greek peasant. The results of Hodja's daily dealings are a triumph of trickery over stupidity, acceptance over rejection. In addition to relying on his old friend Hodja, Bruce found dozens of ways to portray rural life, often in a lyrical vein. Thus he writes, "The attitudes of the more sympathetic city people and foreigners toward the Greek peasant after World War II expressed the paradox of development. They urged him to progress and at the same time expected him to be changeless. They cherished their leisure with him but encouraged him to be more active and organized. They enjoyed the primitiveness of his village but were forever telling him to build a better house and dairy barn and keep his cows cleaner. They wanted to sing and dance with him, yet insisted that he should make better use of his time."[3]

Over Bruce there hovered an atmosphere of spiritual probing; the quest for spirituality mattered to him and seemed to deepen with age. It would be difficult to overstate his influence on the institution and hence on the history of agriculture and rural society, certainly in Northern Greece. Furthermore, he was attended to in the corridors of power in Athens, not merely at the American embassy, but most significantly for the farm school, in parliament, in the ministry of agriculture, and even at the palace before the monarch fled in 1967. In fact, he had about him the aura of celebrity. Indeed, his genius was best exemplified in cultivating warm relations with key members of Greek governments and the country's elites, in his efforts to put them in service to the farm school, while he himself remained, as farm school tradition had prescribed, apart from the overheated world of Greek politics. He was aided by his wife, Tad (Elizabeth Krihak Lansdale), an inspired woman of wit and stamina. To add spice, there was always about her the hint of daredevil as she skied the ungroomed flanks of tall mountains and climbed to Greece's highest peaks. Indefatigably she hosted hundreds of visitors who arrived on campus annually. She completely shared her husband's vision and together they set their own indelible signature on life at the farm school during their thirty-five year tenure.

Even at the early age of twenty-nine, having been selected and prepared by Charlie House for the job, Bruce, who was the brother of a Presbyterian minister, the son of a YMCA director in Greece who later became the YMCA's general secretary for worldwide operations, and the

---

[3] Bruce M. Lansdale, *Master Farmer: Teaching Small Farmers Management* (Boulder CO: Westview Press, 1986) 16.

grandson of another YMCA executive, seemed temperamentally and culturally the rightful heir to the House family's creation. He was installed as the director of the American Farm School on its fiftieth anniversary in a ceremony—attended by King Paul and Queen Frederica of the Hellenes, Crown Prince Constantine, and a throng of Greek and American dignitaries and friends—only six years after the end of the Greek civil war. He took his place at the center of a web of wise and capable men totally dedicated to the school, who were determined that he would succeed.

When Bruce retired in 1990, many questioned whether the American Farm School, so dependant on his iconic stature, could long survive his leaving. In effect, his departure marked the end of a triumvirate, so closely bound was he to the House spirit. "I am following in the footsteps of a legend," was how George Draper, his successor, bemoaned his fate. A major theme of the school's first half century was its survival under dire threat during the Macedonian Struggle, the Balkan Wars, the Asia Minor debacle and the population exchange, World Wars I and II, followed by the Greek civil war, events all world-class and treacherous that pummeled Greece in nearly unbroken sequence. Ironically, once these hazards receded, many dreaded that the farm school would wither under unproven leadership. Others entertained another serious misconception, just as ironic: that a modernized Greece with its European Union orientation, up-to-date institutions and attitudes, would render an American farm school completely redundant.

Next to the presidency came George Draper. He stood apart from the Protestant missionary tradition in that he had not been designated, as was Bruce, for a place in the line of succession. Instead he applied for the job as any candidate might and was chosen by the board of trustees through a selection process. By profession he was an educator and his wife, Charlotte, a librarian. The couple had previous contact with Greece when he served at prestigious Anatolia College in Thessaloniki, first as teacher from 1967 to 1971, then as vice president from 1978 to 1983. In its epilogue, this volume sums up the progress of the school under Charlotte and George, who demonstrated during his decade a brand of leadership and vision new to the farm school, but ultimately bold and successful. He expanded its parameters by embarking on university-level education and applying European Union funds to broaden its national and international reach. The epilogue also singles out the events of the waning twentieth century as they molded Greece and the American Farm School just as the institution was striding toward its 100th birthday in 2004. George and Charlotte were followed in 2000 by David Buck and his wife Patti, who will, by the time the school celebrates its centennial, have etched their own pattern into the farm school's history.

A predominant topic in book 1 was the Hellenization of the American Farm School, which traces the school's evolution from an American outpost in the Balkans, spearheaded by a protestant missionary, to an essentially Greek institution. This subject, which gains momentum in book 2, is intriguing because of its bicultural implications—each culture distinct in its orientation, yet merging to create a practical and uplifting enterprise. Not least of its merit has been the farm school's role in binding together the American and Greek people as partners in a creative, philanthropic undertaking.

Beyond the human factor, during the Cold War the bond served the political purpose of advancing the United States' foreign policy, the American government using generosity as an entering wedge to keep Greece, of value strategically, cemented to the West. As recipients, the Greeks profited by American aid channeled through the school to the benefit of Greek farmers and the agricultural sector in general.

Can such a limited people-to-people operation still be useful in this new age? American-chartered institutions, many knowledgeable Americans hold, are more relevant now than during the Cold War, because American cultural and educational values can temper the pervasive "image of the United States as a rogue superpower."[4] Echoing that same judgment, Paul Stevenson, a former United States Consul General in Thessaloniki, stresses that "American institutions help counterbalance the damaging perception held by numerous Greek people that the United States is a "*planetárchis*." (leader of the planet or, more pointedly, boss of the planet).[5]

All in all, my hope is that the farm school's story will not endure solely as an institutional history but in addition will refract light on the many faces of Greece over the span of a century. Historian J. M. Roberts has written, "It will always remain true that the closer we get to our own

---

[4] Richard Jackson, "The Role of the United States Colleges Abroad: Anatolia As a Case Study," *Mediterranean Quarterly* 10/4 (Fall 1999): 25.

[5] Paul Stevenson, interview with author, Thessaloniki, 17 October 1999, written, AFS Archives.

times, the harder it is to see what is the history that really matters."[6] It seems to me as I write book 2 of the history of the American Farm School what really matters are those events that bind the school to the destiny of Greece. To identify the political, cultural, and historical strands of that country and to explain how they are braided into the history of the American Farm School has been my special goal.

---

[6] J. M. Roberts, *Twentieth Century: The History of the World, 1901–2000* (New York: Viking Press, 1999) xv.

# CHAPTER 1

# THE WAGES OF WAR:
# GREECE SURVIVES

At the end of the civil war (1946–1949), the situation in Greece was calamitous. The statistics, no matter the variations in tabulations, are appalling: from all accounts at least 684,000 people had perished during the German occupation (1941–1944), some 8 percent of the population. A leading Athens newspaper reports even higher casualties: the war and occupation resulted in the death of 1,106,992 people of whom 105,000 died in concentration camps, 56,225 were executed, and 600,000 died of famine and illness.[1] A recent analysis asserts the civil war that followed left more people dead, imprisoned, or uprooted than the German occupation had.[2] Hunger stalked the villages and cities, and antagonism between right and left factions divided people, even in individual families. The Right's gratitude radiated toward the Americans, who had saved them from the communists, while the Left spewed bitterness toward the American intruders, whom they resented for interfering in its struggle to cross into the Soviet sphere, a polarization that was to linger for decades.

All civil amenities lay wasted; schools had been shut for years, leaving a vast proportion of youngsters virtually illiterate, while roads,

---

[1] *Kathimerini*, 21–22 October 2000, 3. All Kathimerini entries are from the newspaper's English edition. Back issues are filed in its archives in Athens.

[2] Mark Mazower, *The Balkans: A Short History* (New York: Random House Inc., 2000) 131. For casualty figures tabulated by the Greek government, see Ministry of Press and Mass Media, Secretariat General Information, *About Greece* (Athens: n.p., 1999) 33–35.

never having constituted a satisfactory network anyway, were now totally degraded, marooning people in isolated villages. Diseases such as malaria, typhoid, and tuberculosis weakened the bodies and will of thousands of survivors. Over 2,000 villages were gutted and the agricultural sector was devastated. The population of Eastern, Central, and Western Macedonia declined by about 180,000 people as a result of war and occupation.[3]

Needless to say, the immediate concerns after the war were elementary subsistence and settlement of 700,000 internal refugees. One-third of the population was dependent in some measure on the government, which itself was bedeviled by hyperinflation—between 1950 and 1955 the annual rate was nearly 10 percent—and a sizable deficit in the balance of payments. It was into this wrecked Greece that the newlyweds, Bruce and Tad Lansdale, stepped in 1949, he at twenty-three already an old hand in Greece, she at twenty-two fresh from happy days at the University of Rochester and graduate work at Smith College, where she trained as a psychiatric social worker.

Bruce commented on the social psychology of the rural population at the time, noting that what dragged people down were their defeated spirits. He heard the villagers carp about their ill-fated lives, betraying a helplessness suggestive of their ancient forebears in the plays of Sophocles. They were convinced they could not prevail against the "*skotinés dinámes*" (dark forces) stacked against them—the government, the politicians, the civil servants, the military, the village president, and the conniving foreigners.[4]

This attitude poisoned all aspects of village life, especially as it related to education. "Peasants lacked self-confidence. Peasant was almost synonymous with loser. They assumed that education was unnecessary for the less talented and for women, whose functions were to bear children and care for the home. Those lacking drive were expected to remain in the village and cultivate the family land." Their mindset was self-defeating.

> Peasants had no management training. They did not know
> how to plan or organize their operations or how to maintain

---

[3] M. B. Sakellariou, general ed. *Macedonia: 4,000 Years of Greek History and Civilization* (Athens: Edotike Athenon, 1983) 516.

[4] Lansdale, *Master Farmer*, 5.

adequate records. Because of custom many resourceful women were unable to express themselves, except in very subtle ways, so their management abilities were only marginally utilized.... Villagers were unwilling to accept advice from outsiders. As most innovative approaches ended in failure and they lived on such limited budgets, they could not afford to take chances.[5]

In the aftermath of the recent catastrophe, educating the Greek farmer would present a colossal challenge. For one, poverty and its attendant fears had historically kept a throttlehold on people in the countryside. Herodotus's cry from ancient days resounded pitch-perfect through the centuries, "Poverty and Greece are twin sisters."[6] As Bruce suggested above, generations of poverty can bow the head in submission. Hand in hand with poverty was the unrelenting problem of the farmers' resistance to change. Victor Davis Hanson, classical scholar and California farmer, writes, "Because of the perilous environment, the farmers' ideology of the need for hard work is not merely reasonable, but vital. Moreover, the threat of catastrophe—man-made, animal, insect, climatological, and meteorological—explains two other characteristic responses of farmers: repetition and a reverence for tradition. Agriculture and its practitioners are—*must* be—conservative. Experimentation in life-style, work ethic, and daily routine can, if proved to be unwise or unsound, cause famine and disaster."[7]

Paired with this ingrained conservatism went an almost pathological mistrust of one's neighbors. This translated into a lack of cooperation, a characteristic woven into the basic fabric of rural life apparently since ancient days, a modus vivendi that prevails even today among farmers in the United States. "To the farmer, all humans, from his closest neighbor to his most bitter enemies, could pose risks. Why, if not to cause harm, would a man ever venture onto another's farm?"[8] This behavior collided

---

[5] Ibid., 5, 6.

[6] Herodotus, 7.102

[7] Victor Davis Hanson, *The Other Greeks: The Family Farm and the Agrarian Roots of Western Civilization* (Berkeley and Los Angeles: University of California Press, 1999) 159.

[8] Ibid., 134. For insight into community behavior, see George Daoutopoulos, "Community Development in Greece," *Community Development Journal* 26/2 (1991): 131–38.

with the spirit the American Farm School's founder worked to instill: a sense of community.

Poised in the midst of this catastrophe, even the most prescient commentator, Greek or American, could not have foreseen Greece's vibrant reinvention of itself from a primitive agrarian society to a service economy by the turn of the twenty-first century. Gradually during the postwar period the struggling nation shed its ignominious position as an American client state, stabilized its democratic system, became an ardent member in the European Union, attuned its economy to facilitate entry into the European Monetary Union, and aspired to play the role of a balance wheel in the Balkans.

Considering the country's postwar devastation, the pace of rehabilitation picked up rather quickly owing to an infusion of funds from the United States: from 1947 to 1967, the United States pumped over $3.7 billion in economic and military aid into Greece.[9] By 1956, hard working Greek farmers responded by producing 40 percent more crops than in prewar years. In fact, the growth rate of the GDP in the 1950s and 1960s registered 7.5 percent. Unfortunately, before 1960 agricultural production represented as much as 90 percent of exports, certainly not a strong enough engine to drive the country to real prosperity. Blessed by extensive agricultural plains and fertile valleys, Eastern, Central, and Western Macedonia ranked among Greece's most productive agricultural regions. And it is in the midst of such agricultural resources in the heart of Northern Greece that the farm school sits.

From 1947 to 1955, American specialists streamed into the country to implement health programs, establish statistical services, improve transportation, initiate electrification and other infrastructure, offer advice on agriculture, education, social welfare, public health, foreign trade, and a broad spectrum of other matters. All official American programs worked through Greek ministries and agencies. Among American specialists the watchword was: "Get in touch with the farm school." For a precedent we have only to look back to the early 1920s, a time of comparable human misery in the country when the school served as a focal point, welcoming aid workers. Rare was the American advisor who had not visited the school or at least heard of its accomplishments.

---

[9] David H. Close, *The Origins of the Greek Civil War* (London: Longman, 1995) 214. The Marshall Plan sent approximately $13 billion after the war to help rebuild Europe, an amount equal to approximately some $88 billion in 2001 dollars.

Owing in part to its position as a center of orientation, the farm school's name spread throughout Greece, and also to interested organizations, both governmental and private, in the United States. Also, workers found it useful to contact the school's graduates throughout the countryside to enlist the graduates' help in furthering aid projects

By 1960, workers witnessed improvement throughout Greece. By June 1950 almost all refugees had returned to their villages and by 1960, 800 agriculturalists and home economists, trained by the ministry of agriculture's extension services, were spreading into the countryside to provide communities with education and training in a variety of subjects. In that year studies indicated that illiteracy rates were tumbling. Eighty percent of the boys enrolled in the first year of elementary school in rural areas would remain all six years, contrasting drastically with the 4 percent who had finished elementary school before the war. In line with that trend, the total number of boys in the countryside who attended high school was double that of the prewar period. As for girls, alas, their education in that decade was substantially more limited than the boys.'[10]

The war had transformed Thessaloniki forever: the farm school's location—some 6 miles from Thessaloniki—put it within reach of Greece's second largest city, a locale of historical distinction. During the Nazi occupation, the city had undergone a major shift in its ethnic composition: well over 48,000 Jews—the city's 1940 population was 278,000—had been carried off by the Germans to Bergen-Belsen and Auschwitz for extermination. As a result of the Holocaust, in the early 1950s the remnants of the Jewish community registered less than 1 percent of Thessaloniki's postwar population of 297,000. Although Jews had been settled there since antiquity, the majority of them had arrived in the fifteenth to sixteenth century after they had been expelled from Spain; through the centuries these Sephardim had colored the city's prewar character.[11] When in 1961 the diminished city experienced a

---

[10] Paul B. Orvis and Basil Moussouros, "Survey and Planning Report of the American Farm School, 1966," AFS Archives. Orvis, then dean of the State University of New York, presented this report, valuable not only for its examination of the farm school, but also for the information on changes in Greece as the result of modernization pressures.

[11] For the tribulations of those Jews who returned to the city after the war, see Bea Lewkowitz, "We Were All Together: Jewish Memories of Post War Thessaloniki," in Mark Mazower, ed., *After the War Was Over: Reconstructing the Family, Nation, and State in Greece, 1943–1960* (Princeton and Oxford: Princeton University Press, 2000)

population influx from the countryside, raising the population to 378,400 in 1961, it was on its way to regaining its cultural status and developing into an industrial center.

Some of its citizens would eventually form a base of support for the farm school. The Biblical entreaty to Saint Paul, "Cross over to Macedonia," has lent to the school a special mythology and has encouraged some American supporters to interpret the farm school in a spiritual light: "During the night a vision came to Paul: a Macedonian stood there appealing to him, 'Cross over to Macedonia and help us.' As soon as he had seen this vision, we set about getting passage to Macedonia, convinced that God had called us to take the good news there."[12]

The prominence of Thessaloniki in Hellenistic, Roman, and Byzantine history, its location on the ancient Via Egnatia, and the extensive ruins within the center and round about the city, have imbued it with a palpable history and a visible legacy. In the first decade of Bruce's tenure as director of the farm school the metropolis must have appeared to farm school students—most of them from primitive, often remote villages—as the navel of the universe. Although the distance to the city from the campus was negligible, lack of transportation right after the war and dire poverty would have precluded frequent jaunts to town. But the commotion at the port—the country's second largest—and the throbbing railhead that gave access to Western Europe would have stimulated a thirst for adventure in any country boy.

In 1955 the train from Thessaloniki carried the first shipments of fresh fruit—grapes, peaches, and apples—to Germany, marking the start of profitable commerce. This innovation excited a vision, shared by a sizable population in Northern Greece, of the country transforming itself into a "new California." Although Greece did not own refrigerated railcars, the fruit was sprayed with preservatives for the long voyage.[13] Also, large foreign investors—Hellenic Steel and Standard Oil—were

---

247–72. For primary source material on the Holocaust see Thanos Veremis and Photini Constantopoulou, eds., Ministry of Foreign Affairs of Greece, *Documents on the History of Greek Jews: Records from the Historical Archives of the Ministry of Foreign Affairs* (Athens: Kastaniotis Editions, 1998).

[12] Acts 16: 9–10. Paul's Work Among the Gentiles, *Revised English Bible* (Oxford University Press and Cambridge University Press, 1989) 119.

[13] William Hardy McNeil, *Greece: American Aid in Action, 1947–1956* (New York: The Twentieth Century Fund, 1957) 171.

about to lift the city to a plane of prosperity. Between 1963 and 1973, Thessaloniki realized a greater increase in industrial employment than did any other area of Greece. The environmental cost of the industrialization has been dear. The over-concentration of industry resulted in severe pollution in the city and the Thermaikos Gulf, and uncontrolled urban sprawl began to degrade the landscape.[14]

To add to its various offerings, Thessaloniki is home to the Aristotle University, founded in the 1920s as the country's second university and for decades regarded as its more progressive. After 1960 the university grew spectacularly, instituting nine technical and professional schools including schools of forestry, veterinary medicine, and agriculture, thus creating a common core of interests shared between those schools and the farm school. In 1965 Aristotle University had nearly 20,000 students and a faculty of 800.[15] By 2001 the university enrolled some 55,000 students.

Not least among the city's attractions is the International Thessaloniki Fair, drawing thousands, and where the farm school was a popular exhibitor. The *Christian Science Monitor* lauded the farm school's 1956 display: "Last September at the International Fair in Salonika, while the United States and the Soviet Union struggled for prominence with expensive pavilions, the exhibit of the American Farm School featuring the function of a model Greek farm was praised more than any other exhibit by both experts and public."[16] Without fail, the farm school's exhibition was compared in the press to the Soviet Union's exhibition, a reflection of East-West competition during the Cold War.

In the larger context, when Bruce's directorship began, Greece stood on the cusp of a social revolution owing to a galaxy of innovations. The spread of electricity and roads pulled cities, towns, and especially villages into the modern world. Thessaloniki's economic takeoff after the war was made possible by electrification on a mass scale. Before 1954 electricity in the city was weak, operating only a few hours in the evening. Among the innumerable benefits of electricity was the powering of radio. The United States Information Service distributed radios to

---

[14] Sakellariou, *4,000 Years*, 518–19.

[15] T. B. Dicks, *The Greeks: How They Live and Work* (New York and Washington: Praeger Publishers, 1971) 145.

[16] John Rigos, "American Farm School Lifts Greek Villages," *Christian Science Monitor* (5 April 1957): 3.

scores of villages around the country hooked up to loud speakers in the *plateía* (village square) for the enlightenment of rural communities. For the first time in their lives, isolated villagers were connected not only to Athens, but the radio also brought them news from the world beyond. Boys coming to the school from those villages that had electricity (many places were not yet on the grid) were influenced and stimulated by this wondrous experience. The students would be astonished by the uses of electricity on the farm, in the workshops, and in their living quarters.

Terrible isolation had been one of the most chilling aspects of Greek rural life; in scores of places mule tracks offered the only way in and out of the villages. In William Hardy McNeill's judgment, "...the construction of second-and third-class-roads, together with substantial improvement of the trunk roads, can be said to constitute the most fundamental of all the great changes wrought by the American aid program...if one considers the social and psychological as well as economic consequences." He observed, "...[w]ith the breakdown of traditional isolation...urban attitudes and an orientation toward market farming began to penetrate regions formerly set apart.... By 1956 profound changes in peasant life and attitudes were already underway, changes, which were largely fed and maintained by the new ease of travel and transport."[17]

Progress took many forms between the mid-1950s through the mid-1960s. In 1956, the average per capita income was $270, but in 1964 it had reached $500, hardly lifting the average Greek out of poverty but nonetheless representing a promising trend. Also, the rampant inflation, so pronounced until the mid-1950s, was checked. The biggest swing was in demographics: migrants began to move out of Greece to Western Europe in search of jobs at the same time that elements of the rural population were flowing into Athens and Thessaloniki as well as provincial towns to escape the drudgery and poverty of life in the fields. By 1961–1962 the urban population balanced the rural population for the first time in modern Greece's history, and in the 1960s manufactured goods did, for the first time, exceed agricultural exports, a barometer showing the economic direction of the country.

Regardless of this trend, the farm school clung to its guiding principle of training young men to return to the land, although finally in

---

[17] McNeil, *American Aid*, 51.

1965 it bowed to an imperative to change its curriculum enabling its graduates to continue their schooling at the next higher level; the new certificate opened the door to non-farming jobs.

In its foreign relations in the first years of Bruce's stewardship, Greece found itself snared in a tangle of alliances and a set of hostilities that tugged its foreign policy in a number of contrary directions. While avoiding the smothering attention of the United States, Greece sought to cope with Turkey, especially over the Cyprus problem, honor its obligations to NATO, which it had joined in 1952, and contend with its communist neighbors to the north where the borders were all but sealed. These issues were particularly relevant to farm school history where they intersected with Greek-American relations. That the farm school survived all political weathers with its mission intact was due to a policy harking back to the founder: that the institution must remain politically neutral. Moreover, Bruce added a second unbreakable code: that the school be ever ready to serve the needs of the Greek people and the ministry of agriculture, and later the ministry of education, a standard from which he never deviated.

Some months before the Marshall Plan, irritants enflamed Greek-American relations. The United States set up the American Mission for Aid to Greece (AMAG), an agency to control the millions of dollars Americans were pumping into the country. The United States had negotiated an agreement then with Greece whereby members of AMAG were granted supervisory powers that led to a system of parallel administration—one Greek, the other American. Greece's total dependence on foreign aid was paid for with her independence. "Greece," admits a US government area handbook, "had become for all intents and purposes a client to the United States."[18]

Then the noose tightened even further. In 1950, the American government was exasperated over the instability of Greek politics. Between 1946 and 1952, the country had sixteen different administrations. Finally sounding a piercing note of displeasure, the United States reduced aid from $225 million to $182 million. In 1952 American Ambassador John Peurifoy, in an egregious act of intervention in the domestic politics of an ally nation, had advised the Greeks in a

---

[18] *Area Handbook for Greece*, ed. Glenn E. Curtis (Washington, DC: Library of Congress, 1994) 65. For a recent evaluation of the Marshall Plan see Peter Grose, ed., "The Marshall Plan and its Legacy," *Foreign Affairs* 76/3 (1997): 157–221.

public statement to change their electoral system in order to strengthen the larger parties and eliminate the smaller ones. The implication of Peurifoy's demand was that aid to Greece would be further reduced if the electoral system were not changed.[19] With the ushering in of the simple majority system in 1952, Greece enjoyed eleven years of relative political tranquility. However since governments during that time were all conservative and basically pro-American, this situation alienated those elements of the population who were convinced the country was held in thrall to a foreign power. In such loam, the roots of anti-Americanism sunk, only to sprout more widely in the 1970s and 1980s.

Anxiety over the northern borders was an old problem, outlined in discussions of the Macedonian Struggle, Balkan Wars, and the civil war in book 1. As a specialist on the subject writes, "...[t]he political, military and diplomatic establishments of the country had been haunted by the unabating concern lest a major confrontation between East and West should once again place the northern provinces of Macedonia and Thrace in jeopardy. It was this concern that had prompted successive Greek governments to seek the safety of Western Security arrangements."[20]

The Cyprus problem has long been a divisive factor in Greek-American relations. Simply stated, in the 1950s the majority population of Cyprus was Greek, the minority Turkish. The Greeks' burning ambition was to drive out the British, who had taken the island in World War I, and in an act of *enosis* (union), unite it with Greece. Under no conditions would Turkey agree to *enosis*, and Greece refused a Turkish proposal for partition. In 1960 Cyprus was granted independence from Britain with the formula that a Greek Cypriot serve as president and a Turkish Cypriot as vice president; both partition and *enosis* were forbidden. Neither side was satisfied. In a series of crises through the years, Greece would heap blame on the United States for favoring Turkey, also a NATO member since 1952. For example in 1964, when President Lyndon B. Johnson presented the Acheson Plan for partitioning

---

[19] For a concise summary of these events, see Richard Clogg, *A Short History of Modern Greece* (Cambridge: Cambridge University Press, 1980) 167.

[20] Evangelos Kofos, "Greece's Macedonian Adventure: The Controversy Over FYROM's Independence and Recognition," in *Greece and the New Balkans: Challenges and Opportunities*, ed. Van Coufoudakis et al (New York: Pella Publishing Company, 1999) 361.

the island, the president is reported to have answered insultingly to the Greeks' negative reaction to that solution, a stance that alienated many Greeks.[21] American foreign policy during the Cold War was based unwaveringly on containing the Soviets from expanding into the Eastern Mediterranean, a position that meant Turkey as a strategic anchor in that area, as well as Greece, had to be appeased. This policy was bound to run counter to Greek ambitions in Cyprus. Greek-American relations would be spoiled further during the 1974 Cyprus crisis, when the Greek junta attempted to unite the island with Greece, resulting in the invasion of Cyprus by Turkish troops, while the United States, to the outrage of the Greeks, engaged only in diplomatic efforts to halt the invasion.

In Greek eyes, membership in NATO at times was viewed as a mixed blessing. Although the country was protected by its membership, especially in reference to its northern borders, at the same time, it resented NATO. Whenever a crisis erupted concerning Turkey, often over Cyprus, or concerning the demarcation of air space, or the continental shelf, or harassment of the Patriarchate, Greece regarded Turkey as the major enemy; yet, NATO, it was clear, would not support Greece's claims against Turkey.

Beside the foreign policy issues, there were certain sociopolitical problems that an American working in Greece had to face. Government legislation was not informed by grass roots groups exerting competing pressures on government, a phenomenon common to the bulk of modern constitutional societies. Greek civil society did not form citizen advocacy groups to channel objective or useful information into government for crafting public policy or for obtaining remedies when problems surfaced. Consequently since "the major concern of individual deputies was primarily the satisfaction of individual demands rather than the enactment of a party platform, general policies are typically riddled with loopholes and contradictions."[22] Since advocacy groups did not cohere toward collective action, the pragmatic approach to handle business was for interested parties to work on the personal level. This meant creating relationships where benefits could accrue to both patron and client.

---

[21] John O. Iatrides, "The United States, Greece and the Balkans," in *New Balkans*, 276.

[22] Keith Legg and John M. Roberts, *Modern Greece: A Country on the Periphery* (Boulder: Westview Press, 1997) 171. For this work's analysis of civil society and its relationship to government in Greece, see also, 163, 170–71, 195–99.

Bruce, the young director, warmed to this feature of the culture and applied his gregarious personality to making friends within the Greek government and society. His talent for working on the personal level gave him the leverage to bring farm school problems directly to the proper government agencies. For decades he was the best-known American in Greece, and as the cliché ran during the Cold War, he was "the United States' best ambassador to Greece and Greece's best ambassador to the United States."[23]

---

[23] This designation of Bruce, as quoted in "The Minutes of the Board of Trustees, March 6, 1958," AFS Archives, was coined by His Eminence Michael, Archbishop of the Greek Orthodox Diocese of North and South America.

# CHAPTER 2

# THE LANSDALES CHART THEIR COURSE:
# THEIR FIRST DECADE

When Bruce McKay Lansdale assumed the directorship of the American Farm School in Thessaloniki in 1955, the small but sturdy institution had already attained a surprising renown throughout the country as well as in certain government and philanthropic circles in the United States. The attention paid by the press, the people, and by both governments to the school's fiftieth anniversary was a testament to its achievements.

Coinciding with the retirement of Charlie House and the installation of Bruce Lansdale as director, the fiftieth anniversary celebration had the markings of an affair of state. Gracing the festivities were none other than King Paul and Queen Frederica of the Hellenes and Crown Prince Constantine. Also present were the deputy prime minister, the minister of Northern Greece, the metropolitan bishop of Thessaloniki, the prefect, the commanding general of the 3rd Corps, the ambassador of the United States, and the head of the American Mission, among a throng of others. King Paul bestowed on Charlie the Decoration of the Commander of the Order of George I as "acknowledgement of the philanthropic and Christian work you have offered to Greece." Minister of Agriculture Petros Levantis declared that "more than 2,000 boys from the rural districts, mainly from Northern Greece, have been trained by this school during its brilliant career…and the school has raised the cultural level of our farmers." He pointed to the amicable relationship between the school and the government: "No trace of disagreement has arisen during these long years between the ministry of agriculture and the American Farm School. And this is important and sure proof that all those who have noble and creative purposes...are going ahead hand in hand...."[1] Adding to the luster, Secretary of State John Foster Dulles sent a message to Charlie, stating in part, "Your school has cultivated many good things. The cereals, vegetables, and fruit are only part of these things. It has...cultivated the proficiency and the qualities of thousands of Greek youths who have brought the up-to-date-methods of cultivation to their villages."[2]

Set against all the pomp were the students' simple rituals of hospitality and proud display of their accomplishments. Theodoros Litsas, the school's talented "auteur" and associate director, created an eye-catching pageantry: in orderly procession past the reviewing stand, boys led a cavalcade of hefty bulls and cows and others drove a series of the latest agricultural machines. A group of students posed atop a phalanx of floats to portray scenes from rural life.[3] Descending from the floats, the youngsters, dressed in traditional costumes, performed the regional Greek dances in front of their majesties and sang a hymn to the countryside. A graduating student presented to fifteen-year-old Crown Prince Constantine a farmer's coverall and straw hat, the daily garb worn by the students. Constantine, who would become king less than ten years later, was made an honorary graduate of the farm school. This royal couple, shaken by the anti-monarchist sentiment during the civil war, had focused particular attention on the people of Northern Greece; they saw the advantages of their appearance on campus, keenly aware of the services the institution had rendered to the rural population.[4]

---

[1] *Nea Alithea* (Thessaloniki), 30 May 1955, AFS Archives.

[2] *To Fos*, 31 May 1955, AFS Archives.

[3] For insight into the deeper political role of such pageantry in the history of Northern Greece, see Anastasia Karakasidou, "Protocol and Pageantry: Celebrating the Nation in Northern Greece," in Mark Mazower, ed., *After the War Was Over: Reconstructing the Family, Nation, and State in Greece, 1943–1960* (Princeton and Oxford: Princeton University Press, 2000) 221–46.

[4] Ibid. 228. Karakasidou offers a view of the royal family's activities in the immediate postwar period and their efforts at cultivating the rural population with special reference to Northern Greece: "It was after the enthronement of King Paul...that the Greek royalty became directly involved on the construction and cultivation of national loyalty among the citizens of Greece. In so doing, of course, they were promoting their own legitimacy as well as the legitimacy of their presence and active involvement in the Greek political arena."

Charlie had enjoyed cordial relations with King George II, and Bruce, too, became known to the royal family when Queen Frederica was drawn to a large diorama of the school. The model demonstrated the campus as it stood in 1904, the year of its founding, an inconsequential mud brick house set on a forsaken plain, and as the school appeared in 1955, over 350 fertile acres enfolding a mosaic of durable buildings. The queen insisted Bruce carry a portable version of the diorama to the United States so the American people could visualize the extended scope of the school. Thus she ordered to be made a smaller folding model, one he could pack conveniently for his trip.

At thirty, Bruce was undeniably young for the directorship of this complex institution with its educational components, an income-producing farm, capital investments, and international ties. To his advantage, though, he was steeped in Greek culture, having spent his childhood in Thessaloniki and Athens, as well as unforgettable summers, making life-long attachments at the YMCA camp in Pelion, on the eastern coast of mainland Greece. He credits the Pelion experience with teaching him the basic ingredients of leadership. His father, Herbert P. Lansdale, Jr., national director of the Greek YMCAs, had followed the career pattern of Bruce's grandfather, also an executive in the YMCA. Bruce imbibed the principles of that organization from the beginning. His fluent Greek—learned autodidactically since he never received any formal Greek schooling—was laced with rustic parables gathered from the rural folk of Northern Greece. Concerning his language ability, he admits to writing and reading Greek with difficulty. Considering this utterly desultory mixture of home schooling, a few years at the German School in Thessaloniki and other scattered educational experiences, it is a wonder that he is so articulate, a persuasive writer of formal reports, a published poet, and a riveting public speaker in both languages. In fact, language is central to his legacy, and his command of it runs parallel to his evolution as a leader.[5]

Ann House described Bruce's early connection to the school: he was "a frequent visitor, he learned the virtues of the Farm School milk, rode on the backs of cows, or in a donkey cart, and always with the lads

---

[5] His publications: Bruce M. Lansdale, *Metamorphosis: Or Why I love Greece* (New Rochelle NY: Caratzas Brothers, 1979); *Master Farmer: Teaching Small Farmers Management* (Boulder CO: Westview Press, 1986); *Cultivating Inspired Leaders, Making Participatory Management Work* (West Hartford CT: Kumarian Press, 2000).

of the School as companions."[6] Born in the United States in 1925, he returned there in 1939 to attend high school. He graduated Phi Beta Kappa from the University of Rochester with a BS in mechanical engineering, and became an ensign in the United States Navy immediately upon graduation. In February 1946, while serving in the Peloponnesos as a member of the Allied Mission for Observing Greek Elections, he heard his "call," beckoning him to devote his life to the farm school. With the overwhelming commitment the calling implied, he would prepare himself professionally, emotionally, and spiritually to assume the directorship. He returned to the school as a volunteer in March 1947, paying for his passage by working on a cattle ship tending forty mules and six stallions. In preparation for his future, he earned his master's degree in rural sociology and agricultural education at Cornell University in 1949 and then served as assistant director under Charlie's tutelage from 1949 to 1954.

But life in Greece, at first, would put his wife Tad to the test. Tad, who grew up on Long Island and in upstate New York, might best be described in high school and college as an "all-American girl." An excellent student in high school, she won a scholarship to the University of Rochester at age sixteen, where she met Bruce, and where she was a college cheerleader while Bruce was captain of the University of Rochester football team. Inclined toward social work and community involvement, she never claimed to have received a "call" in the spiritual sense, but through the years she was always eager for spiritual experience and could share with her husband this common yearning. Brought up in the Lutheran Church (Bruce was a Presbyterian), she attended services also at other churches and felt deeply moved, as he did, by participation in sincere religious observances. After receiving her college degree, she studied psychiatric social work at Smith College. The daughter of parents who respected her choices, she took pride in being fiercely independent. "I had never lived on a farm, and had no idea what

---

[6] *Newsletter*, March 1956. Almost always written by one of the Houses, the *Newsletter* was an occasional letter addressed to friends of the school, conveying school news and expressing opinions and reactions about both the school and the world around them. When the *Sower* began to appear regularly in the 1960s, the Newsletter was discontinued. The *Newsletter* collection is stored in the AFS Archives.

To illustrate how tightly the Lansdales were connected to Greece, it bears noting that the family spread Bruce's mother's ashes on Mount Olympus shortly after her death. Marjorie McKay Lansdale died in Rochester NY in 1950.

to expect or what was expected of me. I only knew I loved and respected Bruce and I was ready to commit my life to him, and to his future, wherever it might be."[7]

A week after their wedding in New York in January 1949, Bruce picked up the newspaper and read aloud the terrifying headlines: "Communist Guerrillas Invade the American Farm School and Kidnap 43 Students." "'This is where you are going to live, my dear,'" said my husband of a week as he handed me the newspaper on our honeymoon," Tad writes years later.[8] By the time the newlyweds arrived in Thessaloniki—Bruce on a series of Fulbright grants from 1949–1952—it was September 1949, and the civil war was in its last throes. Their first venture out into the countryside in October disturbed the young bride. "We took a three-day trip...through the deserted ghost villages of northwestern Macedonia and saw the shells of houses and emaciated old people who have been left in the war's wake...."[9]

Then barely having managed to take in her new surroundings, she suffered from a fever of 104 degrees for four days. She was taken to a British military hospital in Thessaloniki where the doctors were not able to make a diagnosis. At a Greek hospital they dosed her heavily with penicillin every six hours and administered electric shocks for her weakened leg muscles. While she lay powerless to use her legs, she began seeing double. Unable to speak Greek to her sympathetic visitors, unused to the customs of the country, she became discouraged. She had sunk into total dependence on Bruce, she, who prided herself on her independence. One cannot avoid the obvious comparison with Ann House, bride-to-be, who, it will be remembered, was leveled by an attack of appendicitis on her arrival on Corfu in the midst of Mussolini's bombing raid on the island in 1923.

It seemed obvious that Tad was not improving and needed advanced medical attention. However, only when Bruce's mother Marjorie was diagnosed with cancer and on her deathbed in February 1950 did the couple rush from Greece to Rochester, New York. Bruce carried his crippled wife in his arms. The four-engine plane took off from Athens,

---

[7] Tad Lansdale to the author, Thessaloniki, 10 July 1999, AFS Archives.

[8] Tad Lansdale, "Notes," 1960, AFS Archives. The *New York Times* carried the news of the kidnapping: "Guerrilla Surrender Demanded," *New York Times*, 30 January 1949, 5 and "Greeks Flee Abductors," *New York Times*, 31 January 1949, 3.

[9] Ibid.

stopping in Paris, Shannon, Gander, and at last New York. In Rochester, Tad's illness was diagnosed at once as polio-encephalitis and treated successfully. Anyone who reads the following account of her ascent many years later up pitiless Mount Olympus would be convinced that her health was fully restored:

> It had snowed the night before and the pass was blocked. We had quite a time toting packs, which had been packed with mules' backs in mind, through the deep snow.... When we told Queen Frederica at a reception of our exploits tramping though snow to the second peak, with a fifteen minute breakthrough of clouds for a breathtaking view of exquisite snow-topped mountain ranges, the rest in chilly fog, half the group in waterlogged sneakers with frozen feet, our hands too cold to remove gloves so we could eat, the queen said, "We in Greece say only the crazy go to the mountains."[10]

When Bruce became director, he was determined that it should be a Greek school under American influence rather than an American institution shaped by its Greek environment. He was blessed by a mature senior staff, individuals who were strongly influenced by two generations of the House family tradition.

Among the staff, considerable credit has already been given to the contribution of Theodoros Litsas, who had played a leading role in guiding the school through the German occupation and the chaos wrought by the civil war before Ann and Charlie's return to Greece. One of Bruce's first recommendations to the board of trustees was that Litsas be promoted from assistant director to associate director. The overall supervision of instruction as well as of the training farm was under Dimitris Hadjis, a farm school graduate from John Henry House's time who had earned a BS in animal husbandry at Cornell University. Also a farm school graduate, Alekos Andreou had continued his studies at Michigan State University, where he was awarded a master's degree; he returned to the farm school as a leading educator. Joining the group was

---

[10] Tad Lansdale, correspondence, "Dear Family and Friends," November 1963, AFS Archives. The queen's reference "*sta vouná*" (to the mountains) is a double entendre. The people who went "to the mountains" during the civil war were all considered by the monarchy to be communists and hence antimonarchy.

Lee Meyer, an American engineer who had more than fifteen years of Greek experience with the Near East Foundation and later as resident engineer at nearby Anatolia College. Not only was he fluent in Greek but his upbringing in Mennonite Pennsylvania fostered in him a certain understanding for the needs of rural folk. Business manager Prodromos Okkaledes, who had worked several years within the Near East Foundation, was also part of this nucleus, so vital to the young director.

At the next level were supervisors of the various production-training departments, half of which were agricultural and the other half industrial arts, assuring the students instruction in carpentry, plumbing, painting, rural electrification, and farm machinery repair. As a routine part of Bruce's daily schedule, he toured these departments, a fine opportunity to maintain a close personal contact with staff members and students.[11]

At this time, the farm school was one of twelve agricultural schools in Greece: eight state and four private. Over the years, a number of state and private middle-level vocational agricultural schools have operated sporadically; some have continued to function into the new century. Among the older ones still operating from the early twentieth century are schools in Marousi (Athens), Ioannina, and Larissa. These state schools offered a one-or two-year program with limited specialties whereas the farm school offers a four-year course; none possessed full boarding accommodations, enriched with extra curricular programs as did the farm school. The others concentrated on agricultural training while the farm school stressed technical or industrial arts training as well, and none of the others owned the broad range of facilities with which to train their students. What is more, only the farm school was founded with the distinct purpose of molding character. At the time, the farm school was operating in keeping with law 1698, 4 April 1939: "Re: Private Agricultural Practical Schools." The academic curriculum had been set by the ministry of agriculture in 1940 based on an outline provided by the Farm School in cooperation with ministry personnel.[12]

---

[11] Bruce Lansdale to the author, Metamorphosis, 30 August 2001, AFS Archives.

[12] For an informative source on middle-level public and private vocational agricultural schools in Greece, see Alexandros Koutsouris, "*Dierévnisi ton krísimon paragóndon pou sindéontai me tin ekpaidefsi ton neoeiserxoménon sti yeoryía*" ("Identification of the Critical Factors That Relate to the Education of New Professionals in Agriculture") (Ph.D. diss., Agricultural University of Athens, 1994).

In 1955, it cost $300 to educate a boy; one year later inflation had nudged that number up to $425. The students were requested to pay only $125 toward their room and board, a fee few families could afford, given the meager cash income of small Greek farmers, the farm school's target population. Poverty-stricken country people, existing on the margins, could be financially ruined by the slightest stroke of bad luck. And of course the farm school, tied as it was to the fate of the Greek people, was bound to be hurt in turn by the countryside's misfortune. We have only to read Bruce's remarks about a cold weather snap to appreciate that linkage: "For one week it was impossible to move out of Thessaloniki in any direction, which made it impossible to provide food and feed to the villages except by airplane drops. A majority of the villagers were forced to feed their animals what food they had set aside for feeding themselves, so that they will be pretty much on a starvation diet... As a consequence, many parents were put in such a position that it was impossible for them to pay the Farm School the third installment on a fee due after Easter."[13]

In the 1950s, the school's annual budget ranged from $150,000 in 1957 ($90,000 realized from sale of dairy and other farm products and some $60,000 contributed from the United States), to $195,000 in 1959. Although the school suffered crippling damages during the German occupation, its recovery had progressed well thanks to the injection of more than $1,000,000, including equipment from various United States aid programs from 1945 to 1950.[14] By the time Bruce retired in 1990, the school's operating budget had increased to $4,500,000.

Contrasting with the impoverished countryside, and compared to any other agricultural school in Greece, the campus in 1956 was eye-catching, enclosing a plant worth, Sanders reported, somewhat over $1,000,000. But, objectively, there was really nothing luxurious about it; in fact, many visitors from abroad found the arrangements primitive. Standing out among the buildings was Princeton Hall (built 1924), a handsome piece of architecture, home to the boys for sleeping and eating—although the dormitory was virtually without heat and truly freezing in winter and the barren dining room on the main floor, where the food was barely adequate, was described by Sanders as "resembling a barracks." On the second floor of Princeton Hall, the boys shared

---

[13] As quoted in *Newsletter*, 1956.

[14] Paul B. Orvis and Basil Moussouros, "Survey and Planning Report of the American Farm School, 1966," 6, AFS Archives.

crowded sleeping quarters; their double-decker beds were purchased as war surplus from the United States Marine Corps. Close by Princeton Hall, venerable James Hall (built 1906), rebuilt after the Germans had blown it up as they withdrew in 1944, held the classrooms. It was solid in appearance, but its interior was drab and poorly lighted; it sheltered the library that by American standards would be woefully inadequate.[15] Kinnaird House (built in the late 1920s), a tidy, whitewashed three-story structure, served off and on as a home and offices for the House family. The first building for students, Haskell Cottage (built in 1929), was used for staff housing, and the two-story Stone Infirmary was equipped to care for sick students. Next to Haskell Cottage was a red brick structure, the first stable constructed on campus, remodeled, and now called Bush Cottage, to serve as staff housing. Metcalf House (built in 1929), a charming residence made of large hand-cut blocks of local limestone, welcomed guests as humble as village families to heads of state including the royal family. Scattered about campus were a handful of houses inhabited by staff. The sheds for industrial arts classes and the animal barns were somewhat flimsy, but they suited the purpose. Still stark, but renovated, the old barracks, built by the German occupation troops, was refitted to accommodate the short courses run by the farm school in partnership with the ministry of agriculture, a program that by 1966 had hosted over 9,000 adult farmers and extension workers.

The farmlands, more than 350 acres, encircled the little community; the flatlands and rolling hills produced a variety of crops. By 1955, water pumped from deep wells and a storage reservoir irrigated more than 50 acres of corn, cotton, alfalfa, vegetables, and fruit trees. On dry land, olive trees, almonds, and vines flourished. Since such public utilities as water and electricity did not extend from Thessaloniki to the school, wells were indispensable. In 1963, a new earthen reservoir holding 5,000,000 gallons collected during the winter months was completed for summer irrigation; fortunately since then other water sources have been discovered. In early 2001, the farm school was still drawing exclusively from its own store of water, but used public utility electricity made available to the school in 1970.

---

[15] Irwin Sanders, Raymond Miller and Robert Miller, "Report on the American Farm School of Thessaloniki," 1955, 16, AFS Archives. A comprehensive report covering all aspects of farm school life and the relationship of the institution to the economic and social changes in Greek life.

As for the student community, in particular the graduates, we learn from a school survey that from 1927 to 1955 just under 1200 boys had received diplomas: 41 percent had gone into farming; 28 percent found their way into various agricultural services, making a total of 69 percent in farming or related activities. The rest were studying in the United States, or had emigrated, or were in non-farming occupations, or unknown. With rare exceptions, they were, according to the survey, respected by their fellow villagers not only as farmers, but as leading personalities in their communities. For many years they had been among the few farmers in Northern Greece advocating new approaches to agricultural practices.[16] Orvis and Moussouros point out, "whether working on their home farm, in government, cooperatives, or other occupations related to agriculture, Farm School graduates have demonstrated that they have acquired more than just skills and received more than just technical knowledge. The Christian atmosphere of the School, the emphasis placed on the dignity of manual labor, and on the responsibility of leadership, have inspired graduates to contribute to the progress of their communities."[17]

By all accounts boys attending the school were hardworking, serious youngsters, almost all of them coming from desperate circumstances. Take one, Konstantinos Boukouvalis, in 1956 a fourth-year student from Epiros. In his village of 1,000 inhabitants, 876 people were classified indigent, Konstantinos' family among them, meaning each member had at his or her disposal less than $4.00 per month. Konstantinos's family owned 4 acres broken into four plots, fifty sheep, fifty goats, six chickens, a mule, and their house. From these resources they supported the mother and father, five daughters, and two sons. From a visit Tad and Bruce made to Konstantinos's home, they offer a picture of his situation: "Together we went back up to watch the process of milking the flock.... The father sleeps out with the flocks at night to protect them from the wolves. The family lives in two rooms, one with an open hearth, where they do all their cooking squatting before the fire. The other contains a makeshift bed, really a board covered with layers of woven materials, and their loom on which they make their own clothes and blankets after they have shorn the sheep, and spun and dyed the wool...." The Lansdales describe the boy's striving and initiative. "When

---

[16] Ibid. 14–15.
[17] Orvis and Moussouros, "Survey, 1966," 14, AFS Archives.

we arrived Konstantinos was down in the field, miles from his house, harrowing with borrowed oxen their three-quarter acre plot preparing to plant sorghum. The road to his village, which had been made with volunteer village labor, had gone to pot. Konstantinos had asked the village president if they couldn't repair it, but he had shown little interest. After Konstantinos had spent a day by himself working to build up again the stone support for the shoulders of the road, the president of his village sent him two helpers and by the time we got there the road was quite passable."[18]

For a glimpse of the student body in the 1950s, we turn to an anonymous visitor:

> 150 boys were gathered in front of Princeton Hall. They were going home for Easter. Some had crates with calves or sheep; others, sacks with chickens or pigs; still others had bundles of small fruit trees and shrubs. They were taking them to their home farms. Dimitris Popanos carried a bundle of boards and a long level to run lines for contour plowing on his father's farm in Thrace. This scene is repeated at Christmas and at the end of the School year. The School sells the livestock and trees to the boys at a very low cost to improve the quality of stock in the villages, where they have mostly scrub for cattle, water buffalo and low grade sheep and goats.

The boys were deeply immersed in campus life. Aside from hard work in the classroom, barns, and fields, they were active in extracurricular activities. Student government, begun in 1949, introduced them to group dynamics. Like most boys, the students took avidly to athletics. Bruce's years as a star on the University of Rochester football team apparently were serving him well in gaining the students' respect and popularity. The boys tended to place him on an Olympian peak. Kyriakos Spiropoulos, class of 1959, remembers how envious the boys were of their director's fullback physique. "We thought of him as a bull, and when we played ball, if he was near, we played all the harder to

---

[18] "Landales' Report," 5 May 1956.

impress him."[19] The students also remember his Sunday night talks, when he spoke to the boys using a specific format: he would choose a word, define it, and draw on the definition as a springboard to launch a discussion that stressed the moral imperative of how members of society ought to behave toward each other. In a similar vein, at morning assembly every day before class, Bruce always recruited a student to give a five-minute talk on a subject of his or her choosing to faculty and schoolmates after a minute of prayer. This tradition has proved most beneficial to the students. That the gathering of the whole school each day fosters a sense of community and good fellowship is clear, but equally consequential is the practice in public speaking. Almost any farm school graduate will point to that experience on stage as laying the groundwork for his or her self-confidence.

The boarding facilities could accommodate 200 boys, but in 1955 the student body was only 170. The recruitment effort was hindered by circumstances and in some cases by donors' stipulations. In one case, applicants had to be war orphans. Bruce reported, "There are lots of orphans in most villages, but to find one who has finished grammar school and is in line to own property is rather difficult. It is common practice for families to take boys out of school after the fourth or fifth grade to put them to work on the farm."[20] Circumstances did change along with the ministry of agriculture's requirements and within a few years, 200 boys were in attendance.

Given an operation with the goals and ethos of the American Farm School, the quality and dedication of the staff were absolutely key to its success. As we have seen, members of the faculty were often themselves farm school graduates and thus shared a common ethic that they passed on to future generations. Since a significant number of faculty members enjoyed advanced education in the United States, they ably conveyed the spirit of the American educational experience when they returned. For instance in 1965–1966 alone, four teachers were studying in the United States. In the early 1950s, postwar inflation kept the salary scales low in terms of purchasing power. Working against the staff were the

---

[19] Kyriakos Spiropoulos, interview with author, Thessaloniki, 15 May 1998, written, AFS Archives. The Spiropoulos family has deep roots at the farm school. Kyriakos's two uncles and four brothers are graduates, and a few of their progeny have followed.

[20] Bruce Lansdale, "Report to the Pan Macedonian Scholarship Recruitment Trip," 3–6 August 1950, AFS Archives.

extraordinary demands placed on them owing to the boarding component: living on the grounds, being always available, it was not unusual for the men to work from early morning until midnight attending to the needs of students. The opportunity to hold down a second job to amplify their incomes, an option for others among the local population, was not possible for them. The director, and occasionally an English teacher or two, and sometimes a specialist for a certain project, were the only Americans on the staff; the rest were Greek.

The small number of Americans working on campus was only one outward sign of Bruce's overarching philosophy, a credo that dominated his entire tenure from the moment he became director. The American Farm School, according to his lights, could not be alien: it must be Greek, at the service of the ministries of agriculture and education, dedicated to educating the young people of rural Greece, and through the short courses to training adult farmers and rural workers. By fostering this principle, he more closely defined and accelerated the school's process of Hellenization; furthermore, this conviction lent the institution a steadfast direction and accounted for its survival. Bruce's standards were generated by a personal conviction, yet they reflected also a talent for prudent and adroit diplomacy.

The campus church, Aghios Ioannis Chrysostomos, kept even today with limited electricity suggests the piety of an earlier, perhaps Byzantine time: it is another sign of the drive toward Hellenization. Until the construction of the church, the liturgy, since the mid-1930s, had been celebrated in a chapel outfitted in Princeton Hall. The momentum for construction of this first church building started through the energy of the staff during Charlie's stewardship and came to fulfillment when it was dedicated on the day of his departure. The little structure, its outer walls made from rocks of variegated hues, most of them revealing embedded fossils, was constructed stone by stone by students with the aid of other campus residents. The church promptly became the locus for the celebration of life events and religious occasions not only for students and staff but also for their relatives and for townspeople on special occasions. Then there was no resident priest as there is now; a local priest came to the campus for liturgy and other duties as did visiting priests from abroad. In the late 1980s, clergymen appointed by Archbishop Iakovos came from the United States to reside on campus to officiate at liturgy, to attend to pastoral work, and to immerse themselves in the life of the school. Among them were Father George Gallos and his wife Anna, Father John Mamangakis and his wife Stella, and Father John Sarantos and his wife Niki, and the current priest, Father Kyriakos Axarides. Ranking Greek clergy have honored Aghios Ioannis Chrysostomos Church and the campus with their presence, including Archbishop Iakovos (archbishop of the Greek Orthodox Archdiocese of North and South America from 1959 to 1996) and Athenagoras I, archbishop of Constantinople and ecumenical patriarch from 1948 to 1972. Bruce cherished his friendships with many members of the Greek clergy, meanwhile learning about Greek Orthodoxy.

Bruce and Tad moved quickly to bring fundraising in the United States and later in Greece to a heightened level. Fundamental to the farm school's stability during his tenure was Bruce's success at raising money. The Lansdales made whirlwind sweeps, often lasting two months and covering thousands of miles, to introduce the school to a receptive public. With the aid of the board of trustees, who helped with organizing and contacting interested parties, he spoke to groups in the United States, sometimes in churches, other times to organizations, often to smaller gatherings in homes, or at an assortment of events, animating his talks with Hodja stories. Alexander Allport remembers frisking Bruce at the end of his performances to remove the many checks supportive American contributors had tucked into his pockets.[21] Bruce and Tad tried to visit the United States annually and sometimes more often. During Charlie's tenure, Princeton classmates and the board of trustees had donated based solely on their faith in Charlie, whom they trusted to concentrate on "good works," but they lacked true familiarity with the school.

Bruce encouraged the board members and the public, in this more mobile age, to travel to Thessaloniki. In May 1963, for the first time in its history, the board met on campus. Two members, Konstantinos Zannas and Henry Hope Reed, were resident in Thessaloniki, but twelve stateside members out of twenty-nine planned to attend; and lately the frequency of board meetings has been increased to once a year, incorporating the trustees deeply into the life of the School.

---

[21] Alexander Allport, interview with author, Quechee VT, 15 July 1999, written, AFS Archives. Allport served as director of the farm school's New York office from 1960 to 1968 and again from 1986 to 1990. He was also a trustee from 1973 to 1976.

Next, Bruce searched for potential board members among Greeks who would commit themselves to the school. Along with Dimitris Zannas, who had succeeded his father, Konstantinos, came Pavlos Condellis, George Legakis, Stavros Constantinides, and Machi Seferdji, among the first resident board members. To further his plans to include Greek residents in the life of the school, Bruce encouraged fundraising organizations in Thessaloniki and Athens and established development committees in those cities.

To build a stateside organization even before he became director, Bruce had founded the Cincinnati and Rochester committees, organizations that still play a role in supporting the school. He went on to start many more committees across the United States. By 1995, they numbered fourteen.

Deposited in the American Farm School Archives are scores of articles—stimulated by Bruce—in major media in the United States, covering the Farm School, especially during the tense years of the Cold War. Former ambassador Norman Armour's statement typifies the official United States sentiment of the time: "It would be difficult to exaggerate the importance to Greece's future and to American-Greek relations in general of the work being accomplished by the American Farm School...."[22]

Attracted by the school's reputation as a humanitarian agent and Bruce and Tad's hospitality, American visitors wended their way in droves to this still primitive reach of Europe, far off the beaten tourist path. Leonard Bernstein and his wife, actress Felicia Montenegra, came in the late 1950s. United States government officials visiting in 1963 included Vice President Lyndon Johnson and Supreme Court Justice Earl Warren. Historian William Hardy McNeill, who did much valuable and original research in Greece, was a frequent guest along with his wife Elizabeth; they found in the farm school a micro laboratory to test their theories on rural development in Greece.

The Lansdales were sought out as visitors in Washington as well. The couple brought luster to the school when President John F. Kennedy invited them for a luncheon at the White House on the occasion of Prime Minister Konstantinos Karamanlis's official trip to Washington in 1961, the first-ever by a Greek prime minister. The president, upon the

---

[22] As quoted in *Newsletter*, June 1957.

occasion, conveyed to the Lansdales his congratulations for the fine job they were doing at the farm school.

In Greece, Bruce was able to extend the school's outreach in the postwar era as travel became more convenient: the world was opening up for foreigners and Greeks alike. The Community Development Program, under the coordination of the farm school's Andonis Trimis, by 1963 had gained popularity and covered 150 villages including 1,000 leaders representing a population of 160,000. Under this initiative, rural people were finding ways to solve their community problems proactively.[23]

In the process of outreach the farm school continued to draw people from outside Greece, especially from developing countries who arrived on campus for professional development. In just one instance in 1963, twenty-five agriculturalists from twelve Mediterranean countries spent four weeks at the farm school studying new farming methods used in Greece. At a farewell reception, officials from the ministry of agriculture, twenty village girls studying poultry at the Girls School, and other guests enjoyed an evening of international friendship conducted in three languages.[24]

Neighboring villages always remained a priority for the school's outreach programs. In fall 1963, in the evenings (the vehicle was too busy doing other daytime tasks), the school truck would rattle along the country roads, most of them unpaved, distributing Rhode Island Red chickens in cooperation with the ministry of agriculture. The farmers' demand for young chicks, pigs, calves, and lambs to improve their breeds was on the increase.

The graduate follow-up program was another connection to the countryside. Started in 1955 to assist and encourage alumni to apply what they had learned, Nikos Mikos visited graduates on their land. He

---

[23] For an overview of community development efforts in Greece through the years, see George Daoutopoulos, "The Prospects for Community Development in Greece," *Annals of Public and Cooperative Economics* 67/2 (1996): 283–90.

[24] *The Sower* 13/5, 1963. *The Sower*, launched in 1926, was an occasional house medium for the farm school. During the period covered in book 1, the *Sower* began as nothing more than a pamphlet printed on a large sheet of paper and folded into four sides. The publication kept farm school supporters abreast of the progress and news of the school. As the school's paper of record, the information and photographs it contains can be found in no other source. The Greek-language *O sporeas*, introduced in 1976, serves the same purpose as *The Sower*, but addresses Greek residents. Both the *Sower* and *O sporeas* are contained and catalogued in the AFS Archives.

encouraged them to implement innovations and supported them in overcoming opposition to new farming methods. He also steered the young farmers to local agriculturalists who could assist them and helped them navigate the bureaucracy for a loan from the Agricultural Bank of Greece. The follow-up program proved essential to the farm school's educational program, extending training far beyond the secondary school years, creating a lifetime experience for alumni.

In the early 1960s the institution benefited from a variety of United States government aid sources. Special government funds were earmarked for travel grants and study fellowships. The Fulbright Program aided in providing teachers and in return the farm school played host to all Fulbright teachers when they arrived in Northern Greece until they could find permanent housing. It was through the Fulbright Program that Phillip Foote, headmaster of the Cathedral School, first came to campus to teach English. He later became a member of the board of trustees.

When the United States Agency for International Development (USAID)—a tremendous source of funds for capital improvements and special programs—began its support in the early 1960s, it inaugurated an era of major construction early in Bruce's career.[25] A school without student-paid tuition was doomed to be strapped for funds, so USAID helped tremendously. Rochester Hall (named for generous donors from Rochester, New York), a conference center and headquarters for school administration, was finished in 1963. Cincinnati Hall (named to honor Cincinnati's faithful supporters) was partially funded by USAID's Public Law 480 grants and was ready for occupation in that same year. It was a desperately needed facility to lodge visitors, many of whom had previously been put up in the Lansdales' home. At a cost of $20,000, also

---

[25] US/AID/ASHA is an abbreviation for United States Agency for International Development's Office of American Schools and Hospitals Abroad. In late 1991 the abbreviation became FHA/ASHA, when the agency was incorporated into the Bureau of Food and Humanitarian Assistance. All USAID grants were matched by funds raised by the school. From 1958 to 1967 the school received $1 million and from 1968 to 1999, $8,755,244 from USAID for construction, renovations, furnishings, scholarships, staff training, educational programs, and other items. C. William Kontos, "Address to the Board of Trustees of the American Farm School, 1967" and Summary USAID grants, #1199, AFS Archives. Kontos, a USAID official, in his address claimed "the School is only one of a very few schools below college level receiving regular support" from USAID.

footed by USAID, the venerable James Hall was renovated to relieve a grave classroom shortage. Another building constructed with USAID funding, Massachusetts Hall (named to recognize donors from Worcester and other Bay State cities) was dedicated in 1966. Additionally, the school constructed new staff housing, including Cornell House, erected over a span of a few years by volunteer labor from Cornell University; Craig Ritchie Smith Cottage in 1961 (Smith retired in 1965 after twenty-two years of service as chairman of the board of trustees); and Tuxedo Cottage, completed a few years later, thanks to Trustee Dippy Bartow and others from the Tuxedo Park community in Westchester County, New York. The Estella Maresi Portico at Loch Hall on the Girls School campus was completed in 1972, the gift of trustee Pompeo Maresi. Also, Princeton Hall was extended thanks to USAID funds.

From a purely material standpoint Bruce's accomplishments in the first decade of his directorship were hardly trivial. As time wore on and the young director moved through his thirties, he not only charted a practical course, but he composed a moral framework as well. Years later he wrote several lines that define the inner man he aspired to become: "It is not enough for those who seek to cultivate moral authority...to bring their wills into harmony with a set of external commandments prescribed by their faith. They must be prepared, on every occasion, no matter how new or unexpected, to react on the basis of an internal set of moral codes of their own accord. They must be so inspired that their own response is a natural expression of their own ingrained sense of values."[26]

Bruce's reputation was becoming established throughout Greece. Governor of Northern Greece Dionysios Manetis in 1963 awarded him the Gold Cross of the Royal Order of King George I, awarded by King Paul. At the ceremony, Manetis said, "During my years in office I am pleased to state that from personal experience and from many reports of the agricultural services of the government, you and your institution have rendered invaluable service to this country."[27]

Charlie, who had retired to Orient, Long Island, visited the school with Ann in the school year 1960–1961. He died shortly after his return to the United States some ten days before his seventy-fourth birthday. (Ann House died in Hightstown, New Jersey, in 1989 after her 100th

---

[26] Bruce M. Lansdale, *Cultivating Inspired Leaders: Making Participatory Management Work* (West Hartford CT: Kumarian Press, 2000) 109.

[27] As quoted in *The Sower*, 13 April 1963.

birthday). Greek officials, students, and members of the alumni association delivered their special messages on campus at the memorial service in Aghios Ioannis Chrysostomos Church, each with the traditional Greek closing: "Eternal be his memory; lighten the earth of his country which covers him," as they stood in the flickering candlelight. In that era and indeed in years to come, alumni defined themselves even into manhood as graduates of the American Farm School. And in 1961, the majority of them had grown up under Charlie's tutelage. With his passing, Bruce was forever more on his own. But the gigantic shadows of the Houses would always brush him. As he strolled past their bronze busts in front of James Hall, he would stop and ask, not without trepidation, "How're we doing?"

# CHAPTER 3

# SECONDARY EDUCATION IN GREECE: THE HEART OF THE MATTER

For over 2000 years the Greeks have extolled the virtues of education. Menander (343?-291 BCE) in a memorable phrase states its worth quite plainly: Βακτιρια γαρ εστι παιδια βιου(Education is the staff of life).[1] Yet modern Greeks have spent almost their entire history struggling to create a suitable educational system.

Shortly after Greece gained independence from the Turks, the country cobbled together an educational organization in 1833, a derivative based on elements of the French and German systems. Oddly enough, from 1833 until the early mid-1970s, much of this arrangement remained in place despite the variations in political leadership and forms of government, each with its own vision for education.[2]

Until the mid-1970s and even later, education in Greece was aimed primarily at preserving the country's Hellenic-Christian roots, promoting social solidarity, and nurturing the spirit of nationalism, hardly the kind of pedagogy to foster analytical thinking or to convey exciting, fresh knowledge. A scholar at the University of Athens claims, "The Greek educational system attaches particular significance to national history.... In the national narrative reproduced in school, the Greek nation is understood as a natural, unified, eternal, and unchanging entity, not a

---

[1] Menander 1.232.

[2] For example during one of Greece's most unstable periods between 1944 and 1967, thirteen legislative elections were held resulting in more than thirty governments:, see Roy C. Macrides, ed., *Modern Political Systems Europe* (Englewood Cliffs, New Jersey: Prentice Hall, 1987) 334.

product of history."[3] For example, in the 1935 program of study, which remained in effect for some forty years, the ministry of education touted the course in high school physics as a means of strengthening the "religious and moral sentiment" of the students.[4] To underscore this sentiment, the elementary school curriculum offered a heavy load of Greek religion, ancient Greek history, along with geography, some sciences, and arithmetic, adding in the secondary curriculum more hours of ancient language and philosophy to reinforce the static classical curriculum.

Given the nationalistic temper of the country, the system almost entirely excluded technical or agricultural training, making the farm school's emphasis on practical training a rarity. Any program that did not lead to university failed to entice rural youth, because it conferred neither the prized social status nor the academic credentials for government employment, a problem that bedeviled recruitment at the farm school off and on until 1978.[5] Until 1961, when over 62 percent of the population tilled the soil, as mentioned earlier, only approximately eight state-run agricultural schools were operating on the secondary level, and most of them were non-boarding. In contradistinction to the farm school, they did not emphasize practical training, nor did they have facilities to do so. The few other schools teaching manual labor, such as bricklaying and carpentry, did not offer general education courses to enrich the curriculum: in accordance with the traditional European pattern, these youngsters were trained narrowly.

Set up from the beginning under the auspices of the ministry of education and religion (referred to in this book by its shorter designation, ministry of education), the educational system, officially linked to the

---

[3] Efi Avdela, "The Teaching of History in Greece," *Journal of Modern Greek Studies* 18/2 (2000): 239.

[4] Babis Noutsos, "Change and Ideology in the General Lyceum Program: Two Examples," *Journal of the Hellenic Diaspora* 8/1–2 (1981): 53. This volume, with contributions from mostly Greek scholars, constitutes an invaluable source.

[5] For a discussion of technical and vocational education, see Michael Kassotakis, "Technical and Vocational Education in Greece and the Attitudes of Greek Youngsters toward It," *Journal of the Greek Diaspora* 8/1–2 (1981): 81–93. Also A. Papadaki and M. Tzeirani, "*O rólos tis yeorgikís ekpaídefsis kai ton efarmogón stin elláda: Istorikí anadromí kai ekséliksi,*" ("The Role of Agricultural Education and Its Application in Greece"), paper delivered at the international symposium "Agriculture: Contemporary Agricultural Societies." n.d., author's possession.

Greek Orthodox Church and united in the same ministry, has always
been highly centralized. To this day the ministry oversees the drafting of
curricula, hiring of faculty, preparing of examinations, setting of
academic and teachers' qualifications, writing and distribution of
textbooks, managing the facilities and deciding the school-year
calendar—in short every aspect of education of all schools, including the
American Farm School, which, however, has always been granted certain
exceptions. Understandably, the dead weight of centralization, the
staying hand of the church, the chronic lack of funds, and until 1974 the
entrenched status of *katharévousa* (a difficult language that the majority
of farm school students could not handle easily) kept the system
unreformed. *Katharévousa*, a language modified from Ancient Greek
with the addition of modern forms, was used by the more conservative
newspapers, the government, the law courts, the military, and the
University of Athens. The educated classes spoke a modified version of
*katharévousa*, called *kathomilouméni*, while the population at large
tended to communicate in demotic, a language that has been spoken
through the centuries in ordinary discourse.[6]

This fracture in the language took on wider dimensions. Since the
founding of the modern Greek state, the use of *katharévousa* in schools
and in society reflected a political and class division, *katharévousa* being
the language of the conservative and more educated segments, demotic
being the language of the liberal and often rural elements. Taking a
decisive step in 1976, by legislative decree the government designated
demotic as the language to be used throughout the whole educational
system.[7]

Progressive Greek educators recognized that they were saddled with
an educational system rooted in foreign sources and mired in the
nineteenth century, and although many leaders—politicians and
educators alike—had attempted sporadically to harmonize the curricula
with the growing complexities of their modernizing country, their efforts

---

[6] For a comprehensive treatment of the language problem through the ages drawing
on Greek sources, see Peter Bien, *Kazantzakis and the Linguistic Revolution in Greek
Literature* (Princeton: Princeton University Press, 1972) 13–146.

[7] Legislative Decree 309/1976, Vivliothiki Voulis, Athens. For a broad discussion of
the national identity aspects of the language problem, see John Campbell and Philip
Sherrard, *Modern Greece* (London: Ernest Benn, 1969) 38–43.

had foundered.[8] Regarding fiscal implications, it is important to keep in mind that in the early postwar period, Greece's resources were scant and its needs overwhelming, so that from 1957 to 1963 the average expenditure on education was a mere 2.3 percent of public investment.

After the close of the civil war in 1949 until the fall of the junta in 1974, the problems in education at all levels were staggering. Specifically, overcrowded classrooms, insufficient number of schools, and underpaid teachers—deficiencies not uncommon to a degree in almost every country—sorely plagued Greece. A survey prepared by the American embassy in 1965, when Prime Minister George Papandreou was attempting sweeping educational reform, stated that the ministry of education was handicapped by opposition within its ranks from those educators who feared an infiltration of Western techniques, such as the latest methods essential to teaching mathematics and science developed in the United States and Europe. Secondary schools in Greece did not possess even the most rudimentary equipment for scientific experiment; they also lacked audiovisual aids, language labs, and sound administrative practices. The learning process was further handicapped by forces in government who used education as a political tool. Astonishingly, the teaching of Greek and international history stopped with the First World War, so as not to antagonize citizens whose divisive and passionate views polarized them into conservatives and liberals, rightists and leftists, royalists and republicans.

The problems stretched beyond secondary schooling, affecting university study on both the undergraduate and graduate level. Students seeking quality graduate training, then as now, were compelled to study abroad since Greece lacked requisite graduate schools. The brain drain still bleeds the country of many capable young Greeks who secure permanent positions in the United States or Western Europe—the very people who could best assist their country in furthering its educational

---

[8] As early as 1917, Eleftherios Venizelos, Greece's liberal statesman and prime minister, engineered some mild reforms. His government established demotic as the teaching language in both primary and secondary schools; made elementary education compulsory; reduced the number of classical middle-level schools (gymnasiums); and promoted the construction and improvement of agricultural schools. But his reforms were scuttled owing to the worldwide depression. Campbell, *Modern Greece*, 146. For a survey of progress in Greek education,, see *Kathimerini*, 21 October 1999, 3.

and social aspirations.[9] As a side effect, study abroad also causes an outflow of Greek financial assets to support students residing in foreign countries.

Traditionally rural Greeks, like their urban compatriots, placed a high value on education not only because they were ever conscious of their brilliant heritage, but also because education provides the passport to social mobility, some degree of prosperity, or perhaps a position in the civil service—in short various paths leading away from their hardscrabble existence on the land. To that end, indigent farmers hoarded their meager drachmas to send their sons (daughters were rarely educated beyond primary level before the 1970s) to secondary school. Secondary school was not free until 1964. Owing to this mentality, in the early 1960s, 40 percent of the students at the universities in Athens and Thessaloniki reported that their fathers were either laborers or farmers, a figure higher than that of the developed countries in Western Europe. In fact, sociologist Maria Nassiakou concludes that "the decisive motive for children of farmers to enter higher studies was the tendency to leave the Greek countryside."[10]This determination, she explains, is why there are many more children from the "poorer classes" in higher education in Greece.

The inclination to flee the land inspired the founder of the American Farm School to reinforce the institution's fundamental ethos by inscribing in the credo these pointed last two lines, "I am proud to be a farmer and will try to be worthy of the name."[11] Under Charlie and Bruce the school's goal of keeping the graduates on the land was destined to become a point of contention between the school and those graduates who aspired to other opportunities. The psychology of escape is driven home in the following observation: "The young man in his twenties or thirties nursing a two-centimeter nail on the little finger of his left hand is

---

[9] Cultural Affairs Office, American Embassy, Athens, Greece, "Education in Greece: A Report on the U.S. Government and Private Assistance to Greek Education," 20 October 1965, AFS Archives. This report is useful for the itemizing of all funds Greece received from American private and government sources for education and research from 1950 to 1965; for a discussion of Greek government reforms in education; status of US libraries in Greece; and some of the problems faced by Greek educators.

[10] Maria Nassiakou, "The Tendency toward Learning in the Greek Countryside," *Journal of the Hellenic Diaspora* 8/1–2 (1981): 66.

[11] The creed, the farm school's original statement of its belief, has been recited by the students since the early years.

not seen so often now as a few years ago. The nail was the young man's way of telling the world he had escaped manual labor on the land or in industry, and had joined the fortunate few who worked with their brains or wits. Otherwise, of course, the nail would have broken."[12]

In this regard it is important to realize that, until the mid-1970s, the paucity of state technical or vocational secondary schooling channeled rural students, whether they liked it or not, into classical high schools leading to university.[13] Of those families who did remain on their land, many chose to send a child to university to increase his prospects for leaving the farm, and thereby relinquishing his share of an already fragmented land-holding, a tactic sure to inflate the number of university aspirants. The sheer demand for higher education persisted. A few statistics here underline the point: higher education enrollments increased from 28,302 in 1960 to 121,116 in 1980.[14]

In their search of opportunity, thousands of rural folk migrated to cities, provincial towns, or abroad from the 1950s on, claiming their greatest area of discontent lay with the low standard of education in their villages. It was certainly true that in many village schools in 1961, for example, a single teacher often taught the six primary grades crammed together in one classroom. Of the 9,300 primary schools throughout the country, 4,300 had only one teacher and, until 1976, in the big towns there were occasionally 80 students to a classroom.[15]

In contrast to these general inadequacies, the farm school while never luxurious, enjoyed money arriving annually from the USAID for capital projects and scholarships. Thirty-one grants from 1968 to1999 amounted to $8,755,244, averaging $282,500 each—to say nothing of European Union money flowing in from the end of the 1980s onward.

---

[12] "Education: The Problem of the Long Nail," *International Herald Tribune*, October 1978, 6s.

[13] For an understanding of the grip of centralization on vocational and agricultural education, see Nikos Terzis, "Continuing Education, Initial and Continual Training and Adult Training As a Continuum: the Case of Greece, Final Report to CEDEFOP," 1 November 1994.

[14] C. A. Karmas, A. G. Kostakis, and Thalia Dragonas, *Occupation and Educational Demands of Lyceum Students: Development Over Time* (Athens: Centre of Planning and Economic Research, 1990) 24. In 2000 Greece had the highest percentage of university students among all European Union countries according to *Kathimerini*, 21–22 March 2000, 2.

[15] Campbell, *Modern Greece*, 384.

This support helped to position the institution at the center of agricultural education.[16] In addition, World Bank funds and private and corporate donations from Americans, and later Greeks, allowed the institution to establish a campus that at least by Greek standards was attractive and, in most respects, comfortable. As noted earlier, though, the dormitories at the Girls School and in Princeton Hall were not heated, a distinct discomfort since in the winter-gloaming Thessaloniki turns cold and damp, the temperature sometimes sinking to the freezing. What Hesiod reported on the frigid winters in Northern Greece still holds true today: "The animals shudder,/ with tails between their legs; they find/ No help in furry hides, the cold goes through/ Even the shaggy-breasted Boreas/ goes through an ox's hide…".[17] When the boys moved into the new dormitory in 1977, they were comforted by central heating, as were the girls who moved a year later.

Schools in the rural sections were disastrously affected by rapid urbanization. As farmers deserted the countryside, the quality of education there grew poorer: classrooms frequently lacked enough students of the same grade level to create a stimulating learning environment. At the other end of the process in cities and towns, classrooms spilled to overflowing and the number of teachers proved inadequate. The influx into Athens and Thessaloniki is indicative of the flood out of the villages. From 1951 to 1961 the population of Athens experienced a net increase of 330,861, while in the same period Thessaloniki's population expanded 25 percent to 378,400. In 1961 for the first time in modern Greek history, the urban population began to surpass the rural population.

Inadequacies were fundamental and common to the cities as well as the villages. For instance, teaching methods were antique. In secondary school and continuing through university, students were taught to memorize their material and to accept the voice of the teacher uncritically, while their textbooks, devised by the ministry of education, inculcated them with information that the government deemed appropriate, a routine that hardly prepared the graduates for leadership roles in their communities or for coping with the modern imperatives of

---

[16] USAID Grants #1199, 1999, AFS Archives. Also, Cultural Affairs Office, American Embassy, Athens, "Education in Greece: A Report on the U.S. Government and Private Assistance to Greek Education," 20 October 1965, AFS Archives.

[17] Hesiod, *Works and Days*, 496–526.

business management. They entered the world unpracticed in the efficacy of reasoned debate, untutored in the principles of consensus building. Two Greek educators state, "In the context of a Greek educational system that is highly centralized, uniform, and ethnocentric, the main purpose of teaching history appears to be the development of national consciousness rather than the cultivation of critical thinking."[18] That element of critical thinking, which Americans like to think of as the hallmark of their education, is a feature that the farm school has attempted to graft onto the curriculum over the decades, not always with success. The introduction of critical analysis in the farm school's classrooms brought a startling departure, as pointed out by Tad Lansdale in the following anecdote: "You would have to understand the Greek education system to realize how revolutionary it is for teachers to tolerate opinions...much less from students or former students, and how extraordinary our staff is."[19]

Modern language education has proved a weakness in the curriculum. Since World War II, although American and British influence has predominated in Greece, and the English language is the lingua franca internationally, English language study has been inadequate. A Greek schoolteacher writing about state schools in 1981 complained, "...the official Greek state educational curriculum has not responded in the least to the demand for the learning of English. Instruction in English has been introduced only recently, but under terms excluding the desired result. The language is taught for a minimum of hours to classes consisting of approximately forty students, at different levels of knowledge."[20] Today English language instruction in state schools starts in the fourth year of elementary school and continues until the student finishes secondary school, although some of the weaknesses stated above may still remain. English language study at the farm school through the years, especially since 1977 when the ministry of education assumed the supervision of classroom instruction, has on several occasions been the target of criticism leveled by various trustee-directed evaluation committees. In earlier times—before 1977, when the

---

[18] Thalia Dragonas and Anna Frangoudakis, "Introduction," *Journal of Modern Greek Studies* 18/2 (October 2000): 235.

[19] Tad Lansdale, "Dear Friends," 15 September 1963, AFS Archives.

[20] Aloe Sideris, "Some Information About Private Education in Greece," *Journal of the Hellenic Diaspora* 8/1–2 (1981): 59–60.

curriculum was under the more lenient supervision of the ministry of agriculture—many students did manage to attain a degree of fluency. But to the frustration of the able faculty, the hours allocated to language study are now too few to make genuine language fluency possible throughout the student body. The evaluators as well as the English teachers themselves decried insufficient oral practice and inadequate hours devoted to class work at the school, in addition to the students' poor preparation back in their elementary and gymnasium days. Recently, the *Kathimerini* newspaper's comments on the language problem suggested a continuing problem: "As far as language learning is concerned, pupils in the technical education stream are disadvantaged in comparison with those who are in the general stream, which includes a larger foreign language component."[21]

The English language faculty at the farm school work creatively to upgrade the language program by providing extra sessions, supplementing government textbooks with tapes, vocabulary word lists, and enrichment books, and by dividing the class by ability into small groups.[22] In theater presentations such as the annual production on American Thanksgiving Day, students playing Native American and pilgrims, dressed in convincing period costumes, deliver their lines in admittedly accented English, which somehow seems to add to the fervor. Anastasios Apostolides, who came to the school in 1979 as a math teacher and is now principal of the technical school program and vice principal of secondary programs, has fostered the students' interest in drama and in a wide range of other cultural pursuits. He has directed the budding actors in exceptional productions, including ancient drama, rendered in demotic Greek, of course. An enticing exchange program, which sends students to the United States and Western European countries, acts as an incentive for them to improve their English, the program's language of instruction.

For all the reasons discussed above, Greek educators, students, and parents longed for an extensive educational overhaul after World War II, a move advised by international organizations and foreign consultants, and urged by a series of Greek governments and opposition parties. But

---

[21] *Kathimerini*, 21–22 March 2000, 3.

[22] Mary Chism, interview with author, Thessaloniki, 1 March 2000, written, AFS Archives. Mary Chism was hired in 1975 as an English teacher. She also coordinates the student exchange program.

it was not until 1976 under the Karamanlis government that the country finally embarked on sustained reform.

Not to be overlooked, however, in the postwar period was Prime Minister George Papandreou's gallant stab at truly comprehensive reform between 1963 and 1965, although his progress would be wiped out by the military dictatorship (1967–1974).[23] Faithful to his campaign pledges, Papandreou made education a focus of his government's program. Before 1964, public education was free at the primary level only, but in that year the Papandreou government declared schooling free from preprimary through university, and compulsory attendance was extended beyond the six years of primary through the first three years of secondary school (seventh, eighth and ninth grades, called gymnasium). *Katharévousa,* the formal language, was eliminated as the language of instruction below university level, and was replaced by demotic on the secondary level. (Demotic had nearly always been the language of instruction in elementary school until the fourth grade.) An examination at the end of secondary school was instituted instead of the former university entrance examination. To pass the university exams students spend hours at *frontistíria* (private tutoring schools), cramming. The poor as well as the rich still spend thousands on this tedious after-school exercise, an abuse Papandreou long ago aimed to eliminate.

The Papandreou government provided a real boon for rural students. Before 1964, entry into gymnasium had been by examination. Rural students who passed the exams, if they had no gymnasium in their villages (usually the case in the countryside), were obliged to travel to a town where they often had to board. The extra costs of living away from home, of purchasing books, and of paying tuition fees charged for gymnasium, were beyond poor families.[24] The Papandreou government, having made the first three years of secondary school obligatory, ipso facto did away with the examination between elementary and secondary school, and also eliminated the tuition fee, although for poor students the expenses of living in a neighboring town still presented an often insurmountable barrier to further education. Moreover, as a relief for impoverished families in the countryside, his government initiated a free

---

[23] The leading educational reformer was Evangelos Papanoutsos. For a collection of his articles from 1954 to 1965, see *Agónes kai agonía yia tin padeía* (*Struggles and Anxiety in Education*) (Athens: Papadpopoulos Book Store, 1996).

[24] Campbell, *Modern Greece,* 385.

lunch program for primary school children. Also, the curriculum in both primary and secondary school was updated to accent pure and applied sciences and mathematics, and modern languages were introduced.

Realistically, these noble educational reforms as well as other improvements were unrealizable in a country with such limited technical and financial means. In fact, attempts to implement these measures led to budgetary deficits and inflation. For the first time, the expenditure for education jumped more rapidly than the fast-rising total national expenditure.

When the colonels' junta seized power in 1967, it leapt to dismantle the Papandreou reforms and to pervert the essence of learning through forms of indoctrination and regimentation. To that end, school textbooks were rewritten to change the content, returning the general flavor of education to the pre-1964 curriculum. Zealous to foster a strong sense of nationalism in the students, the dictatorship placed more emphasis on the past. Hours of compulsory religious instruction were increased, lessons in modern languages were eliminated, and freshly issued textbooks on history and civics stressed the cult of the army, explaining to the children that the army had saved the country. Continuing in this vein, the new books told eleven-and twelve-year-old children how political parties had pushed Greece to the edge of the abyss.[25] To set the tone of instruction, George Papadopoulos, the colonel who led the junta and set himself up as the minister of education, insisted teachers assign essays based on his ideas—notions of the narrowest sort. To an audience of political authorities in the northern city of Serres on 7 March 1969, he said in pompous *katharévousa*, "In the area of education, we have accomplished much. We have given all that it was possible to give in an effort to improve what was an unacceptable situation we found when we came. We face difficulties and it is important that you know these difficulties are due to the lack of human energy for education."[26]

In the villages, teachers were required to make speeches in support of the regime or lose their jobs, a truly draconian punishment since teachers who were sacked were thereafter forbidden to teach, even privately. The colonels reduced the period of compulsory education to the six years of primary school, replaced demotic with *katharévousa* as the language of instruction in the secondary schools, and also in the last

---

[25] "School Programs Revised in Greece," *New York Times* 12 October 1969, 27.

[26] George Papadopoulos, *To pistévo mas* (our creed) (Athens: n.p., 1969) 79.

three grades of primary school, although children in the first three grades were still taught in demotic. The language switch produced chaos for youngsters, disrupting their linguistic and perhaps, some critics claim, even their intellectual development in the midst of their elementary education.[27] In this respect at least, the junta was supported by the Greek Orthodox Church, a conservative force that vehemently defended the use of *katharévousa*. The dictatorship reversed the Papandreou scheme of using the secondary school graduation exam for entrance to the university and replaced it with the old university entrance exams. For all its talk of improving the educational system, by 1970 the junta had trimmed the expenditure for schools to 9.2 percent of the national budget from 11.3 percent in 1967, even cutting out the free school lunches that Papandreou had provided.[28]

Yet the dictatorship did advance vocational and technical education by planning KATEE: *Kéndra Anotéris Technikís Kai Epangelmatikís Ekpaídefsis* (Centers for Higher Technical Vocational Education). This form of technical junior college began to function only in autumn 1974, after the fall of the junta and the return of democracy in July. The regime also encouraged new agricultural programs in 1971, including a three-year course at the farm school called SEGE: *Scholí Ergodigón Geotechnikón Epangelmáton* (School for Agricultural Foremen). SEGE, which the farm school actually designed and implemented, is regarded by some people as one of its most effective ever in terms of practical training.[29] KATEE proved meaningful for farm school SEGE graduates, who were able to continue their agricultural education at those centers on the post-secondary level.

To prod Greece toward educational improvement, the World Bank granted $14 million in 1971, and $23.5 million in 1972 for, among other substantial projects, construction of three vocational schools. Again in

[27] Richard Clogg and George Yannopoulos, eds., *Greece Under Military Rule* (London: Secker & Warburg, 1972) 131.

[28] Ibid., 138.

[29] For a helpful overview of vocational and technical education, especially for KATEE and its evolution to TEI,, see Konstantinos A. Karmas, Thalia G. Dragona, Anastasia G. Kostaki, *Prosdokíes kai théseis ton spoudastón tou technoloigikoú ekpaideftikoú idrímatos Athinón* (*Expectations and Outcomes of Students of the Technical Vocational Institute of Athens*) (Athens: Ministry of Education and Religion, 1986) 27–290. Also *Epangelmatikí ekpaídefsi stin Elláda* (*Vocational Education in Greece*) (CEDEFOP: Thessaloniki, 1995) an official publication of the European Union.

1975 the World Bank approved a loan of $45 million to develop five accelerated industrial vocational training centers, three higher technical centers, and other institutions. The school also benefited appreciably from the World Bank's funds.

When the junta fell in July 1974, the democratic government, headed by Prime Minister Konstantinos Karamanlis of the New Democracy Party, proclaimed education a top priority, second only to national defense. The government pressed forward to launch the most ambitious school construction campaign heretofore undertaken in Greece and to hire teachers; the laudable goal was to reduce the abysmal student-teacher ratio, the worst in Europe.[30] Education Ministry Undersecretary Vassilios Kontoyanopoulos told reporters, "It is not altogether unjust to describe education in Greece as an urgent problem hanging fire for fifteen decades."[31]

Hastening to revamp the junta's educational revisions, the government reinstituted nine years of compulsory education. Demotic was once again mandated as the language of instruction throughout the system, this time to include university. After completing the compulsory nine grades, gymnasium graduates were now offered three tracks to continue their secondary education; among them, and for farm school hopefuls the most significant track, was the newly created technical vocational lyceum. Graduates of the technical vocational lyceum were admitted into technical junior colleges, KATEE, or in 1983, its later version, TEI: *Technologiká Ekpaideftiká Idrímata* (Technological Educational Institutions). From there, it even was possible, although in practice extremely difficult, for students to take additional courses to enter university.

In an attempt to sharpen students' skills to match employment opportunities in such fields as nursing and tourist services, by the mid-1970s the government had increased the number of KATEE. This system was designed to help erase chronic underemployment, a problem that classical education inflamed rather than dampened. The goal to improve

---

[30] Among recommendations from an OECD report were: eliminating student fees for secondary and higher education; extending compulsory education from six to nine years; raising school leaving-age from twelve to fifteen. See Henry Wasser, "A Survey of Recent Trends in Higher Education," *Journal of the Hellenic Diaspora* 6/1 (1979): 85.

[31] "Education: The Problem of the Long Nail," *International Herald Tribune*, October 1978, 6S.

technical and vocational education was also propelled in large part by Karamanlis's determined quest for full membership in the European Union, since better-educated agriculturists and other specialized personnel would be needed to meet the higher standards imposed by this organization.

More than one scholar has pointed out that "research has shown repeatedly a connection between the education of farmers and their ability to use sources of information, to take up advice, and be willing to make adjustments on their farms and agricultural businesses to respond to the needs of modern agriculture."[32] This truism, seen against the facts that in 1991 81 percent of Greek farmers had completed only elementary school and one out of every three farmers under forty-five had also ended his education with elementary school, offered a tremendous challenge to the Greek government to attract young farmers into the proper educational channels.[33]

In sum, by 1977 a logical technical vocational system had been put in place offering rural youths three options, two of which were implemented at the farm school: TEL: *Technikó Epangelmatikó Líkio* (Technical Vocational Lyceum) and TES: *Techniki Epangelmatiki Scholi* (Technical Vocational School) both secondary school programs. By then, as noted above, the postsecondary centers for technical and vocational education, KATEE, had opened the door to TEL graduates; since TEL was a fully equivalent senior high school, graduates alternatively could take university entrance exams.

Strategically, these important reforms were geared to reduce the overproduction of university students. As a draw to rural students who were not attracted to the vocational and technical domain, the ministries of education and agriculture wisely fashioned a curriculum that extended beyond the confines of narrow agricultural topics, lending the program a certain measure of attraction. Certainly by the end of the 1970s—some thirty years after the close of World War II and the civil war—this policy redrew the map of technical vocational education in Greece. Still to be pinpointed were the exact needs in the economy and then to match them

---

[32] George Archimedes Koulaouzides, "Product Evaluation of the Curriculum of the American Farm School and A Proposal for the Development of a Distance Learning System for Agricultural Adult Education 1998" (MS thesis, University of Surrey, 1998) 14.

[33] http://www.ana.gr/hermes/1999nov/agric/htm.

up with the educational process and to introduce computer technology in the classrooms. Also, such concepts as marketing and management remained outside of education.

The PASOK government, which took office in 1981 under the newly-elected socialist prime minister Andreas Papandreou, displayed its interest in education by increasing the annual budget in that area by almost 50 percent in its first four years, when pressure from the European Union and increased urbanization were pushing Greece toward Western standards. Between 1981 and 1989, 7,980 new classrooms were built for secondary students and three new universities were founded. Nor was vocational education neglected. In that period the number of students studying in TEI trebled as the government pumped money into those institutions. Unfortunately, the economy failed to provide employment for the bulk of the TEI graduates, who fell into competing for civil service appointments or clamoring to enter the university, or agitating to have their diplomas elevated to university level.[34]

In 1982 the Papandreou government was absorbing a startling 70 percent of university graduates into the public sector, a percentage that makes clear that "higher education and employment as a whole functions as an extension of coveted public employment...mainly anti-chambers of non-productive...public employment."[35] No wonder the Houses— prescient as they were—had balked at training students to enter the public sphere decades before this hydrocephalic civil service scheme reached such proportions.

TEL and TES remained the mainstays of the secondary agricultural vocational curriculum until 1998, when the socialist government under Prime Minister Kostas Simitis instituted two new programs of study, EL: *Eniaío Líkio* (Comprehensive Lyceum) and TEE: *Technikó Epangelmatikó Ekpaideftério* (Technical Vocational School).

By that year four universities had been established offering matriculation in degree-granting programs in agriculture open to those who had earned a lyceum diploma and had passed university entrance exams. And to expand opportunities for students seeking higher

---

[34] Constantine Tsoucalas and Roy Panagiotopoulou, "Education in Socialist Greece: Between Modernization and Democratization," in Theodore C. Kariotis, ed., *The Greek Socialist Experiment: Papandreou's Greece, 1981–1989* (New York: Pella Publishing Company, 1992) 328–30.

[35] Ibid., 314.

education in agriculture in 1996 the American Farm School gained the distinction of establishing the first private agricultural two-year college in Greece: the Dimitris Perrotis College of Agricultural Studies, offering two majors: agricultural business and agricultural production.

# CHAPTER 4

# SECONDARY EDUCATION AT THE FARM SCHOOL: A CURRICULUM IN PROGRESS

Under pressure from alumni and students one aim (at times a reluctant aim) of the American Farm School starting in 1965 was to comply with the dictates of the government's curriculum. Compliance ensured that graduates would receive certificates "equivalent" to those issued by state high schools throughout the country. Equivalent certificates permitted the graduates to continue their education on a higher level, opening the way for sinecures in the civil service, a goal toward which the preponderance of the Greek people aspired. Moving into harmony with the government's curriculum brought about warmer relations between the farm school and the ministry of agriculture and later the ministry of education, all three entities routinely sharing information, consulting with each other, and often acting together in the decision making process. In 1965, the farm school took the first step toward equivalency all the while maintaining (with the government's permission) its unique practical program, the breadth of which was available at no other agricultural institution in Greece.

In the early decades, however, the directors of the farm school had, for many reasons, stubbornly avoided instituting a program of study that granted this precious equivalency. By 1925, the farm school was providing a five-year course of study for boys who entered at the ages of twelve to fourteen; many had worked a year or two after on their family farms, gaining a bit of experience and maturity before enrolling. When Charlie succeeded his father as director in 1929, he accepted only those applicants whose families owned land, a fundamental criterion to insure that the students upon graduation would return to tilling the soil. As for educational changes, he modified the program from five to four years and changed the language of instruction from English to Greek. In the first two decades English had been used because Father House and several key staff members did not speak Greek and the textbooks were written in English.

But to the extreme frustration of the boys who had been told by the school and everyone who appreciated its quality that their education was first-rate, in fact the best in Greece, their certificate was not regarded by the government as "equivalent" to the state's gymnasium diploma.[1] Since the farm school was not recognized as equivalent, its graduates were never given the grades or promotions they deserved in the civil service.[2] Without this equivalency the graduates' only recourse, like it or not, was to return to their villages to act as agents of change on the land and leaders in their communities, roles for which they had been groomed by the founder and his son. Another reason to dodge equivalency was the Houses' desire to escape the heavily classical curriculum that the government would impose at the expense of the practical. As mentioned earlier in 1955, the year Charlie retired, a survey of graduates since 1927 demonstrated that 41 percent were engaged in farming and 28 percent were working in various agricultural services, a cause for satisfaction to the retiring director.[3]

Because the certificate was not equivalent, in some circles the institution was ridiculed as the "School for Oxen Keepers," a source of humiliation for alumni and a hindrance to recruitment.[4] No wonder that eventually friction developed between the school and its students, and finally, over time, a substantial rift cropped up between the graduates and their alma mater.

---

[1] To the disappointment of the Houses, many graduates before the mid-1930s had secured positions in the civil service owing to the fact that in earlier days the farm school's five-year program resembled the course of study in state mid-level agricultural schools.

[2] Irwin Sanders, Raymond Miller and Robert Miller, "Report on the American Farm School of Thessaloniki," 1955, 15, AFS Archives. This report also provides an interesting picture of the day-to-day lives of on-campus residents (appendix 9).

[3] Ibid., 14.

[4] Bruce Lansdale, interview with author, Metamorphosis, 13 May 1998, written, AFS Archives.

310 STEWARDS OF THE LAND

Underpinning the school's policy were several intertwined reasons ranging from practical to abstract. Fundamentally the school's stated mission, set forth by the founder, was to train farmers. Success in this well-chosen enterprise had allowed the place to carve out a well-defined niche, one that was clearly understood by the governments and people of the United States and Greece. This lack of ambiguity aided fundraising, making it simple for donors to grasp the needs and worth of the institution. Since the school's clientele was clearly needy and appreciative, its donors were delighted to contribute, especially in the early postwar decades when it was widely publicized that Greece was locked in atrocious poverty. Although Greek officials were generally mistrustful of foreigners' motives and ambitions, by and large they did trust the farm school to carry out policies in the interest of the country. Incontestably, the last thing the country wanted was more farmers drifting off the land into the cities, where lack of employment could lead to political upheaval. As for abstract reasoning, there was, and still is, a strong vein of nineteenth-century romanticism that runs through Europe and the United States imbuing those who work the land with a kind of nobility.

To shield the boys from this opprobrium of "oxen keepers" and to enable them to proceed to the next stage of their lives, in 1965 Bruce took the bold step of setting aside the founder's sticking point and adapted the curriculum to achieve the prized equivalency. He was bolstered by the special report, authored by trustee Irwin Sanders, recommending highly that the school proceed in that direction. For the first time, the *Sower*, the school's official news medium, announced that "some...would be better able to serve their country by receiving further education working for the Ministry of Agriculture."[5] This drastic reversal, which overturned more than fifty years of unwavering policy, was facilitated by the short-lived 1963–1965 George Papandreou reforms, which did much to modernize the curriculum throughout Greek schools. In line with Papandreou's reforms, the farm school was not required, for instance, to teach the classics in ancient Greek, a subject that would have surpassed the language proficiency of most of the boys and delayed other learning.

---

[5] *The Sower*, no. 62, 1965–1966.

Under the new rules for the 1965 gymnasium-level programs, students entered after finishing elementary school. From the school they now received an equivalent gymnasium-level diploma when they completed a four-year course, the fourth year added to the countrywide three-year program to accommodate the school's unique component of practical training. During their fourth year, the youngsters specialized in one of several agricultural fields and took extra training to fit them for a more advanced approach to farming. About 30 percent of the curriculum was devoted to general education, 20 percent to agriculture classroom instruction and, in keeping with the institution's ethos of practical training, 50 percent of the time was given over to work in the fields and shops. Bestowing equivalency on the farm school's diplomas spurred recruitment and, most importantly, opened the way for the boys to progress to the next higher level of schooling.

In 1965, the ministry of agriculture licensed the school and it began operating under a law from the 1930s. The applicants were selected for their interest in farming, their leadership potential and scholastic ability, qualities the school believed necessary if the candidates were to act as leaders among their fellow villagers. The curriculum ran as follows: in the practical segments the students acquired skills in carpentry, masonry, plumbing, and electrical wiring. They were taught how to maintain and repair farm machinery, both modern and traditional, and they studied methods of soil conservation, crop rotation, vegetable gardening, pomology, and viticulture. From hands-on experience they learned such skills as care and breeding of cattle, pigs, sheep, and chickens. During the last year, each boy concentrated on an agricultural or technical specialty. In the classroom, following the state gymnasium curriculum, the students pursued studies in religion, Greek language and history, mathematics, and science, as well as agriculture.

In the Greek context the institution shined as an impressive enterprise. It boasted an enrollment of 200 boys studying, living, and working on over 350 acres of land, a faculty of 35, a selection of modern farm machinery, some 50 buildings, and land valued at more than $1 million. The entire operation had an annual operating budget of $450,000. And the school was expanding. A report issued in 1966 declared, "One indication of the School's growth is the increase in annual operating expenses over the past ten years. These have more than doubled...financed in part from increased income, Greek and United

States government aid and private sources."[6] In 1968 Bruce announced to the trustees that, in terms of quality and quantity at enrollment time, the boys had received better preparation and exhibited more potential than any entering class in recent years. Obviously, the school had entirely escaped its "oxen keeper" repute and was facing the next decade with optimism.

The following rough profile of students in 1968 gives an added dimension to our understanding of the farm school's clientele and exposes the poverty in which these youngsters were raised. The students came from villages ranging in size from 100 to 1500 families. The average number of families was 330. Very few students were from large towns; the average size of their farms was 6 acres. The animals comprised perhaps a goat, one or two sheep, occasionally mules, and a horse, and a few chickens. More fortunate families owned one or two cows. Usually, the land was planted in wheat or tobacco as a single major crop; fruit or small gardens were the main source of income in some sections.[7]

In 1970–1971, just five years after equivalency had been granted, another curricular change occurred when the junta government, conscious at some level of the country's economic and social problems, announced a major reform pertaining to vocational agricultural education: SEGE: *Scholí Ergodigón Geotechnikón Epangelmáton* (School for Agricultural Foremen), which took the form of a three-year curriculum for boys sixteen to nineteen years old who had finished nine years of school. The program provided a more technically-oriented education tailored to older students. The reform was enabled by the increasing availability of elementary and gymnasium education in the countryside, an improvement on a minor scale that had come about during the two decades of postwar reconstruction. Of necessity the government was compelled to modernize the country's agricultural technology to feed a swelling urban population: migration to the cities had succeeded in decreasing the proportion of the farming community from around 70 percent in the early 1950s to less than 40 percent when

---

[6] Paul B. Orvis and Basil Moussouros, "Survey and Planning Report of the American Farm School, 1966," 7, AFS Archives.

[7] *The Sower*, no. 69, 1968.

the new program was implemented.[8] Bruce explained the situation to the *Christian Science Monitor*: "The growing agribusiness needs technicians. This demands distinctive training, people who fill the gap between those with basic training and university graduates."[9] The government had indeed grasped the new reality: with the fading of subsistence farming the call was now for a new type of farmer, a person with management skills who would be capable of operating a modern business. This recognition folded easily into the school's primary mission: to hone agricultural and leadership skills.

To meet the requirements for agribusiness at the time, the farm school, owing to its unique features, was certainly well positioned. Much to the school's gratification, the SEGE program, actually designed by the school staff at the invitation of the minister of education, was considered so useful that the ministry of agriculture employed it as a blueprint, adapting it to other disciplines in state institutions located in Komotini, Ioannina, and Hania.[10]

As the first recruits under the new plan, eleven farm school graduates from the former gymnasium program were selected to continue their studies in SEGE, thus integrating older boys into the student body, a move that would have repercussions in the student revolt of the mid-1970s. From the first, the expectations for the new curriculum were high—at the very least it promised terminal technical training. And at most, since graduates from the new program received a Foreman's Certificate, they were eligible for entrance into the junior college, KATEE. From there, after certain formalities, there was a chance for the best students to enter university.

Under the new curriculum, the first year the boys studied Greek, geography, English, religion, math, general agriculture, and animal sciences, and worked in the fields, barns, and workshops. In the next two years each student chose one area of specialization from the four offered: farm machinery, gardening and floriculture, dairy calf rearing, and minor

---

[8] Albert Brown, "Report for the Schools and Hospitals Program Abroad for the US Agency for International Development," as quoted in *The Sower*, no. 80, 1973.

[9] "Reshaping Greek Village Life," *Christian Science Monitor*, 26 February 1972, 10.

[10] Andonis Stambolides, "Report to the Board of Trustees," October 1998, AFS Archives. Andonis Stambolides came to the farm school in 1964 as teacher of Modern Greek language. Subsequently, he held positions as director of the high school program, and associate director for education until his retirement in 1996.

livestock. The whole program of study was infused with the concepts and application of farm management, a subject in those days rarely encountered in Greece. Included were two summers of on-the-job training related to the students' specialty.[11]

At this crucial period of more sophisticated directions in agricultural education, the farm school, a robust institution with a professional staff of forty-five Greek citizens and eight Americans, was hitting its stride. Of the thirty-nine faculty members, seventeen were proud bearers of degrees from American universities, as were several members of the administrative staff.

To avail his faculty of an American education, Bruce was using every available resource. Culturally, it made better sense to educate Greek faculty in the United States than to import American faculty who might suffer from Greek language deficiency and perhaps culture shock, and thus be ineffective until they adjusted.[12] This tactic enabled the school to retain its Greek personality, while extending to students the spirit of American education via an American-trained faculty.

Any new academic program has its practical ramifications. From the outset it was apparent that the more mature SEGE boys, especially those who had graduated from the old program, some of whom were eighteen years old or older, would be cramped in the dormitory facility in Princeton Hall. At peak enrollment 250 students were housed there, fifty to sixty of them in a series of ward-like sections where they totally lacked privacy, a condition that perhaps had suited the younger boys. But for the new cohort, the atmosphere was spartan. The din precluded quiet conversation among bunkmates and the close quarters denied the students the opportunity to read and to work at desks of their own. In keeping with the farm school ethos, the boarding program was intended to serve as a learning component, encouraging students to develop concern and respect for classmates. Living conditions were no longer appropriate to achieve that goal.

In spring 1974, ground was broken for a new dormitory, thanks in no small measure to the United States government's USAID, which stepped forward with a $500,000 grant. The building was completed in

---

[11] For an explanation of the 1971 SEGE program, see *The Sower*, no 77, 1972; *The Sower*, no. 79, 1972.

[12] Bruce Lansdale, interview with author, Metamorphosis, 16 May 1999, written, AFS Archives.

1977, when the boys moved into new spacious quarters.[13] To some village boys the amenities were so impressive that some were overheard to say: "This is like living in a hotel!"

Catering to older boys, the new program raised the school to a new pitch. Bruce's observation caught the imperatives of the times. "The challenge for the board of trustees as they face the eighth decade of the School's operation is to assure a stable transition from a farm school to a modern technical institution...."[14] While Bruce was perceptive in this observation, he and his staff, as it turned out, did not fully prepare for the more adult behavior and appetite of the older boys. Nor could they presage how the political and social turmoil of the 1970s in Greece would influence the youth culture on campus. The school would later find the students in revolt for what they perceived as the juvenile and restrictive treatment they received on campus from their mentors.

On balance, there was much to admire in the SEGE program. Its emphasis on industrial training and the technical aspects of farming crystallized the ethos of the farm school, preserving the original principles of the founder yet adapting to the exigencies of the modern world. Also, as Bruce suggested above, SEGE did in fact serve as a transition between "a farm school and a modern technical institute." And thanks to various sources of funding, the school possessed the facilities to deliver effective management training. However, that curriculum today still has detractors who claim that classroom subjects were subordinated to the practical program, leaving some graduates with poor skills in writing and other aspects of theoretical education. A balance had not yet been struck.

During the SEGE years, students were afforded an intriguing approach to practical education: student projects. Patterned on the Future Farmers of America and the 4-H model, the projects had been initiated in the mid-1950s but really came into their prime in the mid-to-late-1970s. Managed by a team of about nine students, the projects were small business enterprises. The boys would raise pigs, for example, to sell to local farmers, or perhaps broilers to sell to local restaurants, or flowers, or vegetables. The team financed its undertaking by drawing up a budget and submitting it to the school, which played the part of the Agricultural Bank of Greece by furnishing the capital. The boys failed or succeeded,

---

[13] *The Sower*, no. 89, 1977.

[14] "Director's Report to the Board of Trustees," September 1973, AFS Archives.

we may be sure, according to the attention they paid to management and to the information they had absorbed in the classroom, fields, and workshops. The profits they realized from the projects went into a fund to spend on a class trip their senior year. The majority of the projects succeeded, but now and then one flopped. An administrator relates one grim experiment: "One of the project's groups in pig production found that they could get free garbage from our kitchen. They wanted to know why they couldn't feed the garbage and save on their feed bill, but they had no facilities for boiling it. Their advisers counseled against it, but the boys insisted. Before long, they had a dead pig. We autopsied it and showed them it had been poisoned by spoiled garbage."[15]

SEGE students now found it possible to spend three years at the farm school followed by three years of junior college at KATEE. After they finished junior college, they could take exams, and, if successful, enter the second year of university, the very pattern that Evangelos Vergos, a farm school graduate, followed. For example, he graduated top in his class at KATEE and continued his education at the Aristotle University of Thessaloniki. Upon receiving his degree there he went on to earn his Ph.D. from the University College Dublin. He has been dean of the farm school's Dimitris Perrotis College of Agricultural Studies since its founding.

In 1977 the SEGE program was terminated when the post-junta government of Konstantinos Karamanlis instituted yet another curriculum. A watershed event, this new program of study called TEL: *Techniko Epangelmatiko Likio* (Technical Vocational Lyceum) constituted a transition to a senior high school specializing in agriculture. And in another historical move, girls were enrolled in the fully coeducational enterprise. The first lyceum ever established at the farm school, TEL represented a vast departure from the institution's pattern. Consider that the founder had envisioned a course of education for "the head, the heart, and the hands" suited to young boys, future farmers, who, imbued with a special ethos, would return, he fervently believed, to the land in a spirit of commitment. No longer. Now, the educating of senior high school students, whose lyceum certificate was a launching pad for higher education, including the university, brought to campus

---

[15] Alan Linn, "Thessaloniki Farm School: The American Farm School Is Having a Valuable Impact on Greek Agriculture," *The Farm Quarterly* 21/3 (Summer 1966): 76 (AFS Archives).

youngsters harboring more worldly aspirations. Since the Greek economy was not positioned to absorb into business or industry the vast number of lyceum graduates throughout the country, they would surely turn to civil service or higher education with an eye to settling in the city rather than in their villages. Disquieting, too, to the school was the danger that the homespun values and spiritual emphasis the youngsters found at the school could well be rejected by these more savvy young people.

To implement these changes in 1977, the Karamanlis government passed a bill pertaining to technical vocational education. Under this legislative act, the American Farm School housed the first fully accredited TEL in the country; its curriculum was heavily academic. A second program, running simultaneously, was TES: *Techniki Epangelmatikí Scholí* (Technical Vocational School), emphasizing practical skills. These innovations entailed a serious complication since now for the first time the oversight of the programs came under both the ministry of agriculture and ministry of education, the ministry of agriculture overseeing the practical program in the afternoon and the ministry of education supervising the classroom instruction in the morning. Until this time, for over seventy years, the school had functioned solely under the ministry of agriculture, an office that knew the institution intimately, and with which the director and his staff had enjoyed a warm working relationship.[16] Bruce reasoned that to his detriment he had few close contacts in the ministry of education. On the other hand, the ministry of agriculture had been deeply involved in the life of the school to the extent that in 1929–1930 the ministry was authorized to enter into an eight-year contract with the farm school to grant scholarships. Thanks to this donation, the school had been enabled to increase the number of students and to enhance the quality of education, to say nothing of the cooperative energy ignited by this arrangement between it and the ministry.

As the years rolled on, this relationship only deepened and the government continued its scholarship assistance. In 1984, Bruce warned the board of trustees of the implications of this close association: "The School's role as an extension of the Ministry of Agriculture is one of the most difficult for people to understand and is often considered a threat to

---

[16] Stambolides, "Report 1998."

its independence. In a list of foreign schools given to a local magazine,
the Farm School is not mentioned by the ministry of education among
foreign schools. Obviously, this is a strength as well as a weakness.
Being that closely associated with the Greek government and accepting
$200,000 per year, while at the same time allowing the operation of a
Short Course Center by the Ministry of Agriculture on the campus makes
it possible for the government to exert pressure on the School." Bruce
went on to outline the positive aspect of this intense relationship:
"However, by working closely with the Greek officials, the School is
also able to influence their thinking in terms of innovative approaches to
agricultural education through its example." He admitted it was not an
easy role to play, but in reality the bond was not "one to be avoided if the
School is to fulfill its aims."[17]

A rare phenomenon in a highly centralized government, the
interdependent relationship between the school's educators and officials
in the ministries was a two-way street. The rapport provided the school a
channel through which to funnel its own ideas regarding the country's
educational system. In turn, the ministries benefited from the resources
of the experienced Greek members of the farm school staff, usually
American-educated men, pleased to share with officials their vision of
agricultural education and to work on its implementation. By spreading
its own philosophy of education within these influential circles in
Athens, the institution was able to exercise an influence that traveled
well beyond its campus in Northern Greece. As a result of the growing
synergy, farm school educators were often invited by the ministries to
help draft curricula for state agricultural schools and short courses
throughout the country.

To assure that the school would always be identified with Greek
interests, Bruce throughout his career, steered to the Greek
community—officials and ordinary citizens—for clout and advice, never
blurring this orientation by turning for support to the United States
embassy, although he maintained good relations with many of the
diplomats including the ambassadors. For example, Tad and Bruce
became close personal friends with Henry R. Labouisse.[18] Obviously the

---

[17] "Director's Report to the Board of Trustees," 1984, AFS Archives.
[18] Labouisse was appointed ambassador to Greece in 1962, went on to become
chairman of the board of trustees of the farm school, and head of UNICEF, which under
his direction won the Nobel Peace Prize. His wife, Eve Curie Labouisse, the daughter of

friendship with Ambassador Labouisse gave Bruce a direct connection to American officialdom. He turned to the ambassador for help, he claims, "only when I needed specific access to Greek personalities whom Henry Labouisse had cultivated."[19] In a country where fear of the smothering attention of the superpower was obsessive, Bruce's strategy and statesmanship was effective and brought about a unique association between a foreign personality and the Greek body politic. On the other hand, there were times when his stance became nigh to unworkable. For example, during the tense years of the junta and under the anti-American hue and cry of the government of Andreas Papandreou, relations between the school and some departments of the government were, at times, taut.[20] Be that as it may, the relationship with the ministries under Bruce's direction should be counted as the onward journey in the Hellenization of the school. This deliberate orientation shielded the school from anti-American rage after the fall of the junta and the coming to power of PASOK, when some other private American-chartered schools, like Anatolia College, according to its former president, William McGrew, were torn internally by hostile staff members or buffeted by outside forces.[21]

But in 1977 the worry was whether the ministry of education or the ministry of agriculture would snatch the upper hand in the matter of supervising the school. Would the two ministries become bogged down in a turf war, meanwhile disabling the school? In spite of the frequent and generous remarks voiced by Greek officials, the school was categorized legally, as both private and foreign, in a country that has historically been hostile to such schools to the extent that governments have sometimes threatened to do away with them. In light of this situation it was imperative to retain warm and personal relationships with officials in the ministries.[22]

---

Pierre and Marie Curie and the author of her mother's prize-winning biography, also became a member of the board of trustees after her husband's death in 1987.

[19] Bruce Lansdale, interview with author, Metamorphosis, 10 October 1999.

[20] Bruce Lansdale to the author, Metamorphosis, 17 October 1999, AFS Archives.

[21] William McGrew, interview with author, Thessaloniki, 29 October 2001, written, AFS Archives.

[22] Bruce draws a parallel between Greece and other European countries: he claims it is typical for European ministries of education, who are distanced from farmers, to lean to a theoretical approach, while ministries of agriculture do not have the authority to bestow

Patience and mutual respect seem to have worked wonders. A formula suggested by the farm school was agreed upon: a joint committee comprising both ministries was set up to oversee the school, an equitable arrangement that helped to balance the ministries' interests in the institution. The arrangement also insured that the school, supervised by both ministries, would have all its concerns attended to by the appropriate authorities.[23]

At this critical juncture, the American Farm School offered classrooms, laboratories, a dairy, piggery, poultry houses, and egg hatchery. Additionally, students gained experience tending fruit trees and olive trees, gardens, and field crops. To serve production needs there was a feed mill and milk pasteurization plant, and workshops for farm machinery repair and maintenance. Plumbing, carpentry, blacksmith, and electrical shops were on site for the students' practical experience.[24] Establishing the lyceum entailed a number of serious adjustments: curricular revision, staff and facility upgrading, fresh recruitment techniques, and a slightly altered ethos to accommodate the ambitions of students who were looking forward to a university education.

The TEL program, the three-year technical vocational lyceum, highlighting management throughout, offered specialties in maintenance and operation of farm machinery, plant cultivation and animal husbandry; each specialty offered a course of study valuable for rural boys and girls who would become the future agribusiness people and agricultural technicians of Greece. The program, fully equivalent, afforded its graduates a choice: if they did well in university entrance exams, they could matriculate at a university or alternatively enter TEI without exams.[25]

When in 1978 the girls were enrolled in the lyceum program on a level equal to the boys, they moved into the newly built Konstantinos Zannas wing of the dormitory. One can imagine the jolt of exhilaration

---

an "equivalent" diploma. Bruce Lansdale, interview with author, Metamorphosis, 9 October 1999, written, AFS Archives.

[23] Irwin Sanders, "Comments on the Academic Program of the American Farm School: A Report to the Board of Trustees," November 1986, AFS Archives.

[24] "Report Submitted to the Ministry of Agriculture," 29 January 1981, AFS Archives.

[25] *The Sower*, 1978, no. 93.

experienced by both boys and girls: even the once-hesitant staff members sensed the excitement and succumbed to the new ambiance.

As a means of leavening the theoretical portion, the farm school faculty worked creatively to augment the core curriculum, initiating for the first time in Greece such resources as teaching packages, which the school would later develop for application in the short course centers throughout the country. The faculty also employed audio and visual aids, and organized practical opportunities such as a physics club, where the students could carry out experiments. The students received summer assignments requiring them to collect data from their farms and villages; their records and comments were analyzed in the classroom in the fall semester.

TES, the two-year technical vocational school, ran parallel with TEL, offering a specialization in mechanized agriculture. TES graduates, desiring future training, could enter the second year of the TEL program. TES attracted students who concentrated on the practical aspects of agriculture; they received 70 percent practical and management training as opposed to 30 percent classroom instruction, in contrast to TEL students, who spent 60 percent of their time in the classroom and 40 percent in practical training. For the first ten years of its existence only boys attended TES. Weighted heavily toward farm machinery work, TES did not appeal to girls. However, later when TES offered a specialization in horticulture, young women gladly matriculated.

Even if TEL and TES added still more complications to the dizzying set of curricula that had changed since 1965, these two new courses of study added gusto to life on campus among students as well as staff. And as word of the new program traveled through the villages in 1978, the school received the greatest number of applications in its history. It stands to reason that the switch to coeducation broadened the applicant pool appreciably. Not only did the inclusion that year of the thirty girls in this historically all-boys bastion inspire a wave of euphoria, other positive factors, such as the wide-open opportunity for graduates to continue their education at the university and to secure a civil service post, acted as magnets. With an operating budget now set at $2,156,000, and a solid course of study in place, the American Farm School was playing an ever-increasing role in modernizing Greek agriculture and society.

Under the Andreas Papandreou government, cooperation between the school and the ministries of agriculture and education was still strong, although anti-Americanism, a hallmark of the Andreas Papandreou era, often caused strained relations. But teamwork was evidenced in sundry activities, such as the farm school's developing a training program for farmers at the request of the ministry of agriculture at the KEGE: *Kéndra Epangelmatikís Georgikís Ekpaídefsis* (Centers for Vocational Agricultural Education), a fresh adaptation of the former short courses. In a reciprocal spirit, the ministry of agriculture lent speakers to the American Farm School Alumni Association seminars. The farm school staff participated as needed when the ministry of education sought their opinion on topics related to agricultural education. Over the years the staff sat on various government committees that decided on curricular content, design of labs, specialization areas, and courses taught at state agricultural schools throughout Greece.[26]

Undeniably, the school was still playing a vital and unique role in Greek education. If one considers that a purpose of education was to mold the minds of future Greek leaders, the institution's Hellenization seems even more pronounced. This cultural process becomes all the more exceptional when we read the opinion of William McGrew, for twenty-five years the president of the prestigious American-chartered Anatolia College: "Greek educational law is based on the premise, widely shared by the general public, that education should be public and free. Consequently, private schools, even when tolerated legally and often admired for their high standards, are widely viewed as anomalies that should be bound in conformity to the standard public school model. This reigning assumption, and the system based upon it, obviously come into direct opposition to any private school's reason for existence, namely to provide a different or better education than the public schools."[27]

Once again in 1998 a new set of curricula was instituted, this time largely in an effort to align Greek education with that of the European Union members states. The three-year EL: *Eniaío Líkio* (Comprehensive Lyceum) offers a general course of study accompanied by specialty

---

[26] Andonis Stambolides, "Report to the Board of Trustees," February 1987, AFS Archives.

[27] William McGrew, "President's Report to the Board of Trustees of Anatolia College, 1974–1999," January 1999, AFS Archives.

classes related to agriculture. The diploma, fully equivalent, is accepted for entrance at TEI, or as a qualification to take university entrance exams. The other program, TEE: *Technikó Epangelmatikó Ekpaideftirió* (Technical Vocational School), replaced the previous TES curriculum. TEE offers two cycles of study. After the first two years, students receive a certificate allowing them to continue into the second cycle of TEE, or alternatively, they can join the work force. Graduates of the second cycle of TEE receive a certificate allowing them to obtain additional training at IEK: *Idrímata Epangelmatikís Katártisis* (Institutes of Professional Training), which are post-secondary centers offering specialties ranging from accounting, hairdressing/beauty science, to agriculture, including vineyard management and greenhouse production, among others. Or they can continue their studies at TEI, or join the workforce.

"These are good years for the Farm School," director of the secondary school Socrates Adamides asserts, remembering that "it was hard to recruit students before the establishment of the lyceum in 1977. In years gone by, farming was considered pure drudgery, beneath everyone's dignity, and few wanted to remain in the villages." Now, he maintains, "attitudes have altered. Students today are enthusiastic about their agricultural studies that give them the expertise to work their land as they might any other professional venture.[28]

As for the status of the graduates who have passed through the school's many curricula, a survey tracing them from 1904 through 1996 records data on 2,850 students who were graduated from the various programs as follows:

1904–1965 1,150 graduates, vocational junior high school
1966–1971 250 graduates, three-year vocational junior high school
1972–1979 150 graduates, school for agricultural foremen (SEGE)
1980–1996 1,300 graduates, vocational technical lyceum and vocational technical school (TEL and TES)

As for the career patterns of TEL and TES graduates, we learn from the survey the following:

---

[28] Socrates Adamides, interview with author, Thessaloniki, 24 October 2001, written, AFS Archives. With a degree in philology from the Aristotle University, Adamides came to the farm school in 1971 as a teacher of Greek language and history. He became director of the secondary school in 1997.

Continued their studies: TEL 66 percent TES 42 percent
Are working in their own trades or businesses: TEL 45.8 percent
    TES 66.3 percent
Are farming: TEL 21.25 percent TES 42.4 percent
Found work because of their Farm School education TEL 57.3
    percent TES 65 percent[29]

The ever-changing curricula, when viewed through the lens of the social, political, and economic forces, become fairly intelligible. The SEGE program, followed by the first lyceum, were both incubated during periods of national stress, rapid modernization and, finally, Europeanization. SEGE came into being under the junta, and the lyceum was born during the return to democracy under Konstantinos Karamanlis, just as Andreas Papandreou was building his PASOK party.

---

[29] *"Ekpaideftiká prográmmata dhefterováthmias: Amerikanikís Yeorgikís Scholís: 1980–1997"* ("Educational Programs, Secondary Education: American Farm School: 1980–1997") March 1997, report to George Draper and the board of trustees, AFS Archives.

# CHAPTER 5

# GREEK AGRICULTURE: FROM SUBSISTENCE TO AGRIBUSINESS

The level at which Greece's agricultural sector performs today would have been impossible to envision at mid-century, when Bruce Lansdale assumed leadership of the country's premier agricultural school. Greece has become virtually self-sufficient in food, except for milk, meat products, and feed grains, and the annual per capita income for farmers has risen from a paltry $202 in 1950 to $6,500 in 1992. A robust member of the European Union since 1981, the small Mediterranean country is vitally involved in that organization's agricultural programs and has benefited from them hugely.[1] The nation's progress has attracted people in developing countries in Africa, Asia, and Central America to visit, observe, and take heart from what has been achieved in a land that at midpoint in the twentieth century was suffering from hunger and abject destitution. In the wake of this agricultural development have come seismic sociological changes, giving women and men farmers an ease unknown in the agrarian misery they had historically endured.

---

[1] This book will use the term European Union throughout, although designations changed during the period under study. An official publication traces the name as follows: "The terms European Community and European Union are used to speak of that political entity which was born as the European Economic Community (or Common Market) through the Treaty of Rome, 1957." The European Union milk quota for Greece, set in the late 1980s, came to about 100,000 tons. This quota was still applied in 2002, when "Greece is short of milk and is forced to import huge quantities of powdered milk; by contrast Portugal has a quota of 1.6 million tons with a number of animals smaller than Greece's 300,000." *Kathimerini*, 26 June 2002, 5.

From Greece's liberation from the Ottoman Empire until the 1960s, the country had remained a predominantly agrarian society, a population of subsistence farmers doggedly coping with hand-held tools, wooden plows, and the stone threshing floor. To appreciate the number of people struggling on the land, one can glance at the following figures. The total working population in agriculture remained at about 75 percent until 1962, when it sank below 50 percent, owing to a tide of emigration to urban areas in Greece and later to Western Europe; by 1995, the number plummeted to 20.4 percent.[2] While the number of farmers would continue to dip as the years rolled on toward the new millenium, those who remained, especially after 1981, faced a different threat: an infinitely more complex and competitive arena, typified by rapidly advancing technologies; Greek farmers were entering a zone in which they were hardly trained to navigate. No wonder, then, that a progressive agricultural school in Greece, gifted with American know-how and a knack for reaching out, bolstered by funds from the United States government and from philanthropic organizations and individuals, could assume a position in Greek life out of all proportion to its size. Since an effective system of agricultural education had eluded a succession of Greek governments decade after decade, officials in Greece as well as individual citizens applauded the American Farm School's contribution and spirit of cooperation.

Indeed, the farm school played a particularly important role in the development of agriculture in the early postwar period. Farming, then comprising the most important sector of the Greek economy, fit the school's niche exactly and particularly after the widespread devastation in the countryside caused by nine years of war and occupation. At the end of hostilities, the landscape lay barren and disfigured. Draft animals had nearly all perished; flocks of sheep and goats had been woefully diminished; the land remained uncultivated, owing to the extended absence of the rural population, 700,000 of whom (a tenth of the population) had left their homes. At the end of their homeward trudge after the civil war, it was not unusual for hapless families to discover their houses reduced to ashes, torched in reprisals by the German

---

[2] For insight into the effects of emigration on agriculture after 1962, see William H. McNeil, *The Metamorphosis of Greece since World War II* (Chicago: University of Chicago Press, 1978) 116–17.

occupation forces, or ransacked subsequently during the civil war.[3] This was markedly so in Western Macedonia, an area from which a number of farm school boys hailed. The following cry of despair by student Macheos Tziakos was expressed commonly throughout the ranks of farm school boys: "When the Germans left in August 1944, they caused severe damage. Early in the morning they sprayed the houses with gasoline and put fire to them. They carried away the men and shot them. We lost our house, our livestock, and all our belongings. The destruction was complete."[4]

Throughout Greece, 2,000 villages had been destroyed. After liberation, according to the Paris Reparations Conference, the damage was estimated at $8.5 billion. Vast areas of vineyards, olive groves, and forests had been demolished; hand tools, the chief implements for cultivation, were unobtainable. The population was starving.

Beginning in 1944 the United Nations Relief and Rehabilitation Administration (UNRRA) imported 2,667,500 tons of food into Greece, an inadequate measure considering the magnitude of the crisis. Despite UNRRA's efforts, children in Western Macedonia and elsewhere continued to show signs of acute malnutrition and starvation.[5] UNRRA supplies were not restricted to food; clothes and household goods were distributed, as well as mules and tractors. That undertaking lasted until the end of June 1947, followed by the Truman Doctrine. By April 1948, the Marshall Plan was delivering aid that would substantially affect almost every phase of Greek life. However, the real work of rehabilitating the agricultural sector could not begin until 1949, when the prostrate country returned to peace after the civil war. By 1956, seven years after the end of hostilities, the situation had improved, although Greece, still ensnared in poverty and freighted by structural problems and cultural habits, could hardly be called the breadbasket of Europe.

---

[3] For a graphic description of the Germans burning an area almost in view of the farm school, see Edmund Keeley, *Inventing Paradise: The Greek Journey, 1937–47* (New York: Farrar, Straus and Giroux, 1999) 215–16 and his novel based on the same event, *Some Wine for Remembrance* (Buffalo NY: White Pine Press, 2001).

[4] As quoted in *Newsletter*, 1945.

[5] Norman Gilbertson, "Greece 1945–1953," 34, AFS Archives. Gilbertson came to Greece as a relief worker with UNRRA in 1945. Documenting the conditions of the time, he took a large number of photographs, a portion of which is included in this report. He later served as manager of the Girls School (1947–1950) and as a consultant to the farm school (1986–1991).

Yet, one scholar, seeking a positive note, wrote in 1967, "The agricultural economy is poor. However, it has grown at the surprising average rate of five percent per year in the postwar period."[6] Owing to the infusion of funds from the United States at a per capita rate surpassing that given to any other European country at the time, people no longer trembled on the edge of starvation. The aid was designed to rebuild areas of destruction, to modernize the country's economy, to create an infrastructure more thorough than existed in the prewar period, and to help support the armed forces.

The country has always needed to import materials from abroad, an expense its high-value agricultural exports such as tobacco, raisins, and olive oil had historically helped to defray. Aside from economic and humanitarian considerations, the political implications of supporting the segment that employed well over half the population was obvious to the United States, whose stated objective was to keep Greece in the Western camp. Having aided the Greek nationalist forces to turn the tide against the communists from 1947 to 1949, Washington would exercise the necessary precautions to quash any issue that the Left could exploit, especially a worsening economy. A former director of extension in the ministry of agriculture put it neatly: "We have survived all of our political upheavals only because of our bolstered agricultural economy."[7]

In rural areas poverty was persistent: in 1955, the year that Bruce assumed the directorship, the ministry of welfare reported that 29 percent of the population, some 2,764,017 people, were destitute. Unfortunately, the basic problem was structural: a surplus farming population. It was commonly estimated that 500,000 farmers, or 25 percent of agricultural manpower, could be shifted out of that sector without lowering production. Adding to the list of hindrances was low productivity, underemployment, monoculture (not always a negative), poor agricultural infrastructure, an abysmal road network, and lack of the most basic technology throughout the countryside. Probably the worst and most persistent feature was the small parceled-out production units. These difficulties would take decades of striving to overcome, and some of the worst disadvantages, such as small, separated fields still exist.

---

[6] Pan A. Yotopoulos, "The Greek Farmer and the Use of His Resources," *Balkan Studies* 18/2 (1967): 385.

[7] As quoted in Linn, *Farm Quarterly* (Summer 1966): 149.

In spite of these drawbacks, farmers were making laudable gains in production. By the beginning of 1956, they produced 40 percent more farm crops than before the war. Rice production, for instance, shot up 700 percent thanks to reclaimed salty coastal lands and flood plains, and drainage of former lakes and wetlands. In the mid-1950s, Greece became self-sufficient in wheat. This emphasis on agriculture and the urgency for farmers to develop modern habits made the farm school a primary resource center. Its creative education it extended to youngsters, the short courses it conducted for seasoned men and women farmers, the advice it offered on special projects, and the demonstration units on its farm helped push Greece into the modern world.

Instrumental in agricultural progress were several elements, one of which was the ministry of agriculture's creation of a new extension service in 1950, an organization geared toward winning the confidence of the farmers, who instinctively mistrusted government agents. Going against the custom of the land, where manual labor was considered below the dignity of government officials, the ministry of agriculture made an effort to indoctrinate agents with the efficacy of getting their hands muddied so that they could demonstrate effective farming methods in a practical manner. By 1956, according to William Hardy McNeill, "Agricultural extension agents had become a real influence in the countryside and many villages profited handsomely from their advice, receiving help from their hands as well."[8] The farm school, assuredly the only institution in Greece to stress hands-on practical training, offered extension agents a blueprint to follow. The ministry of agriculture invited the farm school to conduct courses for extension workers. The school can take credit for training dozens in this helpful corps, not least in teaching them to repair and drive vehicles across a jagged landscape, totally lacking at the time a passable road network, gas stations, and repair facilities.

Additionally, making affordable fertilizer available to farmers, on special credit terms they could more easily repay, gave a tremendous boost to production; before the war few farmers used fertilizer at all because it was too expensive, or they were ignorant of its benefits. In an

---

[8] William Hardy McNeil, *Greece: American Aid in Action, 1947–1956* (New York: The Twentieth Century Fund, 1957) 55.

overwhelming number of instances its use often doubled or tripled yields.[9]

Other measures lifted the agricultural sector to even greater heights, including the introduction of new crops such as fruit, vegetables, and cotton. These were all added to the list for export. In fact, by 1977 Greece had become a net exporter not only of olive oil, wines, tobacco, cotton, and raisins, but fresh fruit and vegetables, too. Self-sufficiency in cereals had been attained in 1958; sugar production also surged forward. Moreover, improved seed, the use of pesticides, and the extension of water control and irrigation gradually increased yields. Artificial insemination of cattle and sheep with semen from superior stock raised the quality of indigenous livestock.

As the transportation system, namely the road network, spread across the country agricultural progress zoomed forward. "It was then possible, particularly when allied to the development of the modern refrigerated lorry, to send perishable agricultural products into Western Europe and Middle East markets in first class conditions on a regular basis," one observer comments.[10] Enabling growers to sell products in the markets removed them from the ranks of self-subsistence farmers and placed them in the category of budding businessmen.

The following numbers give an idea of the sector's significance to the overall economy during the first two decades after the war. Specifically, figures for 1956 of $213 million worth of exports show that about $187.5 million, or 88 percent, were in cotton, grapes, tobacco, and raisins. In a spectacular improvement toward the end of the 1970s, Greek agriculture could feed the whole country and provide a net $330 million surplus to offset the yawning trade deficit.[11]

Regrettably, by the end of the 1970s, farming families were still the poor people of Greece, their incomes registering between one-half or one-third of workers in other sectors of the economy. Most officials in the Greek government recognized that they were not doing enough to encourage seasoned farmers or youth to stay on the land, to whittle away bureaucratic problems, to enhance extension services, or to expand

---

[9] McNeil, *Metamorphosis*, 91.

[10] James Pettifer, *The Greeks: The Land and People Since the War* (London: Penguin, 2000) 12.

[11] "Greece: An Ecomonic Survey," *International Herald Tribune*, September 1977, 9s.

educational opportunities for young farmers.[12] Xenophon Zolotas, decrying the deficiency of trained personnel in the country, especially "managerial talent," declared in 1966 that "these weaknesses must be eliminated."[13]

Interestingly, starting in the 1960s, Greek agriculture began drifting into a state of transition, as many farmers began to conceive of their land in a fresh way. As "scientific farming is increasingly practiced, the basis of attachment is changing from the traditional idea of land ownership as a sacred trust closely bound with one's immediate family to a view of land as a source of financial gain," Sanders observes.[14] He notes that those who lived in war-torn zones, such as farmers in Northern Greece, the farm school's region, and likewise those who labored on poor land everywhere, were willing to give up their family plots to move to a safer place that could offer them an easier life. Similarly, on fertile plains where mechanization had succeeded, the proprietor's attitude was more in line with that of farmers in Western Europe or even the United States: there the land constituted a commodity, a business, rather than a sacred heritage. This novel sentiment was to facilitate the farm school's task of instructing students to become managers—keeping accurate books and records, staying abreast of scientific advances, and remaining informed about the changes and opportunities in the field of agriculture as regulated by the relevant ministries in Athens. Of course, this modern perception of the land marked a break with traditional rural values, ushering in a whole new social psychology. In this atmosphere, the farm school's unwavering ethos of encouraging boys to remain on the land ran counter to the swelling trend in Greece.

Modernization, many feel, has come at an exorbitant price. The transformation has wrought dire changes in Greek rural communities and taken a toll on the environment. As Greek rural sociologists and scientists note, "External inputs of machines, fossil fuels, pesticides and fertilizers have displaced farmers, and Greek rural culture has been put under pressure, as more and more people have been forced to migrate in search of work. Local institutions, once strong, have become co-opted by

---

[12] Ibid.

[13] "National Income and Economic Development," *Hellenic Review, International Report* (April 1966): 29. Zolotas also served briefly as prime minister in 1989.

[14] Irwin T. Sanders, *Rainbow in the Rock: The People of Rural Greece* (Cambridge: Harvard University Press, 1962) 60.

the state, while farms have been simplified and some resources, once valued on the farm, have become waste products to be disposed of off the farm. Some external inputs are lost to the environment, thus contaminating water and soil, while the overuse of some pesticides causes pest resistance and leads to pest resurgence, encouraging farmers to apply yet more pesticides."[15]

The farm school's location in Northern Greece (Eastern, Central, and Western Macedonia, and Thrace) encompassed regions of both variety and importance to Greece's agricultural status. Northern Greece, for instance, contributed immensely to the country's agricultural sector and hence to the economy of the nation. The estimated value of its farm production, thanks in part to extensive rolling plains and broad rivers—rare topography in mountainous and littoral Greece—represented in the 1970s as much as 30 percent of the total value of Greece's output.

Valuable as it was, Northern Greece suffered from some of the same disadvantages found in other sections of the country. Farmers' holdings in the three Macedonia regions averaged less than 10 acres; in Epiros, the number slid to 5, while generally throughout Greece the average was between 6 or 7 acres. To make matters worse, even these meager lots were divided into three to twelve separate pieces, owing in large part to inheritance and dowry practices, and were often scattered far apart.[16] As Ann House observed, "When the million and a quarter refugees came from Asia Minor in the early 1920s, land was apportioned by the government according to the size of the families; now children have grown up, married and have children. We found three generations now

---

[15] George Daoutopoulos and Myrto Pyrovesti, E. Petropoulou, "Greek Rural Society and Sustainable Development," in K. Eder, K. and M. Kousis, eds., *Environmental Politics in Southern Europe: Actors, Institutions, and Discourses in a Europeanizing Society* (Dordecht: Kluwer Academic Publishers, 2001) 169.

[16] In his absorbing study of the beginnings of the agrarian society in ancient Greece, Victor Davis Hanson refers now and then to modern Greek farm practices. About the fragmentation problem today in Greece he writes, "...on the Aegean island of Milos survey archeologists found that modern Greek farmers average a mean time of two hours and fifty-five minutes traveling to work each day...many modern Greek farmers claimed three weeks lost per year simply walking to their disparate fields. One farmer spent almost five hours a day in commute time, sometimes to do a fifteen-minute operation on his distant plot. Another agrarian apparently traveled five thousand kilometers in a year to and from his farm" (Victor Hanson Davis, *The Other Greeks: The Family Farm and the Agrarian Roots of Western Civilization* [Berkeley and Los Angeles: University of California Press, 1999] 53).

living on the land, families of ten or twelve on farms of five or ten or fifteen acres."[17] Undoubtedly, land consolidation, then as now, is the optimum solution. But farmers are opposed to the radical concept of shifting around plots, and the difficulty of equitable land distribution is almost insurmountable. Complicating the problem is the absence of a national system of land registration, which to this day has not been instituted. The size of land holdings and their fragmentation render them almost useless to economies of scale, thus impeding farmers from managing their land as a business enterprise. For instance, on those fields that were not mechanized, where lumbering beasts were still pulling plows, farmers needed up to 1.5 acres just to feed the animals. Greece did embark on a voluntary land consolidation program in 1963, as well as compulsory consolidation in connection with large irrigation projects, but consolidation, still essentially voluntary, has remained limited. Thus the burden of success is placed on the only alternative—efficient land management. Management was an approach that had been incorporated into the farm school curriculum, which stressed record keeping, accounting, marketing, packaging, and openness to innovation.

Projects to increase agricultural output included expanding irrigation and erecting flood controls to increase arable land. To that end, important land reclamation projects in the 1950s and 1960s included work on Northern Greece's great rivers: the Axios, Aliakmonas, Nestos, Strimonas and in the east the Evros, on the border with Turkey. By 1958, over 272,000 acres were irrigated in the Thessaloniki plain. These rivers, except for the Aliakmonas, although they course through Northern Greece, have their sources in Yugoslavia or Bulgaria (both antagonistic countries during the Cold War), or flow through Turkey, a situation that brought into play a spate of political problems.[18] In addition, small irrigation projects, works that farmers could accomplish themselves individually or collectively, were encouraged by the Agricultural Bank of Greece and the ministry of agriculture.

Farmers lacked knowledge of irrigation techniques and the amount of water requirements for various crops in different areas. To instruct

---

[17] Ann House, "Family Letter" #41, Thessaloniki, 10 May 1952, AFS Archives.

[18] For agreements between Greece and neighboring countries dealing with the issue of water, see Axel Sotiris Wallden, "Greece and the Balkans: Economic Relations" in *Greece and the New Balkans: Challenges and Opportunities*, ed. Van Coufoudakis et al (New York: Pella Publishing Company, 1999) 113–14.

farmers, the ministry of agriculture established experimental plots to determine precise water requirements. At the farm school, boys were taught how to create and use a variety of irrigation systems. As evidence of their training, travelers could spy, strung across the Macedonian plains here and there, angular inventions of cast-off pipes and fittings concocted by ingenious farm school graduates. In 1955, a resourceful graduate, Dionysios Giungas, rigged an irrigation system from salvage he took from an abandoned German army junk pile. Because annual rainfall in the area of Thessaloniki averages a scant 17 inches, the farm school had long experimented with ways to conserve moisture in the soil.[19]

Northern Greece was plagued by still another vexation: soil erosion. Millions of cubic feet of soil were washed off to the sea. Reforestation, terracing, and construction of small stone and cement dams and dikes were done by individual farmers or initiated by the government to help hold back the washouts. How to terrace hilly slopes and to use contour plowing was taught even before the war at the farm school, in recognition of the difficulties of the mountainous terrain. A *Reader's Digest* article conveys how desperate those mountain villages were and how any improvement might aid them: "A rutty track, dust or mud according to the season, wavering past houses whitewashed but unscreened. Three hundred people drawing water from a single well. A church, of course; a one-room school; a coffeehouse where copper-faced farmers sit talking politics all day, all month, waiting for their single crop to ripen."[20]

Even though small-sized plots were not conducive to optimal use of machinery, mechanization did move forward spectacularly in Northern Greece, mainly on the plains, of course. For example, throughout Greece between 1944 and 1957, tractor ownership jumped from 251 to 6,121 with 43 percent of all tractors found in Northern Greece.[21] Machines clattered side by side with the eternally plodding mules and oxen pulling

---

[19] "The American Farm School," 1955, AFS Archives.

[20] Robert Littell, "They're Helping the Greeks to Help Themselves, "*Reader's Digest* (September 1960): 129. For problems concerning terracing and soil erosion, see J. R. McNeil, *The Mountains of the Mediterranean World: An Environmental History* (Cambridge: Cambridge University Press, 1992) 274.

[21] Paul Vouras, *The Changing Economy of Northern Greece Since World War II* (Thessaloniki: Institute for Balkan Studies, 1962) 37–38, covers most aspects of agriculture in Northern Greece. Sanders, "Report 1955," appendix 2, p. 55, addresses mechanization.

Hesiodian ploughs. Mechanization, because it did a better job of plowing, cultivating, and harvesting, not only increased yields but also made labor more efficient. To help farmers operate, maintain, and repair the machinery, the farm school joined with the ministry of agriculture to offer instruction and in the 1970s initiated a program for students at the Girls School to teach them to drive and maintain tractors, a radical innovation in a gender-prescribed society. Much earlier, in the short courses, at least one course each year was designed to teach village men and women tractor operation.

By 1959, wheat had become a bumper crop, Northern Greece boasting production of 46 percent of the nation's total, an increase of 103.84 percent over the prewar figure. This leap created a state of self-sufficiency, hardly a petty matter for a country hard-put to pay for imports. The boost was due to the use of early maturing varieties, intelligent application of fertilizer, better plowing techniques, and improved use of insecticides.[22]

Before the war, Greece rarely exported fruits or vegetables, and in remote villages rural folk had planted practically no vegetables owing to lack of land, dearth of knowledge about proper nutrition, and a notion that cultivating vegetables was somehow beneath their dignity. After 1950, however, close to Thessaloniki, an area blessed by relatively fertile patches of rolling land, some irrigation infrastructure, and a road network to ship the produce throughout the region, vegetables were enjoyed. After the war, the government launched an extensive campaign to encourage farmers to grow and consume vegetables; agriculturists envisioned a day when Greece would be a substantial exporter of fruits and vegetables, a dream that ultimately materialized. The faculty at the Girls School had understood the need for vegetables in the diet, and when the first students arrived, their teachers taught them how to plant small, easily maintained kitchen gardens.

In the 1950s, fruit became a significant product in Northern Greece, representing 26 percent of the total value of Greek fruit production.[23] But the lack of sorting, grading, packing equipment, ice-making plants, cold storage, and refrigerated trucks to transport the produce to market had to be addressed before Greece could become a prime exporter of fruits. By 1977, however, with improved infrastructure, peaches and apricots were

---

[22] Vouras, *Northern Greece*, 39–42.
[23] Ibid., 48–51.

shipped to Western European countries, and citrus fruit to the Eastern bloc.

Viticulture became another substantial agricultural activity in Northern Greece; in 1959, it represented 13 percent of the country's total vineyards. In time, exportable wines were produced from the excellent grapes of Northern Greece. Yet, as of 1999 only 4 percent of the land throughout the country was given over to viticulture. The farm school began serious viticulture in 1993 and by 1996 was producing its own distinctive wine, a project enabled and guided by trustee Yiannis Boutaris, a distinguished vintner from Northern Greece. Boutaris donated the vines and the time of his experts to train farm school staff to work the vineyard. The first machinery came from one of his own wineries and his people oversaw the installation and start-up of the winery, all gifts in kind.[24] Interest in viticulture at the farm school had begun in the early years when Father House, as noted in book 1, brought phylloxera-resistant vines from the United States to graft to local stock, an effort to combat the vine-killing aphid. Viticulture had been taught off and on to the students, but it was not until 1996 that it would become a focal enterprise at the school to include a winery. At the beginning of the new century, the school was cultivating sixteen varieties of grapes.

Northern Greece, especially on plains washed by the region's rivers, has provided a base for the country's livestock industry, representing 50 percent of Greece's cattle. The depleted herds—sheep, goats, dairy and beef cattle, draft oxen, donkeys, horses, mules, and water buffalo—were slow to rebound after the war; by the early 1960s they had barely reached prewar levels. And the indigenous animals used for meat were generally of poor quality, a factor that spurred the importation of meat and dairy products in the 1960s.[25] To improve the quality of livestock the government initiated an artificial insemination program, which by 1966 had improved 55 percent of the dairy stock in Greece.

Along with the government, the farm school and some foreign organizations assisted in an insemination project. Slowly, farmers, who had been skeptical of the procedure, expecting from it the birth of freaks, became convinced that the improvement of the herds would increase their income and supply Northern Greece with more food. Farmers

---

[24] George Draper, interview with author, Wayland MA, 2 February 2001, written, AFS Archives.

[25] Vouras, *Northern Greece*, 119.

learned, also, to plant crops for forage on fallow fields adding to the amount of feed available for livestock. The farm school had introduced Jersey and Guernsey cows, imported from the United States, beef cattle, swine, silos, balanced feeds, among a galaxy of other innovations designed to increase the productivity of livestock breeders. By stressing livestock production in its practical program, the farm school was spearheading efforts critical to Greece's agricultural development.

Before the war, the country did not possess a poultry industry. According to custom, hens had been treated as gleaners, turned loose to scavenge the villages' unpaved roads, without benefit of nutritious grains. They laid only about sixty small eggs per year. (For a standard: in the US in 1999, the average laying hen produced 250 eggs a year.) In 1958, the school started the Project for the Practical Application of Broiler Poultry Management in Northern Greece, financed by the Rockefeller Foundation. The project started with a base of 170 White Rock hens and 25 Peterson Cornish roosters, their offspring the first of a new breed developed for the project. By the early 1960s, the school had become a showcase for the poultry industry.

No discussion of agriculture in Northern Greece is complete without including the subject of tobacco cultivation. Thrace and Eastern Macedonia, especially the plains of Drama, Serres, Kavala, Rodopi, and Xanthi, include some of the best tobacco growing fields in Europe, yielding more than 50 percent of the country's lucrative crop of Oriental leaf. As a labor intensive but remunerative crop, it drew the whole family's time and energy. Unfortunately, the tobacco farms averaged a mere 1.5 acres and, like most holdings, they were fragmented. In the immediate postwar period Greek tobacco exports tumbled, owing to a shift in taste away from Oriental leaf. The situation added to the misery of war-scarred villagers, who were not prone to crop diversification.[26]

In Halkidiki in the 1950s when the tobacco market slumped, tobacco-growing families were bundled into trucks hours before dawn, driven for almost four hours, where, exhausted, they dug in iron mines.[27]

---

[26] "Raising different crops on the same farm allows the farmer to minimize risks from all sources.... Vines, cereals, and tree fruit...all pollinate, produce fruit and enter dormancy at different times, ensuring that a minute, an hour, a day or a week of climatic misfortune...will likely catch only a portion of the farmer's total produce..." Hanson, *Other Greeks*, 72.

[27] Bayard Stockton, "The American Farm School in Greece," 1972, AFS Archives.

Ann House observed, "Manufactured items which people needed to buy have, since the war, increased from three to six times an average price greater than farmers get for their produce and this is especially true of tobacco, so the suffering where tobacco is grown exclusively is great."[28]

Although the tobacco industry was ebbing in Greece, the government did not discourage its cultivation because of its value as an income for tens of thousands of people, its flourishing on otherwise unyielding land, its ability to endure drought, and its benefits in season, when it provided employment for underemployed non-tobacco workers. The government was looking to the future, when the market would rebound. By 1976, the country was growing both Oriental tobacco (12,200 tons) and Burley tobacco (16,000 tons), a variety appealing to a larger market. In 1989, tobacco regained its status as a major export item for Greece. By1992 farmers in Macedonia were growing the popular Virginia tobacco, which commands high prices, while farmers in Thrace were switching from tobacco to cotton and wheat.[29]

During the 1980s an average of thirty to forty farm school students each year came from tobacco-growing families. For a number of reasons, one harking back to the founder's pious abhorrence of tobacco, the farm school reluctantly included tobacco growing in the curriculum. In 1967, for the first time the school planted tobacco for seed propagation in cooperation with the Greek Tobacco Institute. Later, when a deeper understanding of the relationship between cancer and smoking became clear, doubts about the morality of the school's association with such a program were strengthened. Actually, the debate on the subject of tobacco spread throughout the European Union, which devised a special program alerting people to the health risks of tobacco, at the same time encouraging new varieties of tobacco that some thought to be less carcinogenic. Pinned on the horns of a dilemma, the European Union had in mind the welfare of endangered Greek tobacco growers. In 1986–1987, the Farm School did offer practical training in tobacco growing and processing. Philip Morris had been donating money to the school and pledged to maintain and increase the annual contribution. Yet

---

[28] Ann Kellogg House, "Family Letter, no. 41," Thessaloniki, 10 May 1952, AFS Archives.

[29] In 2002 Greece was the main tobacco grower in the European Union; tobacco growing represents the main and in some cases the sole source of income for 64,000 families. *Kathimerini*, 21 June 2002, 5.

as Andonis Stambolides stated, "We teach our students to avoid smoking and forbid them to smoke and punish those who do."[30]

Northern Greece continues its role as a key agricultural factor. For example, in 1998, Central Macedonia produced some 65 percent of Greece's total yield of stone fruits and apples, and grain cereals (wheat, barley, and corn) covered 58 percent of Central Macedonia's agricultural area. Sugar beet, industrial oil seed, and fiber crops from that region figure largely in the country's output. Add to that the fact that Northern Greece was responsible for 36 percent of the total dairy industry, and the significance of this region becomes apparent.[31]

In the decade of the 1980s, dairy products represented Northern Greece's main growth area as consumer demand for milk products strengthened. Milk and yogurt producers, a new class of entrepreneurs, provided healthy incentives for farmers to stable herds of dairy cows. The inducements proved effective in encouraging dairy farmers to upgrade their stock through purchase of productive French and Dutch cows, and to insure that their animals received follow-up veterinary services. As support for dairymen, the entrepreneurs worked to obtain subsidies for them to construct cowsheds and other facilities. All these arrangements must have seemed like heady finance to Macedonian farmers elevated now to operating in the realm of big business. Tasty yogurt, once produced on small family farms, became a thriving industry, 77,000 tons of it being processed in factories. In addition, high demand for Greek sheep's yogurt in Western Europe is evidenced by the stocks displayed on the shelves of supermarkets there. Despite the emphasis on dairy products, Greece still was forced to import large quantities of milk.[32]

Northern Greece played a substantial role as the nation adjusted to the rules of the European Union. Farmers from the three regions of Macedonia grew 62 percent of cereals, 50 percent of corn, and 80 percent of sugar beets and rice. Furthermore, at least 50 percent of tobacco and a

---

[30] Andonis Stambolides, "Report to the Board of Trustee Education Committee, 1987," AFS Archives.

[31] George Archimedes Koulaouzides, "Product Evaluation of the Curriculum of the American Farm School and A Proposal for the Development of a Distance Learning System for Agricultural Adult Education 1998" (MS thesis, University of Surrey, 1998) 33.

[32] *Area Handbook*, 1994, 117.

large quantity of cotton were cultivated in that region, which since the war had enjoyed an additional 300,000 cultivable acres in the river valleys as a result of completed anti-flood projects. Fruit trees blossomed on the hillsides above the Macedonian plains. Aid workers who remembered with heartache the ailing Greece of the late 1940s and early 1950s, a landscape of eroded hills and desiccated soil, were gladdened years later to behold the exuberant panorama.

Greece's entry into the European Union in 1981 transformed the country's agricultural sector. As a student of Greece remarked, "To say that agriculture was transformed...is to say that in some way Greece was transformed...as agriculture made up seventeen percent of the economy."[33] A decade or so after Greece's accession, the income of the Greek farmer rose dramatically thanks to the European Union's bountiful subsidies, money granted to support prices for agricultural products such as tobacco, fruit, cereals, olive oil, and vegetables, and to renovate and modernize agricultural structures. As a result, Greek farmers received 40 percent of their income from European Union subsidies.[34]

In 1981, the farm school mulled over the formidable task of readying students for the complex world of the European Union. The Greek government, as well as the farm school's director and staff, recognized that the school's proven ability to provide guidance in the cultivation of new crops and new techniques would be invaluable as Greece set itself to compete in the European Union. And it occurred to all that the farm school could contribute even more by reaching beyond the short courses and moving deeper into adult education. The notion of actualizing the school's potential by extending its range of activity was compelling, but it collided with the reality of the institution's limited income and stated mission, which focused primarily on secondary education. There would come a time, however, when the founding of the Dimitris Perrotis College of Agricultural Studies in 1996 and the creation of a Lifelong Learning center on campus would widen the parameters. But for 1981, the major consideration on campus was insuring the effectiveness of the lyceum and scholi programs, the two secondary school curricula, both only three years old at the time of accession. After careful evaluation, the school wisely decided to bide its time and to

---

[33] Pettifer, *The Greeks*, 35.

[34] Stamatis Sekliziotis, "Annual Report," Agricultural Office, American Embassy, Athens, 1997, p. 6, author's possession.

conduct a detailed study of the institution's goals for the 1980s before deciding to broaden its scope. Farm school graduates, most officials agreed, would continue to be the institution's greatest contribution to the country. As the sector adjusted to the conditions of European Union membership, these alumni, working not only on the land but in various agricultural-related occupations and in government positions, could help administer a variety of agricultural services.

To its detriment, one of Greece's weaknesses has been an inability to embrace the cooperative, an organization central to the economics of agriculture. The Greek farmer has found it difficult to accept the very concept of a cooperative. An individualist who considers his family the most important unit of society, he deems it unnatural to subordinate his own immediate needs to the pressure of the group. The reality was that the cooperatives, organized along party lines, were indebted to the government in power for the loans extended to farmers from the Agricultural Bank of Greece, itself a government institution. That meant that loans and benefits could be awarded according to political connections. Considering the farmer's distrust of the central authority, it was no surprise that he would look askance at membership in a cooperative so subject to government manipulation. To add to the reluctance, lack of experience in group membership and insufficient education isolated him from the rules of order and the businesslike give-and-take of discussion as a group drives toward consensus.[35]

Since its earliest days, the farm school underlined the necessity of cooperation. The campus itself—home to school buildings, dormitories, playing fields, farm buildings, a church, and staff housing—was deliberately modeled as a community where each person is meant to feel respected and in turn is expected to contribute to the commonweal. This culture of community, a campus where students live together in a dormitory, eat together in a common room, pray together, and come into constant contact with the resident faculty and staff, equips graduates to serve in such organizations as cooperatives and in other group settings. Predictably, many of them have, in fact, taken the reins of leadership,

---

[35] For further problems experienced by cooperatives, see Ministry of Press and Mass Media, Secretariat General Information, *About Greece* (Athens: n.p., 1999) 192. Also, Bruce M. Lansdale, "On the Special Aspects of Village Life," paper delivered to the Thessaloniki seminar of the International Conference of Social Workers, September 1964, AFS Archives.

participating in any number of community enterprises, cooperatives included. A survey showed that in 1965, twenty-one alumni had been elected to the presidency of cooperatives and thirty-one to boards of cooperatives.[36]

In the Greece of 1981, with the eclipse of the subsistence farmer, the heightened use of farmland as a business, and membership in the European Union, the need to pool resources and to cooperate on every level had become essential. Thus, the slighting of efficient cooperatives, a predicament that lingered at least to the beginning of the new century, hampered agricultural progress.

In the waning years of the twentieth century, Greece's agricultural base was shrinking. By 1999, agricultural products slipped in importance, totaling only 11.2 percent (the figure includes fisheries and forestry) of Greece's GDP; nevertheless, they were undoubtedly more meaningful to the Greek economy than to the economies of other European Union countries, where, on the average, agricultural products represented merely 2.4 percent of the GDP.

Steadily, demographics were changing: when the number of people engaged in agriculture decreased to 18 percent in 1999, the sector became leaner, but certainly healthier. To be sure, those who remained on the land faced more complex situations that required a better educational background, a factor that had been recognized long ago by the Greek government and the farm school.

Another demographic dilemma was developing: in 1995 only 8 percent of people between eighteen and twenty-five years of age worked in agriculture, a low number considering that between 1981 and 1995 there was an actual increase in agricultural income thanks to European Union inflow. In 1995 Greece's agricultural income swelled to $4.5 billion, a figure that one expects might have lured young men and women back to the land, but the trend was unmistakably to move away from the land. The Third Community Support Framework, called "Agenda 2000" (2000–2006), was designed to aid the agricultural sector but as *Hermes Magazine* predicts, "The European Third Community

---

[36] "American Farm School Graduate Survey Report 1965–66," appendix 1, AFS Archives.

Support Framework is expected to be the final call for Greek farmers before they venture unassisted into the global free market."[37]

Another pressure on Greece's agricultural demographics is the European Union itself, which aims to decrease the farming population within its member states to 5 percent. If we take into account that agricultural products in 1954 represented as much as 79 percent of the country's total exports, as opposed to only 29 percent in 1994, we can easily comprehend the decreasing importance of agriculture to the economy. Statistics, which had been pointing over time to an exceedingly decreased role for farmers in the country's social and economic life, held crucial importance for the school. If, as they say, "demographics is destiny," such paring down raises in a direct way the question, "Why an American Farm School in Greece?"

---

[37] http://www.ana.gr/hermes/1999/nov/agric.htm. To demonstrate the magnitude of the European Union's contribution to Greece, the country is projected to receive $5 billion on average per year in the 2002–2006 period, net of Greece's own payment to the European Union. *Kathimerini*, 17 June 2002, 5.

# CHAPTER 6

# AGRICULTURE AT THE FARM SCHOOL: THE VERY KEYSTONE

When building a historical framework for the American Farm School, it is appropriate to regard its farm as the very keystone. John Henry House envisioned the farm as a laboratory, a place where one could adapt modern methods of agriculture to the specifications of Macedonia. But as much as he loved to experiment on the land—he was an accomplished amateur botanist—he harbored a more compelling reason in 1904 to make the farm the keystone of his missionary endeavor; it was the perfect setting to rouse the students' spiritual imagination. As Ralph Waldo Emerson wrote a half century earlier, "Nature always wears the color of the spirit" and "What is a farm but a mute gospel. The chaff and the wheat, weeds and plants, blight, rain, insects and sun—it is a sacred emblem from the first furrow of spring to the last stack which the snow of winter overtakes in the fields."[1] Echoing the New England writer, House wrote: "The man who sows the seed of ideals, who lives in harmony and cooperation with his neighbor, who carries his grain of knowledge, be it knowledge of art, literature, economy or right, is a sower of a seed; and it is sowers of the seed the School hopes to turn out."[2]

For those with heightened sensibilities, the farm acted as a metaphor: its fields, barren at the founding, within a decade stretched

---

[1] Ralph Waldo Emerson, *Selected Essays*, ed. Lazar Ziff (New York: Penguin Books, 1982) 39, 58.

[2] *The Sower*, 1926.

toward the mountains and the sea, rich in corn and other grains, life-giving, hope-giving. To the skinny, war-scarred boys, this fecundity spoke of a life eased by nourishment aplenty. To Dr. House and other pious observers, the harvest shone as a symbol of spiritual growth that all could attain at the school, a manifestation of God's grace shed on this sacred precinct. When Dimitris Zannas, a trustee who has known the school for over eighty years, was asked, "What sustains the original spirit of the School?" He declared without pause, "Why, the spirit exists in the soil!"[3] For those who knew the place well, it seemed a preserve, suffused with mystique; each return to it was a homecoming. During the years of Bruce's tenure visitors were greeted by a group of students who sang a spirited greeting under the main gate, above which arched a sign with the visitors' names and words "*Kálos Orísate*" and "Welcome."

The farm also endowed the school with a vivid sense of place that endures up to our time, at least as a glowing memory, overlaying the present urban sprawl. Until the last two decades or so, when excessive, unplanned development blemished the countryside, travelers journeying toward the campus immediately noted the lay of the land, and commented, sometimes rapturously, on how the naked hills of the surrounding terrain contrasted with the school's flowering fields. One globetrotter wrote in 1934:

> Macedonia is a land of magnificent distances, of towering, rugged plains, unbroken by vegetation except an occasional grove of scrub olive trees or stunted oaks.... Imagine, then, the surprise of a traveler...to come upon a broad area of cultivated farm land which looks as if it might have been transplanted from a flourishing agricultural district in America. All around lies the barren, treeless countryside in which the vivid patch of color made by this cultivated area glows like a well-set jewel.[4]

---

[3] Dimitris Zannas, interview with author, Thessaloniki, 10 June 1998, written, AFS Archives. Zannas, an attorney from Thessaloniki, has a life-long tie to the School. His father, Konstantinos, a devoted friend and adviser to Charlie House, was the first Greek trustee. Dimitris followed in his father's footsteps as trustee.

[4] Margaret Prosser, "A Macedonian Oasis," *National Plant, Flower and Fruit Guild Magazine* (ca. 1934), AFS Archives.

Today the school's landscape is still so distinctive that the knowing passenger circling down into the Thessaloniki airport can pick out the place below; the campus stretches in a neat green pattern contrasting wildly with the encroaching urban welter. The city's nibbling at the farm's borders has vastly disturbed the sense of place and has become a focus of the school's ongoing land management plan. Should or can the farm remain as an enclave in the midst of dense urban settlement is a question that bedevils the board of trustees, the administration, the neighbors, and even the city officials of Thessaloniki.

The farm connects the school to Greece's historical backbone, the rural population. As such, the school's reputation is bound up in large measure with the great strides taken by farmers. The rural community and its survival "is viewed with increasing concern and nostalgic affection the faster it disappears," as former director George Draper put it.[5]

And the ties that bind have been many. For one, the Graduate Follow-Up Program instituted by Bruce under the direction of Nikos Mikos (a farm school graduate with a degree from the University of Arkansas) helped younger farmers, after graduation keep abreast of changing agricultural technology. Also, as discussed in an earlier chapter, the Community Development Program, initiated in 1958, reinforced the school's relationship with the rural people.[6] Basic to the rapport between the farm school and the countryside was the distribution of livestock to graduates and other farmers. The introduction of top quality animals into their herds enabled these dairymen to raise more productive animals at a greater profit. For another, the short courses further linked the institution to the countryside, drawing farmers to campus to learn food preservation, rural sanitation, beekeeping, modern dairying, uses of improved seed and fertilizer, and a range of other emerging advances. All told, helping the government to train

---

[5] George Draper, "Director's Last Report to the Board of Trustees," May 2000, AFS Archives.

[6] Clyde Sanger, *The Unitarian Service Committee Story* (Toronto: Stoddart Publishing Company, 1986) surveys the relationship between the American Farm School and community development. For an explanation of Community Development, see Bruce M. Lansdale, *Master Farmer: Teaching Small Farmers Management* (Boulder CO: Westview Press, 1986) 60–63, and Andonis Trimis, "European Seminar on Training for Community Development," 1961, paper delivered for the European Office of the United Nations, AFS Archives.

agriculturists, government officials, teachers, and priests in agricultural advances, and providing summer courses for extension agents has achieved for the farm school a unique position of influence and friendship among the Greek people as well as government officials.

Aside from the farm's abstraction as "idea," its purpose was totally concrete. It functioned as an indispensable educational tool and income producing operation. According to one authoritative handbook, an agricultural school's farm "should essentially be a practice area for the application of teaching done in the classroom and the laboratory since the institution farm is first and foremost a teaching aid; its crop, livestock, land, and equipment resources are expendable."[7] On the contrary, the farm school's resources, due to financial exigencies, were never expendable. During the early years of the founder's tenure, the farm's produce was intended mostly to feed students and staff. Later the school depended heavily on the farm's income to help defray at least the cost of running the farm, an extremely expensive undertaking.

The farm has served as a demonstration site not only for the school's own students, but also for visiting farmers from Greece and the world over, and for short course trainees. These visitors journey to the campus to observe, say, poultry, a certain breed of cattle, artificial insemination or embryo transfer techniques, the workings of a model milking barn, or an irrigation system. The farm, the only one of its kind in Greece, was designed to demonstrate the methods and economics of production, processing, grading, packaging, and marketing.

Not least of farm functions is the experimentation and development of new technologies for the purpose of breeding animals, applying new seed or inputs to the soil, or adapting new machinery. The students' education in the midst of such innovation, and their actual participation in it, offers them unparalleled experience. Yet often, all these tasks have proven too diverse for an institution so limited in its finances, the size of its faculty and staff, and number of hours students could devote (especially after 1977) in view of their rigorous classroom curriculum set by the ministry of education. Throughout the years, the school has struggled to strike an equilibrium on the farm among these many demands, with varying degrees of success. To its credit, though, the

---

[7] *The Management of Agricultural Schools and Colleges: A Manual for Practical Use* (Rome: Food and Agricultural Organization of the United Nations) 34.

farm's accomplishments have in many ways been extraordinary, an acknowledged contribution to Greece's improved agriculture.

To a large extent, the livestock—cows, chickens, pigs, and more recently, turkeys—has been responsible for fixing the farm school prominently on the agricultural map of Greece. After the war, the school's dairy herd was meaningful in helping to build the stock in Northern Greece, where the livestock had been decimated.

Throughout the postwar years, Bruce and his staff sought to import a steady stream of cattle into Greece. John Henry House founded the farm school's original dairy herd and under Charlie's direction in 1935 the first pasteurized milk in Greece was produced and bottled. Ann House wrote in 1947 that "the first and only certified TB-free herd of cattle belongs to the American Farm School."[8] The farm school herd had been enhanced by the addition of Elsalita and Elsadora, and the bull calf Elmerit, all Jerseys donated by the Borden Company and Duke Farms in 1946. In this, the first of the many airlifts that were to ferry cattle from the United States to Greece, the animals were flown across the Atlantic to Athens, and then trucked over the nearly impassable roads to Thessaloniki. Although the animals had been subject to all manner of physical stress from the flight, the hardy stock not only survived the perilous passage but thrived under the careful ministering of Dimitris Hadjis. To him, shuttling cows on arduous trips was not a novelty. In 1937 this determined cattleman had scoured Yugoslavia in search of animals to improve the school's herd. He marched his find some 30 miles to a railroad station and then traveled with them all the way to Thessaloniki in a crowded, reeking boxcar.

During the administration of Dwight D. Eisenhower, the president himself gave kudos to the farm school for its work when he acted for the Association of Aberdeen Angus Breeders of America in a White House ceremony, presenting to Greek Ambassador George V. Melas a bull calf and three heifers donated to the rural population of Northern Greece through the good offices of the American Farm School. The president called this gesture "a fine example of people to people approach to

---

[8] Ann Kellogg House, *Newsletter*, 1947. The farm school was assured it was receiving healthy animals: the United States government laid down stringent rules for cattle export. The cattle had to be vaccinated, tattooed, and wear identifying tags. The exact amount of space, food, ventilation, and types of stalls was specified.

mutual aid."[9] Especially endearing was the students' response of thank you to the president. Dressed in their straw hats and coveralls, the boys posed for a picture in a formation outlining the word "thanks." The photograph was sent to the president. That the president personally engaged in this activity reflected, possibly, an act of patronage to some important domestic constituency; nonetheless, his participation was an expression of the United States' foreign policy: assistance to populations in areas regarded as strategically vital. These Aberdeens were the first beef cattle imported into Greece.

For Bruce these beef cattle represented the first step in a cherished project: his goal was to discover if the introduction of American beef cattle into the rural Greek economy could substantially increase the farmers' annual incomes, supplement the diet of the rural population, and staunch the outflow of foreign exchange. Also, he recognized that the dairy industry was so deficient that imports of condensed milk and other dairy products would in a few years cost the country $20–30 million dollars annually.[10] His aspiration to improve the country's herds brought him into close harmony with the Greek ministry of agriculture, which in the 1950s had begun to rebuild the nation's livestock through a number of approaches. As part of that strategy the ministry sponsored a joint project with the farm school beginning in 1967: accordingly, 359 three-to five-week old Holstein calves winged their way from the United States to Greece in the company of their only human steward, Bruce Lansdale.[11]

In keeping with the farm school's experiment with ocean-crossing animals, in 1964 four young bulls of the Santa Gertrudis breed of beef cattle, the gift of Dr. Eslie Asbury of Cincinnati, were unloaded at the pier in Thessaloniki to the cheers of a reception committee of 200 students and staff. The animals were the first Santa Gertrudis to set hoof in Greece.

These crossings were all experimental. Calves less than one year old are delicate: sudden variations in temperature or changes in diet cause sickness and sometimes death. On one shipment from Boston in 1969, even with special temperature settings and special diet synchronizations, one calf arrived dead and ten died after arrival in Greece. Still the outcome was considered good, since a 10–15 percent loss was

---

[9] As quoted in *Newsletter*, June 1958.

[10] *The Sower*, 1971–1972, no. 76.

[11] *The Sower*, 1964, no. 14.

considered acceptable. The calves were nourished at the school and after three months sold to farm school graduates.

The cattle program stimulated excitement in a range of communities in the United States. A few people went beyond the call of duty to see that these animals landed in Greece in fine fettle. Trustee Ruth Wells, a dairy farmer herself, flew from Boston to Thessaloniki in a cargo plane personally assuming responsibility for 300 young calves on their way to the farm school.

As it turned out, the farm school's experiment with beef cattle was brief; raising these animals proved prohibitively expensive. But as opposed to the short-lived experiment with beef cattle, milk cows became a primary focal point and have remained so. The school raised the animals scientifically in a manner that local farmers could observe and ultimately apply to their own herds. According to scientific practice at the time, the pasture was rendered obsolete, a boon to Greek farmers who did not possess acreage for pastureland. The dairy cows were tended in a large facility; they entered the world in a birthing room; slept in private stalls; ate in a feeding hall; and gave their milk to a machine in the milk parlor. In 1973 the school constructed the first self-feeding bunker silo in Greece with a capacity of 540 tons, another boon for farmers without pastureland. The silo was a magnet for hundreds of visiting farmers who came to the school annually for demonstrations. Because of their involvement in caring for these cows, the students after graduation became some of the best dairymen in Greece.

Arrival of these cattle in Greece—a highly visible event, the kind of pageantry the institution has always produced in spades—put the farm school in the spotlight and highlighted its philanthropic essence. These celebratory occasions were attended by Greek and American officials who made pointed statements about warm Greek-American relations. Admittedly, the impulse behind the school's importation was purely altruistic, but it dovetailed politically with the type of warm relationship the United States had been trying to cultivate since 1947 among Greek government officials in an effort to enlist their support during the Cold War.

When in 1973 a shipment of Canadian dairy cows landed at the farm school, on campus to welcome them, beside the students and staff, were a cavalcade of officials: the Director of Agriculture Pantelis Pantopoulos and his wife; the inspector of the Agricultural Bank of

Northern Greece; and Father Eugene Pappas, who blessed the animals, the two new barns, and equipment. Mrs. Pantopoulos cut the red ribbon at the gate of the barn and speeches all around dignified the ceremony. Pantopoulos emphasized that in the eyes of the Greek people the American Farm School had grown to be accepted as a Greek institution "working for the betterment of the rural areas in cooperation with Greek agencies."[12] Bruce, whenever appropriate, emphasized the perception that the farm school was accepted throughout the country as a Greek institution: he played on that theme to avoid the taint of "foreign school."[13] But beyond the shrewd diplomacy that underpinned that attitude was Bruce's sincere yearning to identify closely with Greek culture, and by Hellenizing the institution to render it as responsive as possible to the needs of the people.

A comment made by George McGhee, a former assistant secretary of state and ambassador to Turkey (1951–1953) who was a member of the Farm School board of trustees in 1959, typifies the attitude of many American donors and trustees during the Cold War: "One of the principal problems of Greece is the difficulty faced by the farmer. These young men going back to various communities to make the most of their opportunities, present to them the democratic American viewpoint as an antidote to communist pressures."[14] The farm school was regarded as a valuable institution by both governments—by the American to amplify its agenda of friendship with the Greeks, and in turn by the Greek (strongly pro-American in the few decades after the civil war, especially the eleven years of conservative governments from 1952 to 1963) as an instrument through which it could tap into American generosity.

In 1974, a well-known veterinarian, Alexandros Michaelides, assumed the management of the farm school's dairy unit; in 1979 he replaced Panayiotis Rotsios as manager for the whole farm operation. (Michaelides, a graduate in veterinary science from the Aristotle University of Thessaloniki, served as farm manager until his retirement in 1992. Rotsios, who held a graduate degree from the University of West Virginia in animal production, became dean of studies and served

---

[12] *Athens Daily Post*, 27 March 1974, AFS Archives.

[13] For some of the difficulties experienced by other foreign schools in Greece, see William McGrew, "President's Report to the Board of Trustees of Anatolia College, 1974–1999," January 1999, AFS Archives.

[14] As quoted in *Newsletter*, July 1959.

in that position until his retirement in 1990.) Michaelides, who regarded milk production as a major function of the farm, reduced the dairy herd, leaving only Holsteins, a most efficient breed capable of producing large quantities of milk. Some Holstein bull calves were sold to farm school graduates, who used them as sires. A larger number of calves was sent to the ministry of agriculture's Institute of Animal Reproduction, where their sperm was prized for artificial insemination and genetic improvement of local herds.[15] From 1975 to 1982, the school had sold almost 500 head throughout Greece. In light of these sales, a number of school fields that had been planted for forage were now freed to be sown with other kinds of crops; the new plantings featured vegetables and flowers in the interest of strengthening the school's specialization in horticulture and floriculture, sectors becoming strong in Greece and hence offered to students in their practical program.

However, as useful as the importation of live animals had proved, it was not the only means of augmenting and improving the herd. As new technologies became available, Michaelides was quick to exploit them. In 1965, the school had started using semen imported from high quality bulls in the United States for artificial insemination. Then, in 1985, the farm scored a first in Greece when embryo transplants were conducted on farm school animals. Michaelides had weeks of intensive training in the United States and cooperated closely with faculty at the Aristotle University of Thessaloniki to perfect his knowledge of transfers; he gave numerous seminars for farmers, who enthused, as did students, over the new venture.[16] Embryo transplants had the potential to serve as an effective method for Greece to improve the quality of its herd much more rapidly and at a lower cost than would otherwise have been possible. According to the procedure at the farm, the school's top dairy cows were inseminated with frozen semen imported from the States, following which the embryos were transplanted to other cows, freeing the top quality cows to be bred more often.[17]

By this time the school could rightfully boast of owning the best herd in the country, one that compared favorably with the best in the United States and Europe. If all herds in Greece had been of the same

---

[15] Evangelos Vergos, interview with author, Thessaloniki, 10 October 1999, written, AFS Archives.

[16] Ibid.

[17] *The Sower*, 1989, no. 126.

genetic quality as the farm school's and under the same quality management, the country could have reduced its number of cows by one-third and have continued to produce as much milk, claims Konstantinos Evangelou.[18] For example, when in 1992 the farm school's dairy herd was reduced from 170 milking cows to 125, thanks to improvements in breeding and feed, milk production remained at the same level despite the reduction.

Projects for the development of the dairy industry were endless, but one should not assume that this enterprise overshadowed other aspects of animal husbandry and poultry at the farm school. The development of poultry, an undertaking that the farm school gauged to have enormous potential, was targeted right after the war. In 1957, the school expressed its goal of introducing into the countryside day-old chicks that would grow to laying hens producing high quality eggs. "We are selling about 7,000 a month for hatching purposes. Our reputation this year for hatching eggs has gone very high and we have a long waiting list for them."[19]

Furthermore, in 1958, 300 chicks, first of a new breed of broilers, were shipped to farm school graduates throughout Northern Greece as a first step in a project underwritten by the Rockefeller Foundation. The broiler project had two objectives: to increase demand throughout the country for better quality meat birds and to lower the costs and prices through improved methods. The project seemed a good bet since poultry, despite poor quality and high prices, was popular among the public. These chicks, a crossbreed between White Rock hens and Cornish males, left the farm school at five-days old, having been inoculated against Newcastle and other respiratory diseases, and went to farm school graduates for maturing and slaughter. To provide the most advantageous growing conditions for the birds, the poultry farmers constructed model broiler houses from plans furnished by the school. Graduates were spearheading a promising venture. Primed by their farm school training,

---

[18] Konstantinos Evangelou, interview with author, Thessaloniki, 18 October 2000, written, AFS Archives. Evangelou, a farm school graduate, earned a bachelor's of science degree at California Polytechnic University at San Luis Obisbo. He began his career at the farm school in 1961 as an instructor and was promoted to head of the lyceum farm machinery specialization, assistant director for education, and farm manager until his retirement in 1996.

[19] *Newsletter*, 1957.

they grasped the potential of this nascent industry and went on to build successful large-scale businesses that marketed to Greek consumers. By 1962, forty graduates had set themselves up in the poultry business. Alumnus Emanuel Terjis, *The Sower* recounted, "was the first man in his village to go in the poultry business. Today he owns nine modern incubators, each with a capacity of 7,000 eggs. His yearly production exceeds 500,000 chicks. He recently completed a $20,000 hatchery."[20]. Another flourishing entrepreneur, Christos Saramourtsis, class of 1965, started his poultry business in a building he modeled after a farm school prototype and is now an international businessman, proprietor of one of the most technically advanced broiler operations in Greece. He travels the continent, especially to Holland, to keep abreast of the latest scientific trends on poultry raising, production, technology, marketing, and packaging.[21]

To create a modern poultry industry in Greece, housing for laying hens had to be introduced and adapted to suit Greek conditions. In 1972 the Big Dutchman Poultry House was designed and built on the farm, outfitted with American equipment but adjusted to local conditions and most importantly filled with efficient, high producing hens; its organization and layout were such that the whole operation could be handled efficiently by a single farmer. Chief poultry man Athanasios Pantazidis selected birds best suited to local conditions. The extension of the poultry business to include turkeys may not seem like a big leap, yet customarily farmers raised one or two turkeys in empty village lots. In 1969, the farm school flew 1,200 one-day old poults from the United States and the next year 2,200 more, contributed by friends of the school. The prime purpose for the project was to train students in modern turkey production, introducing strictly controlled conditions for the birds. From this initiation into the turkey business, the farm school earned a satisfying income from the sale of turkeys while helping to introduce a delicacy into the Greek Christmas market. Sold in Athens in the 1970s at A. B. Vassilopoulos (the first supermarket chain in Greece), the flavorful birds now adorn many a table at holiday season in both Thessaloniki and Athens. Raising them has continued as a major enterprise for student

---

[20] As quoted in *The Sower*, 1971, no. 75.

[21] Christos Saramourtsis, interview with author, Neohorouda, 15 October 1998, written, AFS Archives.

training, as well as an important revenue source for the school. In 1978–1979, the sale of 8,500 turkeys netted $60,000.

Starting in the early 1960s, pig production was a central enterprise for two decades. In 1977 sows were artificially inseminated with imported semen. The most exciting aspect of pig production was an embryo transfer program run in cooperation with the Aristotle University of Thessaloniki. With only eighteen sows, the farm school was able to improve its breed through this technology. However by 1980, pig production as a commercial enterprise proved economically infeasible, so the facilities were turned over to students to operate as a student project, which continues to this day as an effective learning experience.

Mechanization and innovation continued apace. From 1955 to 1965 a self-propelled harvester, fertilizer spreader, grader, cotton, and corn harvester, and a large-herd automated milking parlor modernized farm labor on campus and demonstrated to visitors the potential of labor-saving equipment. In 1962 one of the first plastic greenhouses in Greece was built on the farm. Farmers came from all over Northern Greece to inspect it, and in 1977 a large greenhouse covering one-eighth of an acre and outfitted with drip irrigation, brought the latest innovations in greenhouse cultivation to Northern Greece.

The practical program has furthered the school's goals in more ways than one. The founder had hammered home to staff and students that there was dignity in the performance of manual labor. He repeated his mantra throughout his life: "To work is to pray," a value that the farm school has perpetuated along with the pedagogical principle that practical work instills self-confidence, a sense of responsibility, and appreciation for working cooperatively with others.

Before 1978, it was school policy for the students to work the farm under supervision and in the process learn to manage a farming enterprise. Their labor had been viewed as a means of earning their board and keep, and the income from the farm had contributed toward or reduced the net cost of running the school. In 1975, for example, approximately one-half of the school's $1.3 million annual budget came from the sale of farm products.[22] A transformation occurred, however,

---

[22] Steven V. Roberts, "U.S. Farm School Is Still Flourishing in Greece," *New York Times,* 6 April 1975, 2. Athanasios Pantazides, a graduate of the University of Arkansas, who came to the farm school in 1936 as an instructor, served as minor livestock supervisor until his retirement in 1978.

when the school switched the emphasis for fear that student labor could lead to, or be perceived as, a form of exploitation, and that the full force of the educational mission was being blunted as the students toiled at the repetitive tasks that farming demands long after they had learned the intended lesson. Also the establishment in 1977–1978 of the TEL program mandated by the ministry of education and weighted toward theoretical lessons, siphoned off much of the students' time formerly allocated to practical work.

Under the new 1978 policy, the farm stood alone as a production unit. Students were assigned to various production departments for the sole purpose of learning the operation, not to perform the day-to-day labor. As orientation, the first-year lyceum students spent three hours a week observing the routine of each department, decidedly a broad-brush treatment compared to the pre-lyceum approach. Then they broke into small groups to rotate through every area of farm operation. In the course of this routine students had limited hours for practical work; as a result they lacked a truly complete perspective of the varied phases entailed in managing a farming business. The danger was that department heads, men who were devoted teachers as well as production experts, might lose enthusiasm for their teaching duties: the whole farm school educational system stood in danger of losing internal logic.[23]

But as a counterweight in this dilemma, students in TES were concentrating intensively on practical farm work and were actually helping the staff in gardening, animal husbandry, farm machinery operation, and maintenance under the careful supervision of department heads. The TES boys, who put in many hours of practical work, got muddied up to their elbows; they had come to learn the fundamentals of farming, and in a direct way, kept the staff members engaged in their teaching mission.

The effect of these changes was transformative. In the school year 1977–1978, the contribution students made toward their own support through their labor on the farm was substantially reduced. When laborers were hired to replace them, the cost of running the farm was greatly increased. Increased costs spurred the school to run its farm more

---

[23] David Willis, interview with author, Thessaloniki, 16 May 1999. David Willis came to the farm school in 1978 and served as associate director of administration and later as operations manager, and director of USAID and capital projects and buildings and grounds.

efficiently along modern lines—in other words, like a business.[24] As a consequence of this restructuring, the school's farm became a kind of production powerhouse selling its eggs and turkeys in Athens. As a spin-off, this new marketing project acted like a blitz public relations campaign spreading the farm school's name among consumers in the capital. And what a producer the farm had become. By 1984 it was producing 4 million eggs, 350,000 chicks, 10,000 turkeys, 150 calves and 350,000 gallons of pasteurized milk a year.

Still another benefit from using the farm as a major production unit was its expanded utility as a demonstration resource, a source of fascination for observers who could scrutinize a high-powered farm in action.

On the other hand, the farm's changed status set in motion a conundrum that continued into the 1990s. When the farm laborers were hired, the students were relegated to the sidelines when it came to the farm's operations. Although TEL students were given plots to carry out their practical education, this arrangement was not as effective as being in close daily contact with the farm.

In 1992, early in George Draper's tenure, the board of trustees and the new director began two initiatives concerning the farm: the reintegration of student education into the farm and the upgrading of the production facilities so that they could, as they had at various times in the past, stand on the cutting edge. It was George's opinion that upgrading the production units was essential, even though farming technology was advancing at breakneck speed, making upgrading an expensive, almost prohibitive, proposition. "It was worth the effort, not only because the production units helped the students, but because they represented an infusion of excitement, ideas, and hope to the adult farmers, too, and because the School's reputation rested in large part on its ability to manage what after all was the *only* large-scale diversified farm in the country."[25]

The reintegration of the farm into the practical program brought about a reexamination of the relationship between the classroom and the needs of the farm. Stating the problems and solutions involved in devising a practical program to work smoothly in tandem with farm functions, a report by a study team in 1993 revealed the perennial

---

[24] Alexander Allport to the author, 16 November 1999, written, AFS Archives.

[25] George Draper to the author, Sorrento ME, 13 June 2001, AFS Archives.

adjustment demanded of the farm school administration and staff. In answer to criticism that the classroom curriculum was not correlating successfully with the practical program, the academic administration noted an improvement whereby students were now at least working "increasingly in the real farm environment rather than being restricted to cultivating small plots set aside for their practical training."[26] This advance was instituted first in the area of farm machinery, one of the specializations developed for the TEL and TES programs.

Inevitably, arranging for students to work on the farm presented almost insurmountable difficulties of coordination and scheduling. The rub came in trying to schedule classroom lessons in parallel with the immediate needs of the farm. If the students were studying about weeds, "the weeds won't wait to surface for the day designated in the curriculum for weed harrowing. For an exercise in weed harrowing to take place, someone must find out which field needs harrowing and arrange for a tractor and harrow for that hour, although the machinery may be vital for use somewhere else by the farm staff." Rendering the students' experiential training even harder to coordinate were their summer vacations that coincided with the peak growth of many crops.[27]

Earlier, in 1981, the administration had reported another more subtle snag: a number of lyceum students were not attracted at all to hands-on-training, an educational component so historically sacrosanct to the school. The institution was faced with the reality that more than a few youngsters harbored ambitions for a university degree or a position in government.

Not to lose perspective, it is worthwhile to hear the comments a Greek official made to the Educational Policy Committee in 1981: "The School's practical training remains unique. The government cannot at this time provide this kind of training, but it views the Farm School as a prototype for future efforts. The School's original and innovative thinking and ability to deal in a practical way with farmers' problems through educational approaches make a most significant contribution to education in Greece."[28]

---

[26] "Report of the Academic Study Team," 1993, AFS Archives.

[27] Ibid.

[28] Unattributed statement as quoted in the "Educational Policy Committee Report, 1981," AFS Archives.

The institution's value becomes all the more conspicuous when we learn that in 1981 the only other technical vocational school in Greece specializing in agriculture and livestock production had closed. That left the American Farm School as the sole secondary institution in Greece with installations and fields capable of promoting practical education in this specialty. While the farm school could take pride in its unique capabilities, its very singularity boded poorly for an agricultural sector in a country that had just gained full admission to the European Union and was joining the most forceful and progressive sectors of Western Europe.

The farm school also bore the distinction of being the first agricultural school to offer computers to students studying agriculture. In 1981 Anastasios Pougouras[29] set up the first computer laboratory chiefly for the students' extracurricular pleasure and for a presence in formal classroom settings. Later, Athanasios Souldouris,[30] head of the dairy department, sharpened his operation by computerizing the dairy's records.

In 1992 when Konstantinos Evangelou succeeded Alexandros Michaelides as farm manager, the conundrum of integrating the farm with the educational goals of the institution was a major challenge. He took the first steps toward a solution for students in the farm mechanization specialization and for those who worked in the greenhouses and on the field crops as part of their practical program. Also, Eustathios Yiannakakis, head of poultry, was dividing his time equally between teaching and the operation of the poultry department, marrying practical skills to theoretical knowledge. (Yiannakakis, a graduate of the farm school and KATEE, came to work at his alma mater as a teacher, and in 1992 became head of the poultry department.) Students specializing in animal husbandry, including practical programs concerned with dairy cows, and laying hens, broilers and turkeys, reaped the benefits of working with Souldouris and Yiannakakis, instructors who could fuse the theoretical and practical lines of their education.

[29] Anastasios Pougoras's first position at the farm school was as an engineering consultant. In 1981 he became full-time librarian and piloted the setting up of computer learning for students. He has participated in information services for the school and worked on European Union information projects.

[30] Souldouris, a farm school graduate, with a bachelor degree in business administration from the American College of Greece in Athens, came to work for the school in 1973. He became head of the dairy department in 1992.

The farm operation had become infinitely more complex when
Dimitris Zannas donated his 145-acre farm in Valtohori to the farm
school in December 1981 as a resource to be passed on to the school
upon his death. His dream was to have the students use his farm on the
Axios River as a laboratory where they could experiment with new crops
and learn stewardship of the land at an early age. At that point Dimitris
Zannas decided to hand over only the management of his farm to the
farm school, letting the school keep any gains from the produce. In 1982,
the school bought 70 acres almost contiguous to the Zannas spread as a
fallback against some future municipal regulation that might force the
school to move the animals from what was rapidly becoming a city
campus. The school hired a farmer to work both pieces of land, but there
was no master plan, no prescription for this rich soil, hard by the river,
until 1996 when the Dimitris Perrotis College of Agricultural Studies
was founded. When the college was born and its mission as an
agriculture institution defined, the Zannas farm immediately became a
centerpiece. At that point Dimitris Zannas deeded over the land entirely,
one of the most significant gifts the farm school ever received. Yiannis
Vezyroglou, a member of the board of trustees, claimed at the time that
you could not have the sort of college envisioned by the school without a
farm like this.[31] The marriage of the college and the Zannas farm was, to
be sure, a defining event in the history of the American Farm School.

As for poultry and dairy operations, they continue to prosper due to
the managers' marketing and scientific expertise. The applied research
project conducted by the poultry department working with the Aristotle
University of Thessaloniki resulted in a successful new product, the
Omega 3 Egg. Housing and feed for laying hens and turkeys has been
constantly upgraded to meet the latest standards. But, production and
marketing aside, in fall 1999 the number of hours students spent in
production units was reduced. The agricultural program with its practical
application, in which the Farm School had so deliberately invested since
Dr. John Henry House arrived in Thessaloniki after the turn of the last
century, had begun to shrivel, at least for the EL students in the 1998
curriculum. Although agriculture for those particular students was still at
the forefront, it was geared more to academic preparation. However, the

---

[31] Yiannis Vezyroglou, interview with author, Thessaloniki, 18 October 1999,
written, AFS Archives. Vezyroglou, himself a farmer, is a resident member of the board
of trustees.

three-year TEE program is more closely allied to practical application and comes closer to reflecting the school's original mission.

# CHAPTER 7

# WOMEN IN GREECE:
# AN ODYSSEY ACROSS THE YEARS

"Enough ink has been spilled in quarreling over feminism and perhaps we should say no more about it," Simone de Beauvoir wrote with mordant sarcasm in the introduction to her book the *Second Sex*, which appeared in France in 1949.[1] Had she foreseen the emergence of women in modern Greece during the last fifty years of the twentieth century, she would no doubt have concluded that 1949 was not the time to become mute on that subject. The struggle had not even really begun, not only in Greece, but also in other countries of Europe and in the United States as well.

The evolution of women's position in rural Greece represents an uphill odyssey across the last half of the twentieth century. From objects of shame and subservience, they advanced to emancipated human beings with full legal rights—truly remarkable progress considering the many centuries of oppressive mores that had been passed down unexamined.

Women were not included in the student body of the Farm School from its beginning, but at the end of World War II a girls school, founded by the British Friends Service Council in 1946 and located on school land, drew women and their concerns into its sphere. Founder Dr. John Henry House's watchword, "When you educate a man you educate an individual, when you educate a woman you educate a whole family," pronounced early in his tenure, echoed through the decades and Mother House vigorously advocated the inclusion of girls as soon as feasible, but

---

[1] Simone de Beauvoir, *The Second Sex*, trans. and ed. H. M. Parshley (New York: Alfred A. Knopf, 1957) xiii.

neither lived long enough to see that dream materialize. Some twenty years after its founding, the Quaker Domestic Training School was handed over to the farm school but the girls remained on their own campus and followed their own separate course of study. Not until 1978 did they move to the main part of the campus and become fully integrated into the school's student life.

From the time the school was founded until the beginning of the 1970s, the position of women in Greece remained essentially unimproved and it has been noted by some observers that traces of women's exploitation and subjugation still exist in certain sectors of Greek society.[2] The disdain for the female child in the family began with the girl's birth, owing to the superstition that the degree of the father's masculinity was measured by his ability to produce boys. Nicholas Gage, in his searing book *Eleni*, the story of his mother's torture and execution by the communists during the civil war in poverty-stricken Epiros, sketches a girl child's beginning during the pre-World War II period in that area of Northern Greece:

> From the moment a female was born in the village, her life was prescribed and ordered by centuries of custom, so deeply etched that no one stopped to question, for a woman was as innocent of self-determination as a member of a beehive. If she survived the first forty days, a girl was taken to the church to be blessed. But she was carried by the priest no farther than the narthex, while an infant boy was shown to God before the Holy Grail leading to the altar.[3]

Eleni's childhood, in the remote village of Lia, where life was especially severe, was deprived of stimulation and joy. Gage charts her fate: "A girl had to put on a kerchief at age eleven so that no wayward curl could invite the lust of a stranger, nor could she utter even a 'Good Morning' to a male outside the family. Only twice a year was an

---

[2] "...women's social inferiority and political marginalization are still paramount: Greek society exhibits one of the lowest levels in the EU of women in positions of power and in socio-economic decision making...while today there are only three women in government...and 19 in a parliament of 300." Ministry of Press and Mass Media, Secretariat General Information, *About Greece* (Athens: n.p., 1999) 278.

[3] Nicolas Gage, *Eleni* (New York: Random House, 1983) 35.

unmarried girl seen by the public, sitting among the women at the Christmas and Easter liturgies. The rest of the time, walls of stone and the vigilance of father and brother surrounded her as she learned a woman's duties."[4]

Severe taboos on relationships were in place all over Greece, although there were regional differences, places where life between the sexes evolved more lightly than in Gage's mother's stony Epiros. Elsewhere, boys and girls did manage ways at least to appraise each other. The *vólta* or Saturday-and Sunday-evening promenade, also referred to in Greek as the *nymphopázaro* (bride's bazaar), still today a popular pastime in some villages, served as a venue for a bit of viewing. In a repetitive choreography, girls in groups of five or six strolled around the main square of their village while the boys, also in groups, passed by, joking boisterously among themselves, all the time surreptitiously eyeing the girls.

Macedonia—Eastern, Central, and Western—the regions from which the largest number of farm school girls hail (the next largest contingent comes from Thrace), was settled in 1922 by over one-half million Greek refugees from Asia Minor. Observers have noted that the women of refugee origin in Macedonia took more initiative, participating more actively in the family decisions, and were less hesitant to express their opinion in the presence of their husbands, than women born of indigenous Greeks.[5] Harold Allen, an official with the Near Eastern Relief who arrived in the area in 1928, noted that "many of these people were cultured and educated folk who had been quite well-to-do before tragic events dispossessed them of their ancestral homes. They included doctors and lawyers as well as merchants..." although among them were masses of uneducated people.[6] These educated people had fled to Greece from large cities, the Hellenic strongholds of Constantinople and Smyrna, where life was sophisticated and attuned to a medley of cultures, unlike the Turkish province of Macedonia, a poor, rural area,

---

[4] Ibid.

[5] Chrysanthi Bien, interview with author, Quechee VT, 29 November 1999, written, AFS Archives. Bien, born in Greece, worked as a teacher and babycare specialist at the Girls School from 1948 to 1955. Since then she has lived in Hanover NH, where she has taught Modern Greek language at Dartmouth College and has co-written a series of books on teaching the Greek language.

[6] Harold Allen, *Come Over into Macedonia* (New Brunswick: Rutgers University Press, 1943) 8

which until 1912 had been a backwater in the decaying Ottoman Empire. Irwin Sanders writes that the women from "refugee families were more sensual,"[7] a conspicuous characteristic that would put them at odds with the women of Northern Greece, who had been liberated from the stifling 400-year-long Turkish occupation only a decade before the newcomers arrived. Farm School demographics indicate that some key faculty, staff, and influential Greek friends of the school and a number of girls came from the refugee population, a fact that could explain a more progressive mentality reigning on campus.

However, among the indigenous rural population, the main business of the family was arranging marriages and honorably setting up the children on the course of their lives; thus, the girl's reputation was key, and her smallest lapse was unforgivable since it would cast a pall over the entire family. For the girl to make a respectable marriage she required good health, a reputation for modest demeanor—meaning eyes downcast, body swathed in a thick homespun dress to hide her sexuality—and a dowry. The dowry, a dominant theme in Greek life, usually consisted of, among prosperous folks, a house, land, and money, or, among the poorest, animals and linens at a minimum. Although the father regarded the dowry as a burden, he also considered it a source of pride and a display of his ambition, since a dowry of substance could attract a reputable husband for his daughter, perhaps even a city boy, thus endowing the marriage contract with a dimension of upward mobility for the whole family.

The dowry could function to the advantage of the girls, too, empowering them in their roles as daughters and wives. Before 1983, when the dowry finally was abolished by the Panhellenic Socialist Movement (PASOK) during the first Andreas Papandreou government (1981–1985), thirty civil code articles had provided a legal base for a daughter to sue her father if he denied her a dowry. Hypothetically, this was a heady right, but in reality probably not a viable action for an uneducated village girl who was since birth confined to the private sphere, often even to the exclusion of school.[8] And the dowry could lend the wife a great deal of clout in dealings with her husband, possibly her

---

[7] Irwin T. Sanders, *Rainbow in the Rock: The People of Rural Greece* (Cambridge: Harvard University Press, 1962 **[1968?]**) 34.

[8] Foreign Area Studies, *Area Handbook for Greece* (Washington, DC: US Government Printing Office, 1977) 77.

only real instrument of power especially if she held land, since by law her husband could not dispose of her property without her consent. Because her dowry in part went to her daughter's dowry, it empowered her to speak out in choosing her daughter's groom. It follows that when the dowry was abolished in the early 1980s women lost some influence but by then life in Greece had changed so significantly that even in the villages a whole set of experiences broadened almost everyone's perspective. But earlier, wives had needed all the leverage they could command.

Traditionally, the wife regarded herself as inferior to her husband and she demonstrated how she appreciated his elevated status by rising when he entered a room. Often she would be seen trudging behind him along the road, her back loaded with firewood for the hearth, with hands free to spin or to knit, as the husband sat astride the family donkey, puffing on a cigarette. But beyond these outward manifestations of subservience, her relationship to her husband and to her father wrought deeper damage; it robbed her of self-confidence and the sense of self-worth. Two scholars who did intensive fieldwork in rural Greece reported: "Early deference to the more kingly male begins at age two or three, and one sees seriousness, reliability, tolerance and sometimes depression in very young girls."[9] Another factor that kept women subordinate was the influence of the Greek Orthodox Church, a socially conservative institution that transmitted values of obedience and authority.[10]

The lack of education clearly operated to keep women in submission. The 1961 census revealed that the compulsory requirement for education through elementary school had been, in many cases, sidestepped: 1,222,000 people over the age of ten were found to be illiterate, by far the largest number of them, nearly a million, women.[11] These statistics highlight the merit of the Quaker Domestic Training School and the enlightenment it brought to those fortunate enough to attend.

---

[9] Richard and Eva Blum, *Health and Healing in Rural Greece* (Stanford: Stanford University Press, 1965) 49.

[10] Adamantia Pollis, "Gender and Social Change: The Role of Women," in. Kariotis, *Socialist Experiment*, 283.

[11] Campbell, *Modern Greece*, 387.

Motherhood was one qualification for promotion to a higher plane, for engrained in the culture was the belief that wives were authenticated as women only when they bore children. The promotion not withstanding, the ideal place for the mother was confinement in the private sphere, in her home caring for little ones, or working in the fields: straying, presumably, could present her with the challenge of keeping her honor intact. Yet responsibility of childcare had its perils. At the time of the founding of the Quaker Domestic Training School (known as the Girls School) in 1946 by the British Friends Relief Service, childcare was still based largely on peasant wisdom. Swaddling, for instance, was one of the worst abuses practiced by rural women.[12] Mothers were convinced these strips of cloth (called *faskiés*) used since Biblical times, which restricted the child's movement and caused sores on the skin, helped to make the baby's body grow straight and tall. And when mothers had to go out to work in the soil, they could more easily secure the infant to a prop some place in the open field if it were swaddled. In Northern Greece, where farming wives toiled unending, exhausting hours in the fields, the child's needs were sometimes neglected as compared with, say, the situation in the Peloponnesos, where women were apt to work fewer hours and then only seasonally at harvesting olives and grain.[13]

A momentary, yet revolutionary, change in women's behavior and outlook occurred during World War II and the Greek civil war. The civil war's dislocation of more than 700,000 people exposed many to urban life, the girls and women blinking in amazement as their world suddenly expanded. In reference to farm school demographics, the greatest resettlement of people occurred in Northern Greece, in the vicinity of border areas where the communist guerillas were particularly active. No doubt many of the students at the Girls School came from families that had been displaced. At the end of the strife they returned reluctantly to their villages, the women tucking away in their minds the prospect of an easier and more exciting life.

An aspect of the occupation that affected females directly was the eagerness of the National Liberation Front (EAM) to welcome them into their ranks. Although life was wretched for everyone during the

---

[12] For an overview of childbirth superstitions, see Ann Peters, "Babies and Bogeymen," *The Athenian Magazine* (June 1991): 14; Blum, *Health*, 77–79.

[13] Sanders, *Rainbow*, 144.

occupation and civil war, the resistance actually provided an invigorating interlude for hundreds of women, allowing them to enter the public sphere, often working and soldiering side by side with the men. Here is how Georgia Pliyannopoulou-Kalini recorded her military service:

> Koula, Thiella and I were chosen for officer training school in Redina. ·Eleven other women age eighteen to twenty were training there to become officers. We trained with the men and in exactly the same manner, instructed by regular army officers... As soon as we completed our training we were assigned the rank of captain as commanding officers of women's companies in various guerrilla divisions. The many battles after we had assumed our duties completed our training and fully tested our courage.[14]

EAM's inclusive attitude toward women was derived from the organization's ideology, which held that the lot of women needed to be improved. Some titles taken from pamphlets that EAM circulated offer an inkling of this professed philosophy: "The Girl and Her Demands" and "In Today's Struggle for Liberty the Mass Participation of Modern Girls Is Especially Impressive."[15] Singling out women, citing their achievements, and bringing them to public notice in the Greece of that era constituted a radical act. EAM acknowledged in these pamphlets that the *andártissa* (female resistance fighter during the occupation) was struggling to "free herself from the foreign yoke and from the bias and superstition of our country."[16] In practice, EAM encouraged women to speak out at public meetings and to play a role in organizations. One of the crack battalions of the Greek People's Liberation Army (ELAS) consisted exclusively of women. Once unshackled, women acted in leadership positions and displayed sharp organizational skills. After the

---

[14] Eleni Fourtouni, *Greek Women in the Resistance: Journals, Oral Histories,* selected, translated, and introduced by Eleni Fourtouni (New Haven: Thelphini Press, 1986) 49. In this unusual source—a book of witness—the women narrate their own experiences during this time, while Fourtouni provides an historical context in which to set their narratives.

[15] Mark Mazower, *Inside Hitler's Greece: The Experience of Occupation, 1941–44* (New Haven: Yale University Press, 1997) 279. Also, see 279–81 for women's activities during the resistance.

[16] Ibid.

civil war, when Greek leftists were exiled to barren islands, places of extreme physical hardship such as Makronisos, the women among them once again assumed responsibilities that required resourcefulness and intelligence.[17]

In Athens during the occupation, girls as well as boys joined the National Panhellenic Youth Organization (EPON), the EAM youth group where young women took part in demonstrations, wrote slogans on walls, and chanted messages at night from rooftops; girls comprised 40 percent of the organization. But this particular form of activism on the part of young women halted at the end of the occupation when the women and men trekked home to their villages to confront the formidable task of rebuilding their devastated country.

As we saw in book 1, the boys at the farm school were victims of kidnapping during the civil war. Throughout Northern Greece, girls were also prey to kidnappers. The *andártes* kidnapped hundreds of girls from their villages and spirited them over the borders to the surrounding communist countries.[18] Norman Gilbertson witnessed the cruelty of the *paidomázoma* (literally the gathering of children, but in this context the kidnapping of children):

> In the morning the girls would go down to the spring with their pitchers, an opportunity to meet with their friends before taking the water for their houses. One morning a group of *andártes* hiding beyond the spring came in and took some of the girls captive and were away before the village knew what happened…people never expected such a daylight raid as attacks were normally under cover of darkness.
>
> The day that I was there one of the mothers had just received a letter through the International Red Cross from her daughter who had been taken to Yugoslavia. I had to read the letter, which was indeed painful and the agony of the girl torn from her family left me speechless in my helplessness.[19]

---

[17] Fourtouni, *Women in the Resistance*, 105–87.

[18] For a description of a kidnapping incident and other episodes during the civil war in Halkidiki, see Joice Loch, *A Fringe of Blue: An Autobiography* (London: John Murray, 1968) 213–42.

[19] Norman Gilbertson, "Greece, 1945–1953," 14, AFS Archives.

Then gradually over the decades conditions in Greece began to change. The old customs were wearing away as fresh ideas were aired; striking new modes of behavior enticed young people and rankled some of the old. In rural Greece, women's work was shifting away from the drudgery of laboring in the fields. By the second postwar decade, the government had made great strides in constructing roads to connect remote villages to larger centers, enabling the farmer to sell his produce and to collect some cash to improve his home and make the wife's life a bit easier. Tractors, those marvelous catapults into the modern world, took the place of beasts, relieving the women, whose responsibilities had typically included caring for animals and driving them, women and beasts tilling the soil together. Perched jauntily on the *trakteráki*, families could now motor to a neighboring village or even shop in the nearest provincial town. In the 1960s, electrification of the countryside was well under way allowing for the eventual installation in the home of laborsaving appliances. Also a network of government extension workers educated the farmer and his wife regarding modern methods and appropriate division of labor.

After the war, families at first refused to allow women to learn how to drive the tractors, when lessons started at the farm school in the short courses, run in conjunction with the ministry of agriculture. For a woman to be at the controls of a vehicle was considered unseemly.

Television was available by the early 1970s, transmitting into village homes from Western countries a host of novel propositions, among them images of youngsters choosing their own mates, a liberty unheard of in rural Greece where arranged marriages had always been the unchallenged rule. Boys began to select their own brides, but a girl dared not display such initiative, although in some regions she was not forced to marry someone she objected to unless there were overriding family concerns or economic considerations, and then of course, she was sacrificed.[20]

In the 1960s another influential phenomenon surfaced, when Greek workers flocked to Western Europe, mainly to Germany. At first the majority of migrants were men but by 1974 roughly 42 percent of those who took up permanent residence abroad were women, a trend related to scant employment possibilities in Greece and also to their hobbled lives

---

[20] Sanders, *Rainbow*, 161.

in the villages. The first group to leave consisted of mainly rural folk from Northern Greece, a movement that affected the farm school and the Girls School directly by increasing enrollment, since many of the migrants' children were packed off to their grandparents. The old people, freighted with their own responsibilities and poverty, found the boarding facilities and tuition-free educational opportunities at the farm school a solution for their abandoned grandchildren. The female migrants traveling back and forth between Germany and Greece carried home pointed tales of a free urban existence and examples of material wonders from the *Wirtschaftswunder* (economic miracle) that epitomized the lives of their former enemies.

By the middle of the 1970s the Greek rural world was being propelled forward by improved agriculture, pressures of urbanization and industrialization, expansion of the middle class, influences from the women's movement, and closer contact with Western Europe and the United States. The influx of tourists exhibiting libidinous behavior, and such popular entertainment provided by transistor radios and cinemas, also opened their eyes to the wider postwar world. Despite the military dictatorship (1967–1974), which found ways to apply its iron clasp to social advancement, those larger forces continued to urge Greek society onward.

However the military junta, with its backward glance and ultra-conservative impulses, dealt a temporary setback to women's aspirations. The regime tried to impose such archaic rules as a dress code for women, prohibited mixed social events for students, and repressed women's rights groups, although enforcement was haphazard. All the same, young women braved the danger of arrest and entered the fray to oppose the colonels' regime. The autumn before the junta's takeover was a landmark for girls at the Girls School. The farm school, at the Quakers' request, assumed the direction of the Girls School, quickening its pace and bringing it in line with the modern trends that were springing up throughout the country.

Almost ten years after the fall of the junta, during the first Andreas Papandreou government, pressure brought about by feminist groups, economic realities, and a realization of the importance of a modern social outlook, motivated PASOK to promulgate the Family Law of 1983. With the founding of the modern Greek nation in the nineteenth century, the new state had adopted European criminal and in 1941 the civil code.

However, the family code was based on the restrictive traditions of rural Greece, thus institutionalizing the patriarchal system; through it, the subordination of women was enshrined in the judicial process. One scholar comments, "The presumed incompetence of women, for example, was given as grounds for denying them standing as credible witnesses in legal proceedings."[21] Causing a seismic shift in Greek society, the 1983 legislation granted equality of the sexes, draining away the husband's role as the sole representative of the family to the outside world. Now the wife was empowered to offer equal representation. Greek society had traditionally looked on marriage as an arrangement negotiated by the families, who rarely considered the compatibility of the bride and groom as a primary ingredient. PASOK's family law reforms rejected this notion and instead treated the two parties as autonomous individuals, equal partners in the union.

Under the old scheme the husband was granted the legal right to make all decisions concerning the family including the education of the children and the family's place of residence; the wife could not register the children in school without her husband's written permission. Likewise, she was not allowed to travel without his written permission. Her presumed incompetence was articulated in the legal requirement that upon her husband's death the responsibility for the children's property was handed over to the court and a family council, appointed by the court. However PASOK's intention was that the husband and wife should share jointly in family decisions, which obliterated the traditional ideology and officially—that is legally—trumpeted gender equality.[22]

Decriminalization of adultery was part of the reform. For that offense the statute had called for fines or imprisonment; punishment had fallen primarily on women, reflecting their weak position in society and within the legal system. In 1986 the government enacted legalization of abortion on demand, paid for by the state. Needless to say, the Greek Orthodox Church was vehemently opposed to PASOK's reform of family law.[23]

[21] Pollis, "Gender," 295.

[22] Ibid., 298–99.

[23] For a discussion of a spectrum of Greek views on abortion, see Alexandra Halkias, "Give Birth for Greece! Abortion and Nation in Letters to the Editor of the Mainstream Greek Press," *Journal of Modern Greek Studies* 16/1 (May 1998): 111–38.

As noted, the Papandreou reforms struck down the dowry. Actually the tradition had started to crumble in the 1960s when girls began working in factories or in tourist facilities, where they could earn their own money to contribute to their dowries. Their wages, flowing into the dowry, enhanced their chances for a finer husband. Also store-bought clothes with a Western European flair came into vogue in Greece and could be purchased with this extra money; the new mode allowed young women to escape the image of the heavily draped peasant girl of yore. In some instances the daughter's contribution simply enlarged the dowry but in another more subtle way it also undermined the authority of the father in his role as sole provider.[24] The heavy emphasis on handwork at the Girls' School after1966 when it came under the direction of the farm school, was an effort to give the girls a means of earning money, which would lend them prestige in their communities and grant them a measure of independence.

Decades earlier another basic right had been won by women. As early as 1930, women had been given the right to vote in municipal elections if they were over thirty years old and could read and write, a factor that eliminated most rural women. In 1949 women were enfranchised for municipal elections even if they were illiterate and ultimately in 1952 they were given the right to vote in national elections, not bad considering that French women were not enfranchised until 1945 and Swiss women had to wait until 1971 for the right to vote. In this regard, though, in 1944 the leftist provisional government in areas of Greece no longer under enemy control had conducted elections in the village of Viniane in the prefecture of Evrytania. For the first time Greek women had participated as voters and had stood as candidates. The provisional government had drafted legislation that included articles on women's rights. Some of these rights would eventually wend their way into the Papandreou family reforms of the 1980s.[25] In the main, Greek

---

[24] The father was not only sole provider, but, according to Jennifer Cavounidis, ("Capitalist Development and Women's Work in Greece," *Journal of Modern Greek Studies* 1/2 [October 1983]: 324) "as the recognized head of the agricultural unit, the husband/father typically organizes the production process and markets the products. Women may influence the allocation of income to which they have contributed, but this does not constitute a recognized right." Also, for a study of girls who worked in factories, their behavior, outlook and situation, see Irwin Sanders, "Greek Society in Transition," *Balkan Studies* 8/2 (October 1967): 322.

[25] Fourntouni, *Women*, 48.

men conceded that women should have that right but claimed their lack of education would impede them from voting judiciously.

Yet with all of the gains for feminism, there was an irony. Each decade fewer women were entering the workforce. As women left their farms and migrated to the city, they found it difficult to obtain jobs in their urban settings. Demographics show that their participation in the labor force decreased from 33.9 percent in 1961 to 22.8 percent in 1981. Another trend, though, altered the course of their careers: a large increase in the number of women who were not only continuing their education beyond the compulsory nine grades but were also persisting on to higher education.[26] In 1970 about 30 percent of university students were women, while in 1989 that number mushroomed to 53 percent. Although this does not tell us the number of countrywomen who were able to vault the disadvantages of a rural life, it does indicate an improved climate for many who struggled to progress. In spite of these encouraging statistics, in 1990 the rate of illiteracy for females was four times higher than for males. Of course, we should take into consideration that this high percentage includes older women brought up in sterner times: this illiteracy rate will drastically reduce in the next decade or so.

A year before PASOK's sweeping changes, the Papandreou government enacted two laws that pertained directly to women working in the agricultural sector. These acts did improve the lives of farm wives. One law enabled them to gain experience in cooperatives—mainly as owners and managers of small hotels and handicraft shops. The cooperatives thrust countrywomen into the public domain, granting them a previously unachieved status outside the home and familiarity in the business world.[27] Through the cooperatives the government hoped to provide families with sufficient income and satisfaction to remain in their native, areas away from the lure of the cities. The other important law was the extending of the pension to farm wives, an entitlement that recognized the economic value of their labor—their status as productive farmers worthy of the same benefits as their husbands. This law applied to women who did not possess land but worked in agriculture as their main "secondary" occupation (a suggestion that still bore the hallmarks of traditional thinking that a woman was first and foremost a housewife,

---

[26] *Kathimerini*, 20 March 2001, 1.

[27] Keith Legg and John M. Roberts, *Modern Greece: A Country on the Periphery* (Boulder CO: Westview Press, 1997) 80.

a worker in the private domain, even though she probably was out toiling in the fields.

Although Greece has received its share of criticism for its lack of inclusion, it is fair to note that in this once patriarchal society a number of women have been included in government. Each of the PASOK governments of 1981 and 1985 had a female minister of culture (Melina Mercouri, who aside from her career as a brilliant actor is remembered internationally for demanding the return to Greece of the Elgin Marbles) and several deputy ministers. In 1993 the minister of justice in Konstantinos Mitsotakis's New Democracy government was also female. Additionally, 11 women out of a total of 300 members of parliament won seats in the 1985 election, and by 1996, 19 women held parliamentary seats, the same number holding for 2000.[28] Two out of twenty-four representatives of Greece's contingent at the European Parliament in 1985 were women. The number of women at the ministry level has averaged 12 percent over the decade of the 1990s. Again, this does not identify any woman whose origins were rural Greece, but it does indicate that women reached the very centers of power and that little girls like Eleni, wars aside, have an opportunity to grow up as valued human beings, educated, and legally empowered.

It was in this social and legal context, then, that first the British Quakers and then the American Farm School seized the opportunity to create a program that would inspire these striving country women to

---

[28] For a comparison, at that time in the United States Congress there were seventy-five women. In terms of percentages, 6.3 percent of the seats in the Greek parliament were held by women while in the United States Congress 14 percent of the seats were held by women. By 2002, France and Italy ranked at the bottom of the European countries, with women counting for some 10 or 11 percent of their parliamentarians. Alan Cowell, "French Politics Find Little Room for Women," *International Herald Tribune*, 7 June 2002, 3.

develop their potential to the fullest. For both the farm school and its alumnae, the educational experiment could not have been more fulfilling.

# CHAPTER 8

# WOMEN AT THE SCHOOL:
# A NEW COSMOS

In the 1930s the American Farm School purchased a few acres of land adjacent to the campus to realize a dream of Dr. John Henry House and his wife, Susan Adeline House: the establishment of a school for village girls, a parallel educational opportunity of the same caliber enjoyed by the boys. But because of the worldwide depression and the prospect of war the project was shunted aside.

Ironically, a decade and a half later it was the ruin wreaked by World War II and the Greek civil war that spurred the plan along toward realization. In 1945, the resulting social chaos prompted the British Society of Friends, which had worked in Greece on and off since 1922, to found a school to shelter village girls and to teach them how to cope with and improve their embattled lives. Norman Gilbertson describes the conditions under which youngsters were being raised in Northern Greece: "Refugee movements from mountain villages to the towns had begun during the war as a consequence of the burning of villages by the German forces as reprisal raids. Although there were such reprisals all over Greece they were outstandingly heavy in Western Macedonia and there can have been few unburned villages to the west of the Kozani-Kastoria line.... Many people returned home after the Germans left but as the hold of the *andártes* built up again in the mountains more and more people became refugees from the high villages and then from the plains near the mountains and readily open to attack."[1]

---

[1] Norman Gilberston, "Greece, 1945–1953," 8.

The Quakers, who had a long history of cooperation with the farm school doing relief work among the Asia Minor refugees in Northern Greece after 1922, grasped immediately that settling the girls close by the farm school would increase their security against incursions by communist rebels. The farm school had organized a few watchmen and desultory patrols, although this protection, to be sure, later proved porous. Supportive of the Quaker project, the farm school made available some 2 acres lying one-quarter mile from the heart of its campus. Of no small matter were the two wooden barracks the Germans had built on that land in 1943 to house the farm school staff. (The occupiers had requisitioned staff housing to billet their own troops.) As dingy and rat-infested as these hovels were, the German barracks provided ready-made shelter for the girls and their teachers. The necessity of housing young women in these huts furnishes testimony to the appalling conditions that prevailed in Northern Greece. The new institution officially named the Quaker Domestic Training School, became known popularly as the Girls School but the Greeks put it in their own idiom, dubbing it the *Scholí Quakéron*, adding the Greek genitive plural ending to the English word Quaker.

Unique in Greece and perhaps the only one of its kind founded anywhere by Quakers with the same intention and learning program, this was the tenuous beginning of an exceptional institution. It offered a practical and comprehensive educational experience to impoverished village girls who had seen their elementary schooling riven by war and were struggling in a broken world where their natural curiosity and desire to learn could only be smothered. How the girls developed under the Quakers and later the farm school is indeed a story that illuminates how the Greek people can prosper and deepen their lives and in the process recast their whole social psychology.

Considering the extreme cultural and financial poverty and the subordinate position of girls in this era, their attendance at the Girls School must have represented for them entry into another cosmos. Some girls were gathered from villages near the northern border, areas of Greece undergoing the worst turmoil, but the majority came from the prefecture of Halkidiki, not far from the farm school, where by 1946 a growing band of *andártes* was threatening them from Holomondas, a mountain in the heart of the Halkidiki peninsula.[2] There the Quakers were well known from relief activities they had carried out earlier; otherwise the families would hardly have relinquished their daughters to attend a boarding school—a kind of establishment unheard of in Greece—no matter how dire the conditions at home. The parents appreciated that under the Quakers' care their girls would be fed and protected. A number of girls, though, were orphaned, caught in the same predicament as the first boys who, more than forty years earlier, had been gathered into the farm school by Dr. John Henry House.

Many girls were malnourished. Almost all suffered from bad health at a time when tuberculosis was rampant all over Greece. The mayhem had also maimed them in spirit; fear, anger, and anxiety dogged them. That first year they ranged from fifteen to twenty years old. Their education did not begin until the second year because the Quakers devoted that first year to relief and rescue. Since the families could not afford to pay, the Friends Service Council in London raised the funds for the students' room and board.

One of the founders of the Girls School and its first director was an Australian, Joice Nankivell Loch, who had arrived in Greece in 1923 with her husband, Sydney Loch. Both were veteran relief workers (usually functioning under the direction of the Quakers) and writers, with finely honed artistic sensibilities. She set down the school's principles and devised its curriculum. Earlier, in the late 1920s, almost single-handedly, she had revived the art of rug knotting among the Asia Minor refugees, using her own designs. These famous *Pírgos* rugs were knotted exclusively in her adopted village of Ouranoupolis with its landmark Byzantine tower, or *pírgos,* where the couple had taken up residence. The designs were based on photos her husband, Sydney, took of Byzantine motifs from the all-male monasteries of nearby Mount Athos, mostly from books and illuminated parchments.[3] She was a gifted narrator and her books, *A Life for the Balkans* (at once an early history of the American Farm School and a biography of Dr. John Henry House) and *A Fringe of Blue: An Autobiography*, are not only beautifully crafted, but stand as singular sources for background reading on the farm

---

[2] Ibid., 9.

[3] Joice Loch, *Prosforion-Ouranoupolis: Rugs and Dyes* (Istanbul: American Board Publication Department, 1964) A unique history of the rug industry that she revived in Greece for the benefit of the Asia Minor refugees.

school in its early historical context. To honor her in 1972, the main building of the Girls School was christened Loch Hall.

After the first year, the curriculum was extended to two years, for the Quakers realized that the girls needed more time to profit from all the advantages the school was able to offer them and to benefit psychically from the restorative atmosphere. Although more girls were keen to come, the enrollment was limited by the amount of dormitory bedroom space available in the barracks—forty beds.

The purpose of the Girls School was to educate young women who would return to their villages, eventually finding their way as future wives and mothers and, in the manner of farm school graduates, assuming the role of community leaders. The students admitted during the first year or so had finished the six years of elementary education (such as it was) two or more years earlier; thus the long absence away from the classroom had weakened their study habits and sense of organization. As time passed, the curriculum broadened, the students carrying home with them improved methods of cultivation of small garden plots, raising of poultry, small animal care, food preparation, homemaking, childcare, diet, personal hygiene, gymnastics, and sanitation. Their academic studies comprised lessons in Greek language and history, arithmetic, writing, and geography. Progress was slow since most of the instruction had to begin at the most elementary level. Later, when enrichment courses became possible, handcrafts, music, dancing, drama, English language study, excursions, athletics, and activities with farm school boys were added. The Girls School could boast of an extraordinarily rich education. On the other hand, no equipment was used that the girls would not find in their own simple homes. Like the farm school, the Girls School was committed to the principle of a practical approach to learning.

The Quakers had an effective recruitment strategy. Two or three girls were selected from some of the same villages year after year, so that when they returned home they could look to fellow graduates for reinforcement, achieving cooperatively what they could hardly have managed individually. Also, the recruitment took place in clusters of neighboring villages, so that a wider community of reinforcement was formed. One is reminded here of Plato's words: λαμβδια χηοντεσ διαδοσουσιν αλλελοιϛ(Having the torches of light they hand them on to one another.).[4] Although in the beginning parents were hesitant to send their young girls to a distant boarding school, the excellent results from a critical mass of youngsters returning to their own villages convinced the anxious mothers and fathers that the experience was indeed positive. It was not too many years before more students applied than could be accepted.

The young women were recruited not according to their intelligence or school grades but chiefly according to their needs. It was important, the Quakers believed, to help the girls and the villages, and so they sought out the poorest places, choosing those living in deprivation. Bruce conveys in one line the out-and-out poverty: "The girls from the Girls School lost an average of four pounds while they were home for Easter."[5]

In a brief description of her village, a student provides a glimpse into the barrenness of the countryside even though this account, written in 1966, comes sixteen years after the termination of the devastating civil war, when conditions in Greece generally were on the mend.

There is no road and no bus to my village of Koupa in Macedonia. It is the last village before the Yugoslavian border. Once a week a truck takes people to Axioupolis, the nearest town. Koupa has fifty houses, a church, a store, and a coffeehouse. The village telephone is in the coffeehouse. On August 15, Assumption Day, our church has its holy day, but there is no celebration. The people go to church, nothing else. The village is too poor. Our house has two rooms and a little kitchen. There is an outdoor oven in the courtyard where we bake our bread. The richer people have sheep or cows. We have a goat. We also have two mules to help with the heavy work, and five chickens. Our farm is ten stremmata [2.5 acres], but they are not all in one place. And they are not flat, but go right up the sides of mountains. On our land we grow oats for the animals and potatoes and beans for the family.[6]

The girl's personal situation was as follows:

---

[4] Plato, *Republic*, 1.328A.

[5] Bruce Lansdale to Craig Smith, Thessaloniki, 19 May 1951, AFS Archives.

[6] Triandafilia Kiti, as quoted in *The Sower* no. 8, 1968.

Everyone wants to leave our village to go study. In the village there is just nothing to learn. My father wanted me to leave so that I could learn something and be clever. When I first heard about the Farm School I wanted to go right away, but I was too young: so my father arranged it so that I could stay with my grandmother and aunt and uncles in Axioupolis. My father wanted me to begin some kind of work. My uncle owns a tractor. They have a little more money than we do. I first heard about the Girls School at the American Farm School from a girl who lives near me. She is married now and has two children, but she studied for a year at the Girls School. Even though she stayed only a year she knows much more than the others here.[7]

In the early days, the accommodations in the German barracks at school were, in most instances, at least a minor improvement over the students' homes. A woodstove kept the frost from forming on walls of the communal study room, but unless one was hunched close by, it hardly warmed the body. The students enjoyed electricity, but the generator turned off at 10:00 P.M. Food during the civil war was exceedingly scarce: the teachers remember meager amounts—"never enough, just adequate to function"—mostly olives, potatoes, and spinach.[8] The girls wore black uniforms fashioned from blackout material sent from England, adorned with white collars to make the dresses appear less austere.[9] As for the ethos of the Girls School, the Quakers, it seems certain, did not attempt to draw the students away from their Orthodox roots. As all members of the teaching staff were Greek, of course, they exerted a predominant influence. The only non-Greeks were the school heads, and they ensured that the girls walked to the farm school every Sunday to attend Greek Orthodox Church services. The institution, claim former faculty members who taught at that time, was intended purely to provide aid, relief, and education.[10] Photographer-

---

[7] Ibid.

[8] Chrysanthi Bien, interview with author, 27 November 1999, Queechee VT.

[9] Nancy Crawshaw, "Quaker Experiment in Peasant Education: The Domestic Training School in Salonika," *Sunday Times* (London), 17 August 1951, Education Supplement, 1. Crawshaw came to Greece first in 1947 to photograph the Girls School and returned to take pictures of the civil war in Greece.

[10] Vouli Prousali, interview with author, Thessaloniki, 11 November 1998, written, AFS Archives. She served as teacher at the Girls School from 1948 to 1952. She returned

journalist Nancy Crawshaw wrote in 1947, "Political discussions are forbidden and girls from embittered rival sectors of the population work peacefully together in the stable atmosphere of Quaker reconciliation."[11] "We taught the girls to be strong, to stand up for their rights, not to back down if they thought they were right, but always to remain calm and to respect the other person," former teacher Vouli Prousali stressed.[12] When they returned home, their confidence, strength, and poise beckoned to other women as worth emulating, facilitating the Girls School's influence to permeate a cluster of villages. And it appears that generally the future husbands, occasionally Farm School graduates themselves, and the fathers, were proud of their educated wives and daughters, accepting their enlightened ways.

The girls grew close to the teachers, were eager to learn from them, and were influenced greatly by their instructors' demeanor. "They watched everything we did," Chrysanthi Bien remembers, "listened acutely to every word we said. They were so eager to learn every single thing they could. And, strangely enough, through all of these hardships—the cold, the hunger—we staff members worked from morning to night without a thought to fatigue, giving everything we could: the spirit was so great."[13]

Until 1949 the staff and girls felt secure on their campus, protected from guerrilla raids by watchmen and farm school patrols. But when the *andártes* kidnapped the farm school's senior class in January, the girls were profoundly shaken, and no one lived without trepidation until the civil war ended in September of that year.[14] The kidnapping, which received wide coverage in the American and Greek press, was perpetrated by a group of *andártes* who descended from their hideout on Mount Holomondas. They took control of the dormitory, cut the

---

in 1969 to the Farm School as a teacher with additional duties as head of instruction and assistant to the principal; later she taught at the Farm School until retirement.

[11] Crawshaw, "Quaker Experiment."

[12] Prousali, interview, 11 November 1998.

[13] Bien, interview, 27 November 1999.

[14] The kidnapping incident is also covered in book 1, chapter 12. Recently, coverage in the *Ellinikos Vorras* (1 February 1949, 1) reporting the national forces account of the skirmish has come to my attention For a richly detailed unpublished narrative of the event by one of the last five boys to escape see Nicholas Hadjimankos, "An Incredible Adventure: The True Story of My Capture and Escape from the Communists," October 1986, AFS Archives.

telephone lines, and ordered thirty-seven of the older boys and four staff members outside.

The kidnappers disappeared with their captives into the night, each boy clutching a blanket and a loaf of bread. Under the cover of darkness some boys scattered and by dawn half had escaped from the *andártes'* control. The national forces pursued, having been notified by Heracles Iasonides, a farm school alumnus and employee, who drove across a road (probably strewn with mines) to notify the military authorities. In an ensuing skirmish between the army and the *andártes*, all boys escaped except five who finally made it back to the school in time for graduation in June. For a while after the kidnapping, the girls were escorted at dusk from their own isolated school to sleep at Hastings House on the farm school campus where they could be better protected.

One of the most innovative features of the Girls School curriculum was the childcare program, the only one of its kind in Greece. As country girls thought of themselves then principally as future wives and mothers, they were eager to master the elements of childcare. After the Germans withdrew in 1944, abandoned infants were found throughout the country, even on the streets, in trashcans and other throwaway situations. It is important to keep this sorrowful fact in mind when learning that each year the Girls School obtained from the local state foundling home an infant approximately six weeks old. "We would pick the sickest looking baby," says former childcare specialist Chrysanthi Bien, "one that we could nurse back to health."[15] The infant was fed as nutritiously as possible under prevailing conditions, whereas in the village or foundling home, owing to a worse state of affairs, the baby would have received only supplementary rice water. The little one was cared for in a separate cottage by a staff member. In addition, three students assumed the roles of mother, father, and sister, on a rotating basis. The "mother" was responsible for the cooking, shopping, account keeping, and general care of the infant. The "father" managed the farming, milked the cow, fed the livestock and chickens, and kept a small vegetable garden. The "sister," a first-year student, learned by doing, assisting in household tasks. After a year the school returned the now chubby, rosy-cheeked baby to the authorities who made it available for adoption. People at the Girls School never saw the child again. There are no follow-up studies to indicate how

---

[15] Bien, interview.

this experience—a team of people acting as surrogate parents—affected the personality of these foundlings.

A raft of superstitions associated with child rearing in Greece, such as belief in the evil eye, can still be detected among Greek women. Sickness, the girls were taught at the Girls School, does not originate in the child's being bewitched by someone with the evil eye, nor by the same token could the baby be healed by superstitious treatment. Instructed in good hygiene, proper nutrition, and the effectiveness of logical thinking, the girls were advised to make a trip to the doctor if their infants became ill. Observers have noted that sound infant care is still not practiced in some places by Greek mothers owing to lack of proper education and the grip of old folkways throughout the country.[16]

A great improvement in the lives of the school girls occurred when in 1960 Mr. and Mrs. Noel Jones, serving as joint principals, raised $50,000 in England to construct a facility that could suitably house the staff and fifty students to replace the dank barracks. Much to everyone's excitement, the building was formally opened by the princesses of Greece, Irene and Sophia. (Sophia later married King Juan Carlos of Spain and today is queen of that country.)

Not surprisingly, the improved living environment transformed the students' and faculty's lives. Set on a rise in the midst of wheat fields, the two-story whitewashed building faced a spectacular view: the Thermaikos Gulf with storied Mount Olympus in the background. The bright sun of Northern Greece, streaming through the windows, set the rooms aglow. With a look toward the future, the administration installed some modern appointments in the kitchen and introduced sewing machines for the practical reason that these appliances would soon be entering Greek homes.

In 1965, the British Friends Service Council notified the trustees of the American Farm School that they intended to terminate their lease of the property in 1966. For years the Quakers had been discussing such an eventuality with the farm school; Quaker policy followed a tradition of starting institutions abroad and, when propitious, turning them over to local agencies. The Joneses submitted their resignations at a crucial

---

[16] Ministry of Press and Mass Media, Secretariat General Information, *About Greece* (Athens: n.p., 1999) 159. For a revealing study of health practices in mid-century rural Greece, see Richard and Eva Blum, *Health and Healing in Rural Greece* (Stanford: Stanford University Press, 1965). Also, Prousali, interview, 11 November 1988.

moment when the curriculum needed revision to meet modern trends in Greece. An even more persuasive factor was the necessity for the Girls School to coordinate its activities with the Greek government since its status had never been truly clarified, the result being that officials regarded the school with suspicion. These measures, the Quakers agreed, could best be handled by the American Farm School. Weighing heavily in the negotiations was the Quakers' kinship with the school: they had always felt at home with the farm school's moral core, so similar to their own. At the same time, Greek extension and home economic agents reported a dearth of farm women sufficiently trained and motivated for introducing new ideas to improve rural life, making the Girls School with its singular initiatives all the more prized. With all this in mind, in June 1966 the American Farm School's board of trustees voted to assume responsibility for the Girls' School.

In academic year 1966–1967, truly a watershed in the history of women at the farm school, Tad Lansdale took over the direction of the Girls School while a search began for a new headmistress. Change breeds it own uneasiness; thus the move was not without difficulties. First of all some members of the farm school faculty did not readily accept the melding of girls with boys within the same institution. The very concept seemed foreign to them even though they had been accustomed to the girls visiting the boys' campus each Saturday night for joint events—skits, songs, dances, movies, or a speaker—and for church services every Sunday followed by lunch together.

The more basic problem, a quagmire that had troubled the farm school for almost its whole existence, was funding. Like the boys, the girls received full tuition scholarships, and the fifty new students represented no small expense. With as much interest as Bruce, Tad followed Greek village life intently and had studied how she might steer the girls toward new possibilities. She herself possessed a keen sensibility for textiles. She realized that textile handicrafts, an industry that had become widespread in Greece since the war, could be introduced into the curriculum for a double purpose: to earn money for the Girls School and to teach the girls a craft they could apply after graduation. Now that the girls had access to radio and magazines, and contact with a more modern world, simple domestic techniques that once had to be spelled out had become the furniture of everyday life. The time was ripe for a turn away from domestic basics to intensified training in vocational skills.

Tad was well acquainted with Joice Loch's rug operation and approached her to help with the project. Loch offered her own designs based on the monasteries of Mount Athos. She sent her master rug maker to the school to advise the staff not only on the knotting technique but also on the use of vegetable dyes. The National Handicraft Association donated looms, material, and a teacher. The rug knotting and other crafts took off quickly and the Girls School was soon holding special exhibitions and selling the creative wares to an interested public. To draw people to the campus and to introduce the girls to the commercial process, their goods were sold at a boutique set up at the school. The crafts project served a fundamental purpose: it stoked the pride and enthusiasm of young women who never before had stood in the limelight as professionals.

Tad also formed a volunteer auxiliary group of women from Thessaloniki—reconstituted later by Girls School headmistress Janey Hamilton, and today called Group for Student Services. The women invited the girls to their homes for lunch or tea, took them to the theater and other cultural events, arranged for speakers to talk on campus, and interested the students in volunteer community work such as visiting old-age homes and orphanages. The group continues to devote efforts to fundraising and helping to enrich the boys' and girls' extracurricular lives at the farm school. Its steadfast contribution through the years has been inestimable. Also, Tad persuaded some English and American women living in the area to teach English to the girls three times a week on a volunteer basis, a meaningful addition to their course of study in a country where the command of foreign languages can mean career and social advancement and where such instruction would otherwise be too costly for village girls. As tentative as their newly honed talents might have been, their grasp of English, their artistic skills, and their brush with the world of commerce lent them a cachet when they returned to their home villages. The students' newfound poise and eloquence and confident personalities traits were striking and, over time, became the very signature of farm school graduates, both boys and girls.

At Tad's invitation in 1972, Elsa Regensteiner, one of the world's foremost weaver-designers, visited the Girls School and pledged to lend a hand, with an eye to nurturing a new approach to contemporary Greek design.[17] During successive visits she guided the students until they were creating credible textiles and tapestries using folk motifs adapted to modern designs. Local journalist Litsa Fokidou wrote in a Thessaloniki newspaper: "She opens the eyes of these young girls to the world of art. She teaches them to look at nature and to borrow from its colors and shapes, which they then transform into personal creations by combining warp and weft on the loom."[18] Under Regensteiner's instruction, the girls were flying high on an artistic venture. Her real gift to them was immeasurable, a donation that many of the schools' graduates claim has lasted a lifetime: to cultivate confidence in their own ideas and to stimulate their own creative gifts. In 1972, the school established the Crafts Center for the preservation and promotion of textile arts.[19] The center closed in 1978 when the girls were integrated into the lyceum program with the boys on the farm school campus in accordance with the country's new vocational curriculum.

In the early 1970s an ambitious, high profile fundraising campaign was orchestrated by Dippy Bartow, head of the American Farm School trustee committee for the Girls School. A piano recital at Lincoln Center raised $12,000. TWA donated $3,000 and a pledge of $20,000 a year for the next three years was received from a foundation. In 1971 the famous singer and social activist Nana Mouskouri gave a benefit performance at Carnegie Hall, which she followed by a benefit at the Kennedy Memorial Center in Washington, DC. Additionally, the Wallace Fund and the Sigma Kappa sorority took a serious interest in the girls and donated generously. Artist Elli Trimi helped the project by working with the girls on their designs.

The goals the Lansdales established for the girls corresponded with those the Quakers had set down. Bruce and Tad encouraged the graduates to return to their villages to act as leaders in their communities and in the process to contribute to raising the standard of living in rural

---

[17] Regensteiner, head of the weaving department at the University of Chicago, was a recipient of a citation of merit from the Museum of Modern Art in New York and the Institute of Design in Chicago. Her work has been shown in more than fifty exhibits and included in several museum collections.

[18] Quoted in *The Sower*, no. 82, 1974.

[19] For traveling exhibitions promoted by the Crafts Center, see *O sporéas*, no. 6, 1977. In 1976, the farm school began publishing the Greek-language *O sporéas* in Greek, reporting local news to local residents. Like *The Sower*, this is a small publication with photos.

Greece. That they would be respected in their own communities for their character, education, and special skills was Tad's hope. She was mindful of the woman's position in Greece and cognizant of the farmer's idiosyncratic expression used decades earlier when chatting about a lowly creature such as a donkey, "Excuse me for mentioning my donkey." But often he would also apply this idiom thus: "Excuse me for mentioning my wife." This startling expression drove home to Tad the importance of having the girls esteemed in their own villages. During the educational process, the girls, she is convinced, "experienced the euphoria of a great adventure," an observation seconded by many alumnae.[20] She devoted herself to the acting directorship for only one year—a creative and stimulating tenure—at the same time performing her demanding duties as wife of the director of the American Farm School, and nurturing her four children, who were still young. Under her stewardship the foundation was laid for an ambitious course of study with a host of enrichments one might expect at a fine preparatory school in the United States. She was succeeded by Helen McCune, a home economics trainer in the ministry of agricultural extension service, who had spent over twenty years doing aid work in Greece.

To ensure that the girls were being trained to take advantage of the broadening opportunities that were surfacing for women, the farm school took an audacious step in 1977 by inviting them to learn the mechanized aspects of farming, an opportunity that had formerly been open only to boys. (However, the short courses held at the farm school had been instructing young adults—both men and women—in tractor driving for years.) A class in tractor driving was started only after intense discussion among director, staff, and parents in consideration of the parents' possible objections. To participate in this program the student was required to submit written permission from her parents; some of the girls were not successful in persuading their mothers and fathers to agree to such a revolution. The school's stated rationale for initiating this instruction was that the girl's competency in driving a tractor could prove vital to her family should her husband or father become ill, disabled, or mobilized. Along with preparation for the license, the students learned to make the standard five-point check before turning on the engine—oil, water, battery, fuel, and tires. Once again, this education

---

[20] Tad Lansdale, interview with author, Metamorphosis, written 26 June 1998.

did more than teach girls just to drive a tractor; it was a means of elevating young women to operate on a level equal to young men, thereby bolstering their self-assurance. An issue of *The Sower* records the expression of an elated student—one of eight to win her license—as she gained a sense of herself: "Eleni Panitho climbed atop the bright red tractor, hesitantly shifted into first, released the brake with both hands, and with a broad smile lurched off down the Girls School driveway with her instructor clutching at the handrail behind her."[21]

Because of social taboos, tractor lessons had not been offered in the 1960s or early 1970s at the Girls School. One alumna, Fifi Teneki, class of 1967, who announces herself proudly as an *agrótissa* (farmer), is a grateful and enthusiastic graduate, one who would share Tad's labeling of her education as an adventure. "I learned so many, many things. One thing they didn't teach, though, at the Girls School was a real mistake. I didn't learn to drive the tractor. Then when I needed to operate one, just when I was caring for my baby and I had to take care of my house and work in the fields, I was forced to take time out to learn the tractor. It made it very hard for me, a real strain. If only they had given me lessons when I was a young girl at the school."[22]

As noted in chapter 4, change shifted into fast gear in 1977, completely altering the course of education for farm school students, when the school initiated two programs for boys and girls who had finished the mandatory ninth grade. The girls joined the boys in the classroom, desk by desk, and in the fields, shoulder to shoulder, taking a leap into the future together. The old Girls School curriculum and crafts program were terminated. In the most revolutionary change of all, the girls moved into the same dormitory as the boys, in a separate wing of course. After the Girls School building was damaged by the 1978 earthquake, the Kostas Zannas Wing of the Charles and Ann House Dormitory was built specifically to house the girls. The cost of repairing the Girls School structure would have been prohibitive; the savings gained by maintaining the boys and girls together were significant. The move was fiscally sound, and the parents, although conservative, rural folk, were ready to accept this degree of cohabitation as one of the

---

[21] *The Sower*, no. 89, 1977.
[22] Fifi Teneki, interview with author, Thessaloniki, 26 February 2000, written, AFS Archives.

innumerable social modifications of the second half of the twentieth century.

The recommendation for integrating the girls into the lyceum program came from Janey Hamilton, the director of the Girls School from 1972 to 1978, widow of the former American consul general in Thessaloniki, and an old hand in Greece. She noticed that the concept of boys and girls going to school and boarding together in provincial towns had recently gained acceptance in the countryside, since many youngsters had to live away from home, unprotected, in order to attend junior high school and high school. Hamilton and her staff reasoned that the parents would accept their boys and girls living close together at the farm school, where they certainly would be well supervised. The government agreed and passed the enabling legislation. The plan was appreciated by the parents, who by this time had broadened their outlook, and had come to realize the necessity of their daughters' receiving an education equivalent to that of the boys. Bruce notes, "As it turned out, as the girls tried harder to keep up with the boys, so they challenged the boys to work harder. And the girls were no longer objects to be observed in church on Sunday; it was a natural step in the development of the School."[23]

A 1993 profile of one female student illustrates how far Greek women had progressed from the days when their parents were miffed that their daughters yearned to master driving a tractor. Elsa Kafantari came from the island of Thasos and was one of seventy girls studying at the farm school in 1993. Her ambition was to set up a greenhouse vegetable business on the island after graduation. "Thasos produces no vegetables. It has to import all from the mainland," she said. At the farm school she specialized in horticulture to learn how to grow greenhouse vegetables. An enterprising young woman, she sold the results of her project—tomatoes, eggplants, peppers, onions, garlic, and lettuce—to faculty and staff who lived on campus or to the school's numerous visitors. As part of her study program, she learned farm maintenance, mechanics, carpentry, and tractor driving. In addition, she raised poultry, slaughtered and dressed hens, acted as a midwife to a sow, raised pigs, reared turkeys, and learned the whole process of milking.[24]

---

[23] Bruce Lansdale to the author, Metamorphosis, 17 December 1999, AFS Archives.

[24] Ann Elder, "Farming the Future," *The Athenian Magazine* (July 1993): 16–19.

Underlying the TEL program was an emphasis on management training. The farm school insisted that for any family—particularly for the rural families in which wives were intimately involved in myriad tasks—management was indeed a dual effort. Thus the joint training of both men and women in management and social development, as well as in technical fields, had by 1978 become an imperative.

Hearing the women's stories lends immediacy to their experiences at the school, permitting us to view the variety of outcomes as the graduates go about their lives. Because of her fidelity to family and her return to the village, the ardent farmer, Fifi Teneki, is certainly a person the founder and his son would have counted as a model graduate. Since she is a product of the Girls School, class of 1967, her education was basically home economics and handcrafts. For her, farming has not been so much a profession as a passion. "I fell in love with farming; it was on the land that I felt that I was creating." As a young child she followed her father around the fields. Naturally, the *Scholí Quakéron*, as she still calls her alma mater, was the perfect atmosphere for her with its proximity to the farm school fields and the sense of organization and empowerment she experienced. An excellent student, she received a prize in English. For her dowry, she was given 3.75 acres, and she inherited another 5 acres; her husband had 5 acres, and together they bought 2.5 acres, adding up to about 16 acres, a large holding for Greece. Only when she had her first baby did she stay at home until the child was two years old; then she took him with her to the fields. "Now after so many years at manual labor I can't do strenuous work any more," she admits. But she's still on the farm. She brings her obvious intelligence and energy to the task of managing the 16 acres outside the village of Agios Pavlos near Thessaloniki.[25]

Yet while farming has been a passion for Fifi, other alumnae are discouraged by agriculture. Katerina Alopoudi graduated from the farm school in 1993, having completed the lyceum program, which had a strong agriculture emphasis. She returned to her village, Vasilika, home to dozens of other farm school graduates. When she entered the school, her intention was to continue on to university. "Anything to escape the drudgery of farming. I liked my classes and found that mathematics really interested me. Also, my plots of onions and cabbages and the like

[25] Teneki, interview, 26 February 2000.

were always cited for excellence." At the school, she made good friends and was impressed by the way everyone learned to live together and develop a sense of self-discipline and order. After she received her farm school diploma, she found that upon entering TEI that she liked science as well as mathematics. Then she married, had a child, and she and her husband decided to stay in Vasilika, she to farm the half acre her father had given her. She found that life away from the school seemed like chaos. Asked if her education did not teach her how to manage chaos, she smiles and answers with a question, "How can you ever learn to manage chaos?" Then she says emphatically, "Farming is very hard work, and there is no real money in these small plots, and we can't afford to buy more land. My sister and I work our land together—a total of one acre, half hers, the other half mine." We sit in her humid greenhouse, chat, and inspect row upon row of cucumbers. Although she is grateful for her education, which she puts to use in her greenhouse to grow her crops and make her business decisions, for all her education and intelligence the holdings are too small for the sisters to really profit from farming.[26]

Irini Dalou is another 1993 graduate of the farm school lyceum program, but her career has taken a decidedly different turn from classmate Katerina's. She works for an accounting office in her village of Galatista, home also to numerous farm school alumni. She found the farm school through her brother, Manolis, an earlier graduate. She is proud that she learned how to solve problems away from home, away from the guidance of her family: "You sometimes meet some other students who are not the kind of people you ought to take up with. These are choices you have to make on your own. And without parents around to guide you, you have to develop judgment and initiative." She found the level of education at the farm school very high, and in fact had to take tutoring classes to keep up. She also liked the opportunity to meet other students from all over: from Epiros to Cyprus. After she received her diploma, she went to IEK, where she found it challenging to strengthen her computer skills. She is poised, professional but cordial in her office setting. Although she comes from a farming family, she seems to have discovered her niche in office work.[27]

---

[26] Katerina Alopoudi, interview with author, Vassilika, 8 March 2000, written, AFS Archives.

[27] Irini Dalou, interview with author, Galatista, 8 March 2000, written, AFS Archives.

The two alumnae of the lyceum program agreed that life at the farm school was as satisfying for girls as for boys; they reported an atmosphere of inclusion. George Draper's opinion is that "...the girls were as fully integrated into School life as women were in Greek society, and maybe more so, since not many Greek women had experienced the joy of welding and tractor driving, and mastery of such skills gave you a self-confidence you might not already have had. All that had been achieved by the staff long before 1990 when we arrived."[28]

In an effective social experiment, the Girls School and the American Farm School were working to liberate young women from the tyranny of ignorance. An additional inroad leading to the very heart of rural society was, and still is, the bond with parents—people who truly want their children to prosper and who feel an affection for the two schools. It goes without saying that the students' families are in a position to contribute incalculably to modernizing village mores and attitudes. Without realizing it, these two institutions proved to be forerunners, on a miniature scale, of the women's movement in Greece. What the students absorbed in knowledge, self-confidence, and leadership capabilities was empowerment. That, after all, was the goal of the women's revolution that swept across Europe and the United States.

---

[28] George Draper to the author, Sorrento ME, 16 May 2000, AFS Archives.

# CHAPTER 9

# THE JUNTA IN GREECE AND THE ROAD BACK

For Greece the period from 1967 to 1974 was once again a time of distress. The country's political instability from 1965 to 1967 was brought to a rude conclusion when a triumvirate of unworldly military officers performed an amazingly successful coup and then managed to hold power for seven years. Throughout the twentieth century the democratic system in Greece was punctuated by five military takeovers, but this one survived the longest in its modern history. This junta—intrusive, arbitrary, and bullying—engendered an unnerving and surreal context for citizens and for foreigners who made their home in Greece. The country would welcome the restoration of democracy in 1974 and the return of Konstantinos Karamanlis, prime minister from 1974 to 1980.

Unmistakably the junta put the times out of joint. A group of army officers led by Colonel George Papadopoulos, Colonel Nikolaos Makarezos, and Brigadier Stylianos Pattakos took over the country on 21 April 1967 without any immediate resistance from the rest of the army. In their manifesto the colonels, with Papadopoulos acknowledged as the predominant force, decried the corrupted political situation and offered a period of stability during which political institutions could be reformed and new parties formed. They also asserted they were saving the country from the communists who, they claimed, had been scheming to take over. Ideologically, they intended to restore a Helleno-Christian civilization.[1] The junta's policy toward Cyprus was to insure *énosis*

---

[1] For commentary on the first five years of the military government, see Richard Clogg and George Yannopoulos, eds., *Greece Under Military Rule* (London: Secker & Warburg, 1972).

(union) with Greece.[2] Arrests were carried out throughout the country as targeted leaders and members of political parties, prominent newspaper publishers, and people from the cultural world were hustled off to jail or exile on some distant island, or placed under house arrest. King Constantine failed in an attempt to reestablish civilian control in December 1967 and fled to Rome.

In 1968, Amnesty International reported the regime's widespread use of torture. This documentation convinced the Council of Europe that Greece was abusing human rights, which in turn persuaded the junta to quit the council to save face before almost certain expulsion.[3] In 1969, the European Union froze some aspects of its association agreement with Greece, relegating the country to near-pariah status in Europe. But the Greek government's relations with NATO were always punctilious, a factor pleasing to the United States, whose strategic interests in southeastern Europe were served by a Greece that remained strongly attached to the alliance.

The breaking point came when students rose up against the colonels at the Polytechnic in Athens on 17 November 1973. As a result of student opposition and other serious problems covering a wide range of issues, the colonels were overthrown by Brigadier Dimitris Ioannidis, a hard-line military police officer who had become critical of what he considered Papadopoulos's misguided approach to ruling Greece. Next, Ioannidis's unsuccessful attempt to assassinate Cypriot president Archbishop Makarios triggered an invasion of the island by the Turks on 20 July; in full spate Turkey occupied the northern section. Ioannidis's reign imploded on 24 July 1974 as a result of his display of hubris on Cyprus. Fortunately, internal chaos was avoided when on 24 July Konstantinos Karamanlis was called back from self-imposed exile to head a national government; he imposed order immediately. His most urgent mission included subduing the army; restoring democracy; preparing the country for free elections; and coping with the perilous situation caused by provoking the Turks. The Greeks blamed the

---

[2] For a view of the Cyprus problem by a former American ambassador to Greece, see Monteagle Stearns, *Entangled Allies: US Policy toward Greece, Turkey and Cyprus* (New York: Council on Foreign Relations Press, 1992).

[3] It is worth noting that the council is strictly a European organization and that the United States is not a member, although at that time nine council countries were members of NATO.

Americans for allowing Turkey, a member of NATO, to invade and occupy one-third of the island.

Throughout the junta period, American foreign policy toward Greece became the object of scrutiny, rumor, and bitter accusation in the beleaguered country. In the United States, too, Congressional hearings and newspaper comment had questioned—and in more than a few cases objected to—American's foreign policy regarding the junta. Much as Lyndon Johnson's had been, Richard Nixon's foreign policy toward the junta was dedicated to maintaining good relations with the Greek government. One reason was for the sake of homeporting the American Sixth Fleet in Piraeus, considered a necessary deterrent to the build-up of Soviet naval forces in the Mediterranean. If the Greeks pulled out of NATO, the withdrawal would obviously leave a gaping hole in the alliance's southern flank.

Outbursts of violent anti-Americanism came to international attention in July 1970, when the car of a United States Army attaché and seven other American-owned vehicles were damaged by bombs in the Athens area. Bombs also exploded in Thessaloniki. The perpetrators were thought to be opponents of the regime who alleged that the United States was behind the colonels' coup. In fact the opposition held that the United States, if it so desired, could restore democracy, so much sway did this super power exert over Greece. It was no secret that since the Greek civil war there had been a close relationship between the Greek and American intelligence agencies and that American officers had served on the Greek general staff and in units as far down as the divisional level.[4] Many Greek people assumed that their military top brass were protégés of Washington because of a tight working relationship between the two armies spanning decades.[5]

When the junta seized power, the United States suspended almost half of its military aid. In 1968 the United States partly lifted the embargo and in 1970 resumed deliveries of heavy weapons and equipment. For its part, the junta was irked that the United States was

---

[4] Barbara Jelavitch, *History of the Balkans: Twentieth Century,* vol. 2 (Cambridge: Cambridge University Press, 1983) 407.

[5] For an analysis of Greek-American relations, see John O. Iatrides, "American Attitudes toward the Political System of Postwar Greece," in *Greek American Relations: A Critical Review*, ed. Theodore Couloumbis and John O. Iatrides (New York: Pella Publishing Co., 1980) 49–73.

dragging its feet in naming a new ambassador. The State Department had purposely left the Athens post vacant for eleven months to signal displeasure with the regime or, just as likely, to dampen the criticism leveled at the United States for its lack of censure toward the dictatorship, a complaint voiced by NATO allies and anti-junta forces in the United States. Finally, under the new Nixon administration, Henry J. Tasca was appointed ambassador to this sensitive post and arrived in country in mid-January 1970, a boost for the colonels, who had recently been humiliated by their treatment at the Council of Europe

While European countries gave Greece the cold shoulder, the United States sent a stream of high-ranking personalities to visit Greece, some to pose for photos with Papadopoulos: General Andrew J. Goodpaster, commander of NATO forces, visited in 1967, Vice President Spiro Agnew called on Papadopoulos in 1971, as did Secretary of Defense Melvin Laird and Secretary of Commerce Maurice Stans.

When the 1973 coup leader Brigadier Ioannidis, in a drive for *énosis*, instigated the overthrow of Cypriot president Makarios (who wanted to retain independence for Cyprus, not *énosis*), and the Turks invaded the island, many Greeks converted their anti-American antipathy into outrage. "Why didn't the Americans stop the invasion? Why didn't the Americans compel the Turkish forces to withdraw after the fact?" were the blazing questions. Still in a state of ire, the post-junta government under Prime Minister Konstantinos Karamanlis pulled out of NATO's military command to return to the organization only in 1980.

As might be expected, life for Americans and American institutions was no longer as sweet as in the pre-junta days. But even if the atmosphere was blustery, the protecting buffer for the farm school was that Greeks, rural folk and politicians alike, considered the institution to be *dikiá mas* (ours), thus safeguarding it from what could have been disruptive or even destructive forces.

As a foreign and private body, the American Farm School— dependent on the government's good will, cooperation, and a degree of likemindedness when it came to the process of educating rural youth—resolved to remain vital for the sake of the agricultural community it meant to serve. The trick lay in finding a way to evade the tentacles of government control, all the while seeking to carve out areas of mutual interest wherever genuine teamwork with the dictatorship was possible. Cooperative efforts during the educational reforms of

1970–1971, as outlined earlier in a discussion of the SEGE program, and partnership in the short courses, evidenced in the construction of a KEGE in 1973, were examples of mutuality.

The excesses of the junta regime in the broader context, compounded with student agitation and labor demands on campus, make this period one of the most dramatic and challenging in the history of the American Farm School. Ironically, only after the return of democracy in 1974, as new political and social forces emerged from the eclipse of the junta, did farm school life take a problematic turn. In 1975, for the first time in the institution's history, a student strike disturbed the old-fashioned—but benevolent—paternalistic system that had existed between administration and students since the founding. And beginning in 1976, labor demands unsettled the harmony between administration and staff. These discontents were, to an extent, connected to the junta and its collapse, much as student unrest in the United States was catalyzed by the Vietnam War. In the farm school's case, complaints on campus were the first spasms of social modernization.

The junta's coming to power marked a turnabout in Greek-American relations, an important factor in the history of the school. Relations with future governments would remain cooperative, but it would take much more effort on the part of the farm school to maintain its special status. There is little doubt that the close cooperation between the United States and Greece in the two and a half decades after the end of the civil war had enabled Bruce and his Greek staff to embed the farm school deeply in Greek society. Through a kind of transference, many people considered the School to be Greek. "The whole educational effort has contributed to the love which the people of Northern Greece feel towards the Farm School. They see the Farm School as their own educational institution," was the sentiment of one highly-placed individual, but it reflected the attitude of thousands.[6]

While the dictatorship was unbearable or dangerous to those individuals who defied it, or to those who stood somehow in harm's way, generally in most rural communities people were not personally tormented or alarmed. To be sure, there were instances in the provinces where individuals were arrested for criticizing government representatives or condemning the dictatorship. But there is much truth

---

[6] Nicholas Martis, minister for Northern Greece, to George Rallis, minister of foreign affairs, Thessaloniki, 23 January 1979, AFS Archives.

in the observation that in a multitude of villages, where thousands had been executed by the communists during the civil war, villagers were only hounded now and then by the military regime. In comparison to the abuse visited on other segments of the population, the colonels' treatment of farmers was restrained.

In fact, the dictatorship was at pains to gain and hold the loyalty of the rural population. The colonels curried favor by increasing the farmers' pensions by 70 percent (they had been the lowest in Greece) and in 1968 by canceling their outstanding loans, an act, however, that many economists have criticized. The value of the canceled loans was equal to approximately 20 percent of the total revenue in the 1967 government budget. But the loans had been onerous, secured from the Agricultural Bank of Greece at an interest rate of 20 percent. As one reporter comments, by 1967 farmers "owed the government bank 10 billion drachmas, which was about one-quarter of what they could produce in a year."[7] By granting free medical care to farmers, the government again gratified them. Also, the authorities strengthened agricultural education at the high school and also at the junior college level by planning the junior college, KATEE. And the junta pleased the rural population by continuing electrification of the villages and roadwork to connect them with the outside world, practical gestures toward a segment of the population that had felt neglected in favor of city people.

There were country folk who were attracted to the colonels. The three ringleaders sprang from the rural sector themselves, and Brigadier Pattakos, especially, spoke like a rustic. They loved to preach the traditional pieties, to praise church-going, and to spout *zíto o stratós*, (Long live the army.). Not given to quoting famous ancient aphorisms, they resorted instead to slogans. Pattakos ordered plastered all over the countryside, *Kathariótita eínai politismós* (Cleanliness is civilization), a message he composed in a moment of inspiration and directed at people who littered.

Although the government touted sympathy for the farmers, the reality was, according to economist Ioannis Pesmazoglou, that "the rising rate of growth of income…significantly declined in 1967–1970" and in 1968–1971 was half the 1963–1966 rate in the agricultural sector. Pesmazoglou claims that subsidies to promote changes in the structure of

---

[7] John Corry, "Greece: Death of Liberty," *Harper's* 238/1433 (October 1969): 80.

crops declined under the dictatorship and that the share of government investment in agricultural development was adjusted down from 28 percent to 24 percent. Pesmazoglou also denigrates the cancellation of farmers' debts, "which was not associated with any systematic policy to promote agricultural modernizing or growth." To add to the country's woes, emigration from farmlands and a slowdown in agricultural production were "becoming mutually determined processes, leading to stagnation in the countryside."[8] This phenomenon was most pronounced in the tobacco growing areas. "Although expenditure for large-scale reclamation or irrigation projects continued to rise, there was a decline in essential supplementary spending to induce the necessary shifts and adjustments in production," Pesmazoglou notes.[9] The *Economist* reported that "One clear statistical failure during the rule of the junta was agriculture. Growth in the form of output was planned to rise by 52 percent a year between 1968 and 1972. In fact, it grew only 2.2 percent."[10] These conditions went far to discourage families from staying on the land, as a result seriously hurting the recruitment effort at the school. Farmers as well as the general population were trying to cope with inflation that had reached 30 percent in 1973. One distinct advantage for farmers during the junta period was the expansion of the tourist industry. Since farmers were typically underemployed, their superfluous energies were invested in a variety of ways in serving the substantial influx of foreigners.

Among the organizations the dictatorship disrupted were the cooperatives, so vital to agricultural modernization. Many of the more successful, where the farmers were beginning to work in the spirit of cooperation, had their councilors arbitrarily fired and replaced with people loyal to the government. In an especially aggressive move against the cooperatives, the regime stripped the highest decision-making body of the cooperative organization of all its power and transferred that power to regime-appointed councils. In an even more crippling action, as a gesture of patronage to big money, the government privatized a number of cooperatives.

---

[8] John Pesmazoglou, "The Greek Economy," in Clogg and Yannopoulos, *Military Rule*, 81–82.

[9] Ibid., 82.

[10] Jonathan Rodice and Michael Wall, "The Gods Smiled at Last: A Survey," *Economist* (20 September 1975): 23.

That the rural population, the farm school's clientele, was not unduly ruffled by the regime spared the school the brunt of anti-American venom, which otherwise could have poisoned its kinship with students, alumni and parents, and blighted its decades-old relationship with communities throughout Northern Greece. Although the countryside remained passive, the universities, features of the urban scene, became centers of resistance, and in the end were among the major players in toppling the regime.

When Papadopoulos acquired for himself the ministry of education it was expected that every aspect of education from elementary school through university would come under special scrutiny. The junta's harassment of university students ran from the petty to the serious; it stretched from imposing a loyalty test on students and seeding classrooms with spies to banning—but not necessarily enforcing—the sporting of miniskirts and beards, and insisting on short haircuts for boys. The regime abolished the National Union of Greek Students and reconstituted it under the watchful eye of the police, who nominated the union's officers. Furthermore, the junta culled from university faculties scholars whom they regarded as anti-regime and replaced them with junta supporters. At the Aristotle University of Thessaloniki, possibly as many as one-third of the faculty had been dismissed by 1969. Some private primary and secondary schools were also purged. Since the teachers in private schools were not state employees, the schools themselves were required by the government to fire the alleged rogue teacher or risk losing their licenses to operate. This regulation would cause anxiety for Bruce when a farm school employee was subjected to government harassment.

In another restrictive move, the dictatorship created the post of commissar for each university. These officials were responsible for attending senate and faculty meetings and reviewing all university correspondence. In each case the commissar was a military officer.

In January 1969, the regime issued Decree 93, the Student Code, listing the penalties for students who were convicted of political offenses. They could be expelled from the university and also be barred permanently from further education if their ideas were incompatible with "national ideals." At the time this code was issued, these sanctions had already been applied to dozens of students who, as a consequence, had been exiled or sentenced to prison. The disciplinary boards were composed of faculty loyal to the junta and the whole process was overseen by the commissar.[11]

Such repression goaded the students to rebellion in 1973. As a prelude to the culminating act of student revolt, which would erupt at the Polytechnic in Athens on 17 November, the general public joined with students on the occasion of former Prime Minister George Papandreou's memorial service on 11 November 1973 to demonstrate against the junta. Pent up hostility against the regime manifested itself when hundreds of mourners, including students, clashed with police. On 12 November students at the Polytechnic demanded certain rights. In a militant temper, they called for the government's resignation. In an act of extreme defiance, the Polytechnic students occupied their school building. After a few days, the regime grew alarmed at the possibility of the occupation's assuming wider dimensions when the students appealed via radio for workers to join them. On 17 November—a still highly celebrated date that includes a national school holiday—the government called in tanks and troops to squelch the mutiny. In the ensuing fracas, probably some thirty-four students were killed and hundreds injured and arrested. The exact number of casualties have still not been verified. The public was generally appalled by the brutal reaction of the troops. Also appalled was Brigadier Dimitris Ioannidis—appalled, however, from his particular standpoint—because Papadopoulos had allowed the students and the general political situation to explode.

University students, owing to their age and a touch of sophistication acquired by living in the capital and other cities, were obviously not farm school boys. At issue in the universities were grave political problems, high matters of state, not the simple (although perhaps valid) demands of adolescent boys rebelling for a slice of personal freedom, although some boys at the farm school did adopt a more serious agenda. As might be expected, the rebellious mood in the cities trickled down to high school students in the provincial towns and circulated throughout the provinces. When the ferment finally jarred the farm school, it was not until fall semester 1975, over a year after the fall of the junta. Democracy that year was the watchword; the social climate had become permissive toward students as a reaction to the junta's oppression; and the farm

---

[11] For a summary of how the universities were structured under the colonels, see *Inside the Colonels' Greece*, trans. and intoduction by Richard Clogg (London: Chatto & Windus, 1972) 90–92.

school boys, in touch with the tenor of the times, were asserting themselves.

Yet another force was at work. The Cyprus crisis had unexpected consequences for the farm school. In October 1974, a group of seventeen bedraggled refugee boys, who had their agricultural training interrupted when their school was closed owing to the hostilities in Cyprus, arrived at the farm school to continue their studies. Traumatized by war and dispossessed, the young Cypriots naturally were grateful for this haven in Northern Greece.[12] Given their situation, a clique did bear a grudge against Americans, and bristled with discontent at being wrenched from home and hearth. Their sentiments, mixing with the disgruntled disposition of the farm school students eventually soured into student strikes.[13]

Meanwhile, in a remarkable feat of statesmanship, Konstantinos Karamanlis, called back from self-imposed exile to head a provisional government, paced himself shrewdly.[14] He set about establishing a government apparatus that would ensure a solid democracy, positioning the country on the road to prosperity and healing some of the social divisions. Toward these ends, he conducted elections on the commemorative date of 17 November 1974: his New Democracy party won 54 percent of the vote. To decide the fate of the monarchy, a plebiscite was held in December 1974; the people pronounced the country a republic. Then he guided the drafting of a new constitution. In a series of trials, after elections, the junta perpetrators were meted out death sentences that were quickly commuted to life in prison, thus sparing the country further bloodshed and turmoil. Some of the junta's most notorious torturers were also sent to jail. To bring a measure of

---

[12] A depiction of the Cypriot students' arrival at the farm school is reported by Brenda L. Marder, "A Unique Institution," *Athenian Magazine* (May 1975): 22–27.

[13] Evangelos Vergos, interview with author, Thesaloniki, 21 October 1999, written, AFS Archives.

[14] Karamanlis is considered one of Greece's finest statesmen. Elected to parliament twelve times, he served as prime minister for a total of fourteen years, and state president for ten. He helped shape Greece from his six ministerial posts and as prime minister from 1955 to 1963, and again from 1974 until 1980. He then became president of the republic, a position he held until 1985, when PASOK forced him out. He was reelected to the presidency in 1990, serving for five years before retirement. For an analysis of Karamanlis's career, see C. M. Woodhouse, *Karamanlis: The Restorer of Greek Democracy* (New York: Oxford University Press, 1982).

healing to a socially and politically divided society, the communist party was legalized.

To appease and insure the support of his countrymen, who were seething against the United States over Cyprus, Karamanlis declared that the status of the heretofore sacrosanct American military bases scattered throughout the country was up for review. The people, eager to cut these bonds, understood this as an act of independence. However, a terrorist group calling itself 17 November in honor of the Polytechnic student revolt could not be placated. In December 1975, the terrorists, adding murder to what had previously been merely a sentiment of anti-Americanism, shot and killed the CIA station chief in Athens.[15]

That year the Turks declared the northern third of the much-contested island the Turkish Federated State of Cyprus. It was the establishment of the partition for which they had yearned. Regarding the Turkish occupation of Cyprus, Karamanlis turned away from a showdown with the historical enemy, Turkey, by adhering resolutely to the foundation stone of his foreign policy: membership in the European Union, a process that had begun in 1961 with Greece's signing the Treaty of Association. As a member of a powerful European organization, Greece would cease to be considered merely a client state of the United States, and assume the role of a respected, independent nation in step with Western Europe. Strife with Turkey could block Greece's accession.

What interests us in particular about the Karamanlis government from 1975 through 1980 are those events that might have colored behavior at the farm school. To a large extent, the general mood wafting through the Greek atmosphere, like the *néfos* (in this connotation the cloud of pollution) that blankets Athens, lay outside of his control; a large portion of the national anger was simply human reaction to living seven years under the boot of tyranny.

In the realm of education Karamanlis set in progress meaningful improvements, especially in the technical vocational domain. What his

---

[15] For insight into how Greece's history of political upheavals spawned the terrorist organization 17 November, see George Kassimeris, *Europe's Last Red Terrorists: The Revolutionary Organization 17 November* (London: Hurst & Co. 2001). The terrorists were not apprehended until June 2002. They have been accused of twenty-three killings, among them four American officials, a British attaché, two Turkish diplomats, and Greek businessmen and government officials.

educational blueprint demonstrated was a turning away from the classical curriculum toward a more practical course of study bent on modernizing the country and preparing its people to meet the pressures of European Union membership. The impressive statistics speak for themselves: the number of teachers in secondary schools increased from 17,932 in 1974 to 22,916 in 1977, and from 570 to 3,027 in secondary technical education. Public education projects scheduled for the early 1980s included five higher educational centers, five vocational training centers, and three tourist personnel training centers.[16] Other projects, partially funded by the World Bank's grant of $60 million dollars, included equipment for vocational facilities at forty high schools and the construction, equipment, and furnishings for ten combined vocational technical centers. It bears repeating that the legislative decree mandating the use of the demotic throughout the school systems obliterated the problems associated with *katharévousa* once and for all.

The course of *apohountopoíisi*, or "dejuntification," was especially exacting in the universities, as the government catered to students who, as might be expected, had become radicalized and now demanded a say—and power—in university affairs. In student elections held ten days before the general elections in November 1974, four-fifths of the students voted for left wing candidates from the new Panhellenic Socialist Movement (PASOK), from the communist parties, and from various Trotskyist and Maoist parties, a forecast indicating a core of strong opponents to Karamanlis's New Democracy Party. The dejuntification process had as a major goal the drumming out of the universities faculty members who had supported the junta. Generally, students were demanding their "acquired rights" for all they had suffered under the junta and all they had contributed toward its collapse. By fall 1975 students had pressed for the investigation and suspension of ninety-eight faculty members. As a consequence, a shortage of professors was causing a void in the lecture halls. Beyond investigating the faculty, the students called for changes in the curriculum and examination system. Many of their leaders were radicals affiliated with the Communist Party of the Exterior.[17] Since the collapse of the junta, political movements in the universities were on the rise, distracting the students from their

---

[16] "Education: The Problem of the Long Nail," *International Herald Tribune*, October 1978, 6s.

[17] Rodice and Wall, "The Gods Smiled," *Economist*, 23.

studies. Success rates on exams as low as 15 to 20 percent seemed to corroborate how distracted the students really were. Naturally this agitation filtered down to secondary schools, the farm school included, encouraging those younger students that they, too, should demand their rights, "acquired" or not.

Also gathered in the net of the dejuntification process were the junta's union leaders, who truly had been puppets of the dictatorship. The new, elected union leadership felt itself empowered; a fresh esprit animated these organizations. They were encouraged by the emerging PASOK party to demand their rights, a call heard also on the farm school campus. Legislation laying out the rights of labor unions now included the right to strike for wages and conditions of employment, activities that had been forbidden under the junta. While the government was drafting this legislation, labor unrest exploded for the first time since the return of democracy when some 150,000 workers marched on parliament to protest a provision that would have put a ban on wildcat strikes. Violence broke out when the workers met head-on with police; one death was reported. The incident showed the mettle of the unions. When the bill was passed, apparently it was less restrictive than originally drafted.

Greek people were entering a new and more sophisticated phase in their social and work relations. This change influenced how the farm school staff and workers defined themselves in the work world. A less provincial, more worldly Greece, was in the making.

Unrest in the country at times like these is understandable, although in many ways the Greek people were making an admirably smooth transition from dictatorship to democracy even as outside forces impinged on the course of Greek events. A grave economic recession struck the industrialized world in 1973. Inflation was rampant. Stagflation set in during 1974 and a second oil shock hit in 1979–1980. Unemployment in European Union countries doubled. In the midst of these broader problems, Greece's own economic woes were exacerbated by the Cyprus imbroglio. Tourists who usually spent billions of drachmas cruising the Aegean and lounging on the beaches each summer were scared off by tension between Greece and Turkey. Anxious over monetary problems as they watched the GNP drop in 1974 for the first time in twenty-five years, Greeks were confronted with large outlays for arms purchases as a deterrent to the Turks. Unemployment, an ugly specter that had diminished somewhat under the junta, rose once again to worry an already beleaguered people.[18]

Rare was the Greek institution or individual that was completely shielded from the general disorder. As it turned out, though, the farm school would weather this period rather well. The administration learned, sooner rather than later, to accept labor issues as part of the normal give and take of institutional life, and it promptly adjusted the rules and regulations of student life to the extent it deemed prudent.

---

[18] Ibid.

# CHAPTER 10

# THE FARM SCHOOL:
## DICTATORSHIP AND AFTERMATH

During the seven years of the junta and six-year aftermath when Greece was working to democratize itself under Karamanlis, the stress of these times was bound to take a toll on the American Farm School. A cluster of difficulties that surfaced for the school in its relations with government, and also on campus with students and staff, would play out over this period from 1967 to 1980. Like the country, the institution was being swept along in a swift current of modernization.

In the beginning, communication with the junta government loomed as a major hurdle. Although Bruce and his staff had persevered at fine-tuning relations within the ministry of agriculture under governments preceding the dictatorship, those precious lines of communication were snapped when the dictatorship restaffed the ministry from the top down with personnel loyal to the military regime. For the first time—at least since the end of the civil war, and probably since the Venizelos government of 1929—the school was left totally bereft of official friends. To confront this regime, a force threatening in its tactics and intent, the farm school would muster up an uncommon amount of patience, stamina, and courage, and no small measure of deftness.

At stake was the morale of the school. As Bruce puts it, "We devised a series of projects to help the ministry of agriculture beside our core educational curriculum for youths, such as short courses and the Community Development Program and others, all part of an effort to create projects that the government could take over and run itself. By distancing itself from us we felt the government was saying: We don't

believe in you, we don't trust you, we don't want you."[1] According to
Bruce's calculations, 50 percent of the institution's effort was directed
toward the school component, the remainder toward helping the ministry
with its projects. When the government revoked the school's tax
exemptions, a privilege extended since early times, Bruce not only felt
the financial squeeze, but the measure reinforced the school's low
morale.

Exacerbating relations with the government were the kinds of
deliberately heavy-handed harassment the junta levied on the institution.
Bruce held firm to the policy that the school would remain apolitical,
neutrality being the axis upon which the institution had revolved since its
founding. He adhered to that position even though he recognized he was
standing on the rim of a volcano when the government ordered its
totemic image—the risen phoenix, a large bird pictured against a
background of flames with a soldier standing guard—to be shown
prominently at all institutions. No matter the cost, in outright defiance of
the directive, Bruce postponed indefinitely the decision to exhibit the
dictators' symbol, which he regarded as a badge of political affiliation.[2]

Undeniably, the administration's delay was an act of defiance, a
dangerous script in a theater where people were frequently carted off and
detained with no justification. Beyond Bruce's immediate reaction lay
his strategy of taking an early stand against the colonels lest they
perceive the institution as lacking spine and therefore open to their
exploitation. In the end, the phoenix was never displayed on campus and
the school never suffered for it.

Another alarm to the school's independence sounded when the
government pointed a finger at the assistant to the director of education
and charged him with being an *ilistís* (materialist.). This veiled
accusation implied that the farm school employee was "politically
unreliable," an allegation that warranted his termination. Since the farm
school was private, the government by law could not itself dismiss the
accused faculty member. That left the school responsible for the sacking.
Again, responding to a spurious accusation would signal a capitulation to
government authority as well as collaboration in a political act. Bruce
and Charlie before him, irrespective of political considerations, had hired

---

[1] Bruce Lansdale, interview with author, Metamorphosis, 2 June 1999, written, AFS
Archives.
[2] Ibid.

excellent men, even people who had belonged to leftist organizations during the civil war, a mark of Cain that rendered a whole category of people unemployable until the liberalization of recent decades; yet no government had before demanded the firing of a school employee.[3]

To ease this taut situation, Bruce arranged through an intermediary to call on Brigadier Stylianos Pattakos, who as a member of the triumvirate stood at the pinnacle of government. The brigadier averred that the accused farm school employee's case was in the process of investigation. Hearing that, Bruce proposed that the employee be allowed to work at least until the investigation was completed. To this request the brigadier first demurred but after a few moments agreed. Time passed and presumably the investigation dragged on; but the farm school employee who had attracted the junta's enmity reached retirement age, took his pension, and left, more than a trifle scathed by the hazard that lay behind the accusation.[4]

During the colonels' sway, there were other points of tension. Once a government official arrived at the school demanding Bruce turn over the short course center and the school's workshops—when not in use by farm school students—to the government to train the adult students it planned to lodge on campus. Bruce explained to the official that installing mature students on the campus would disturb the moral environment that the school had painstakingly fostered through the years. Although on the face of it the official's proposition might seem tactless yet harmless, it smacked of confiscation. Presenting his case in the strongest way, Bruce pressed the official to find another solution. The official, apparently in light of Bruce's obstinacy and procrastination, failed to pursue the matter. "I learned that putting off an undesirable decision was a solution in itself," Bruce concluded.[5]

Running countercurrent to these harassments, the government was driven by a worthwhile ambition: to bring agricultural education up to a standard attained in Western Europe, where countries had succeeded in upgrading their agricultural sectors with the latest agricultural technology and management skills and in improving the quality of agricultural training. Fortunately, Evangelos Economidis, director of education in the

---

[3] Ibid.
[4] Bruce Lansdale, interview with author, Metamorphosis, 26 September 1999, written, AFS Archives.
[5] Ibid.

ministry of agriculture, became convinced that to work in tandem with the farm school, not against it, was in the best interests of Greece. Through him and others in the ministry, the farm school found a formula for cooperating with the government in a spirit of strict political neutrality: the goal of their joint endeavors, to serve the needs of agriculture and education, was respected on both sides.

Economidis admired the farm school's SEGE program. Since that curriculum concentrated on technological as well as practical farming, it seemed a direct answer to many of Greece's educational and agricultural needs. Considering the state of its morale, the American Farm School was heartened to play a meaningful role in Greek agricultural education and also to know that through Economidis it was extending its reach to other institutions. Encouraging, too, was the ministerial decree of 1973 whereby through financing by the World Bank's Second Educational Project for Greece, the school received the new short course training center, KEGE, a swine unit, a greenhouse, and funds for renovating the old dormitory toward the more mature SEGE students As an act of good will, the government agreed to pay 50 percent of the annual tuition for the SEGE students.[6]

In a dictatorship there is always scope for informers and probably almost all educational institutions during the 1967–1974 period were unwilling or unknowing hosts to these spies. The farm school no doubt had its fair share. The duplicitous actors were, Bruce thinks, "disgruntled members of the staff who found in the junta their moment to be heroic."[7] Whether their reports did any real harm is unclear. Bruce suspects, though, that informers may have been guilty of calling the government's attention to the administrator on the farm school's staff—the so-called "materialist." The informers hoped to forge a gap between the school and the government. Bruce's own shield was his reputation as a technocrat willing to cooperate toward aiding the rural population of the country. That Dr. House some fifty years earlier had been seized and jailed overnight by the Greek government when Greece went to war on the side of the Allies in World War I must have occurred more than once to Bruce. (Dr. House, it will be remembered, had worked in Bulgaria from

---

[6] Andonis Trimis, "My Relationship with the American Farm School of Thessaloniki, Greece Since 1954," 17 July 1977, 9, AFS Archives.

[7] Bruce Lansdale, interview with author, Metamorphosis, 19 May 1998, written, AFS Archives.

1872 until 1902, when he moved to Thessaloniki. The Greek government was suspicious that he might have been pro-Bulgarian, a dangerous posture since Bulgaria was an ally of the Central Powers.) But in both cases the principle of non-immunity was clearly operative and under a capricious junta government could have been reactivated to threaten Bruce.

Just as disturbing as threats from the dictatorship was the upsurge of anti-Americanism. The resentment that had festered in leftist circles after the civil war spread to other pockets during the junta and after 1974 continued to infect the population. Due to this strain, Bruce was placed in a relentless state of tension having to justify both Greek and American positions. He found himself persuading those Greeks who were antagonistic to the United States (those who were convinced the Americans had put the junta in power or at the very least were maintaining it) that he had no insider's knowledge of the State Department's policy and was not in favor of dictatorships. On the other side, he was at pains to explain to American donors that the Greek people themselves were still friends of the United States, that generally the Greek people did not uphold the concept of dictatorship and hence were worthy recipients of American largesse. An example of his equipoise: "A number of friends have arrived from the United States in a state of great distress, wondering how they would be treated. They have soon discovered that you shouldn't be too concerned about all you read in the newspapers. The Greeks have a wonderful way of separating people from politics. There have been incidents involving cars with US government license plates, and this expresses disappointment over the whole Cyprus matter. But at a personal level, they emphasize that their affection for the American people continues at a high level." He urged Americans to come to Greece "because there is a great need for personal contact—for Greeks and Americans to look each other in the eye and see that the affection they always held for each other has not changed."[8] Despite the anti-Americanism, friends of the American Farm School remained faithful and generous. In 1972, gifts to the school from donors in Greece amounted to $33,000 dollars, well over the mark of previous years, in addition to Greek government grants of $13,000.

---

[8] As quoted in *The Sower*, no. 83, 1974.

Curiously, Bruce, like other philhellenes who had spent years in Greece, was accused of being a CIA agent, often by those who knew him rather well, so far had mistrust taken possession of the Greek people. Tad, too, picked up her pen to give, as she put it, "a housewife's point of view." She wrote, "the School means many things to many people. Our trustee Stavros Constantinides says he feels at peace with the world when he comes inside the gate, while our other Thessaloniki trustee, Machi Seferdji says that now more than ever, the School must continue to play a leadership role in building Greek-American bridges."[9]

A more subtle form of anti-Americanism was reported by *New York Times* correspondent Steven V. Roberts from some remarks heard on a visit to campus: "Some tensions are unavoidable. The school recently received new milk processing equipment from the United States. A staff member, describing the students' reaction said, 'I was expecting real enthusiasm, but something about the fact that this stuff came from the States dampened their feelings. Since Cyprus they are more suspicious. They keep asking: "Why is this coming from the States? Why are they giving this to us?"'"[10]

Certainly the calf project mentioned earlier, when the farm school flew in American calves at the Greek government's expense, was an important project to demonstrate to the government the real possibility for both parties to interact in a professional manner without the farm school's becoming politically entangled. If we look at these accomplishments and others, we can conclude that despite occasional showdowns with the junta, and a generally tense atmosphere, the director and his staff kept the school faithful to its mission throughout this difficult period.

The return to democracy in 1974 made official contacts easier for the school. Personnel in the ministry of agriculture changed: pro-junta people were drummed out and replaced by technocrats. Among them, however, were officials who were galled at Americans, blaming them for the junta. Some had harbored a grudge against the United States since the early 1950s when Ambassador Peurifoy had meddled in Greek internal affairs. Flashes of tension between the United States and Greece were apparent when the Karamanlis government exhibited a posture of

---

[9] Tad Lansdale, "Farm Notes," 14 October 1974, AFS Archives.

[10] Steven V. Roberts, "U.S. Farm School Is Still Flourishing," *New York Times*, 6 April, 1975, 2.

independence not only as a matter of foreign policy but also to align policy with the sentiment of the people. Once again, the farm school's cultivation of the ministry of agriculture began in earnest to convince the officials that the institution would continue to help the Greek people. And with the rule of law once more in force, the threat of personal harm that obtained under the junta disappeared, enabling the school's director and staff to approach officials more directly.

As to be expected life had altered. If the seas were rough outside the campus, this once tranquil island with its happy, obedient boys and devoted staff was itself now storm-tossed. Keeping in mind that the ethos of the school had always been one of concord with the land and among the members of the farm school community, of respect for the faculty, of purposeful work, and of spiritual preoccupation, it is easy to comprehend how a student strike hit the administration with gale force.

In 1975, the second anniversary of the crushing of the Polytechnic revolt, when university students, still raging over the junta oppression, were demanding a larger voice in the governance of their institutions, and high school students in Thessaloniki were restive, the farm school students, too, were edgy. Merging with the regular student body on campus, as discussed earlier, were the seventeen scholarship students from Cyprus. Adolescent, embittered at the United States, and depressed over their situation, the Cypriots injected a note of rebellion into the tumultuous mood of the times. However, the farm school boys needed no prompting to proclaim their own unhappiness. They had taken stock of the alluring youth culture outside the campus—long hair, rollicking rock concerts, freedom to travel—and they resented the narrow limits imposed by the farm school on their personal freedom: the short hair, the march to church every Sunday, Bruce's moralistic talks to them every Sunday night, restriction to campus every weekend except for the one Sunday a month off, obligatory study hall, and room inspections. They griped, too, about the food, which they found scant and of poor quality. On another scale entirely, they railed against a matter that impinged directly on their future: the school was accredited only by the ministry of agriculture, which meant they could not be accepted directly into university.[11] The agitators were students in the SEGE program, among them mature young

---

[11] This information is based on a series of interviews by the author in Thessaloniki (from 1998 to 2001) with former students and staff: Nikolaos Papaconstantinou, Konstantinos Evangelou, Evangelos Vergos, Stathis Yianakakis, and John Borovilas.

men, some almost twenty years old, graduates of the previous program, who returned now to earn the new, more prestigious SEGE diploma. That the administration and faculty had apparently failed to come to grips with their age and ambitions is a crucial detail. And the school had not gauged the velocity of social modernization shaking Greece since the fall of the junta. As a visitor during the upheaval writes to George Post, chairman of the board of trustees, "...the older student needed a different level of instruction and a more mature relationship between teacher and learner."[12] The Cypriot students trailed along with the protestors, more or less in the role of fellow travelers.

Some SEGE students resented the intensity of the practical program and voiced their complaints, claiming their labor was a form of exploitation. According to Stathis Yiannakakis, then a student, the boys were stressed by the obligations imposed on them by the work program. Since church attendance was obligatory, the students could not enjoy a shred of leisure even on Sunday morning.[13]

Evangelos Vergos characterizes the student strikers as "inexperienced village boys, over our heads. We really knew nothing about how to strike or how to protest. We marched up and down the campus, just up and down, [and] we cut classes." On these occasions, although the students refused to go to class, most did continue to work in their practical program. Student labor, especially in the livestock area, was all-important to the school. In fact, on weekends during this problematic fall semester, the students were managing the animals on their own, being paid by the school for their work on Saturday and Sundays; their independent weekend management responsibilities were considered a vital component of their practical experience.[14]

At one point the administration invited the parents to a meeting on campus; the school explained to them an array of issues, a helpful orientation for confounded mothers and fathers, a few of whose boys had

---

[12] Alan Haas to George Post, Thessaloniki, 8 December 1975, as quoted in Trimis, "My Relationship," 31.

[13] Stathis Yiannakakis, interview with author, Thessaloniki, 29 October 2001, written, AFS Archives. Yiannakakis, a farm school graduate, received his degree from KATEE. At the farm school he began his career as a teacher and is now head of the poultry department.

[14] Evangelos Vergos, interview with author, Thessaloniki, 22 March 2000, written, AFS Archives.

actually quit the school. Vergos recalls the sign the students had erected over the front gate. "Father, They Are Exploiting Us!"[15]

In a fit of independence, eight students, including Stathis Yiannakakis, EvangelosVergos, and John Borovilas, found they could no longer resist the siren song of the "rock scene." One Sunday evening, they stole off to Thessaloniki, some of them to see a rock concert starring Cat Stevens. There they tarried, returning some two hours after curfew. The next day, they were summoned by Andonis Stambolides, who gave them a two-day suspension for tardiness. Since the students had spent their last drachmas on the concert, they were left with no money to join the high-life in Thessaloniki. They returned home for the hiatus.

To play hooky from Bruce's Sunday night talks was another way for students to display contempt for the enforced rituals of the school. He and Tad had made it a practice early in his tenure to come together every Sunday evening with the students. They would all read a portion of the Bible together in both Greek and English and sing a Greek hymn. Bruce would then deliver a talk always with the purpose of reinforcing the basic values of the school, a practice he continued until the early 1980s, when he switched to once a month. Before this era of social change, students generally had enjoyed the directors' presentations, the warmth it generated between them and their director, and the wisdom they felt they had received. But in the more liberated decade of the 1970s, the boys felt put upon.

A truly major complaint, one that had gnawed at students and parents for years, was the value of the farm school certificate. As a graduate of the school, a student could enter KATEE, without exams. Then it was possible for exceptional students after securing a diploma at KATEE to take exams and, if they excelled, to continue on to university. But that degree of success, obviously, was not attainable by ordinary farm school graduates, an issue that helped trigger the student strikes and one that peristed until the lyceum program was started in 1977. John Borovilas, a leader in the 1975 student strikes, claims that due to the extreme pressures exerted during the student revolt, the farm school hastened to grasp the opportunity in 1977 to become an equivalent lyceum.[16]

---

[15] Ibid.

[16]John Borovilas, interview with author, Thessaloniki, 15 October 2000, written, AFS Archives. Borovilas, a 1976 graduate of the farm school, was the third alumnus elected to

Stunned by the student upheaval but nevertheless treating the demonstrators with all seriousness, the administration opened an honest dialogue with them. The staff formed committees to study the students' grievances. Among the remedies were the following: students received permission to wear long hair, to travel off campus once a week after church on Sunday (not just once a month as before), and to have use of a meeting place set aside in the cafeteria. The food improved, although not to the extent they had hoped. When the new dormitory opened in 1977, the administration made adjustments in residential life to appeal to students.

Thus the student strikes and agitation on campus, which had erupted in the fall, were not long-lived or truly militant. By Christmas vacation 1975, the flareup had ended: no material damage was inflicted on the school. The revolt is memorable first because it was the only rupture in the social contract since the school's founding, and secondly because it brought about a more modern understanding of relations between the school and the student body on the part of the administration.

Like the student agitation, the formation of a labor union with its subsequent demands can best be interpreted, too, as a breach in the social contract and an abrupt crossing over into the modern world. Since the beginning faculty and staff, largely an undemanding, devoted lot, had held the director of the school in reverence. Dr. John Henry House, patriarchal, yet approachable, unquestionably inspired that kind of awe. Here was an educated man with a command of at least three languages. He gave off the aura of supreme confidence in contrast to many members of the staff who were unlettered subjects of a decayed Ottoman Empire and who teetered on the brink of unemployment. During Charlie's tenure (1929–1955), the gap had closed somewhat. His immediate staff, men like Theodoros Litsas and Eleftherios Theoharides (who taught English), recruited from the ranks of educated Asia Minor refugees, while they respected him, did not stand in awe of him; generally the relationship among them was collegial, although surely the rank and file were still humble. Yet loyalty to the school and its director was the all-encompassing value.

However with the return of democracy in 1974, the times were radically different. Labor unions in Greece had been interfered with

---

the farm school's board of trustees. He holds a degree from Cornell University and works in finance in New York.

historically by governments. Reacting to the new freedoms, pressures of inflation, and a more modern dynamic in the work place, the workers began to organize. The PASOK party, a growing force in Greek politics, was a catalyst bent on wooing rural communities and urban workers, encouraging them to voice their demands. Union militancy became a force in Greece (and Western Europe as well) in the 1970s. Muffled by the junta for seven years, workers at the school were eager to master their own fate. Not too long after the student rumblings ceased, the workers on 1 January 1976 organized themselves into a union. In the first instance, the union members' sole demand was to obtain from the school about 3 acres by the sea for their recreation, which the founder had at some point apparently designated for that purpose.[17] This demand was not met. Dimitris Litsas, then acting personnel officer, remembers the union members at that point as being "sincere, simple people with modest demands."[18] Interestingly enough, the leaders emerged from middle management and the lower ranks, and were mostly farm school graduates. The union had seventy to eighty members, including teachers. However, that number diminished when the teachers left after the first year to join their own national teachers union.

By 1977, the union on campus was better organized and able to shape its agenda more cogently. Inflation in Greece, despite its drop from 30 percent in 1973, was still double-digit, at 24 percent, presenting a grave financial hardship for the school. Naturally, the workers demanded higher salaries and better working conditions. At the time, since the American Farm School was non-profit and private, it was not required by law to comply with the collective agreements and was not represented in the national labor organization; nonetheless, farm school salaries were pegged at some 10 percent above the collective agreements. Actually, in 1975–1976 two increases were paid to the staff members in an effort to relieve some of their hardship. After 1981, when the PASOK

<hr/>

[17] Harilambros Grafiadelis, interview with author, Thessaloniki, 20 March 2000, written, AFS Archives. A farm school graduate, Grafiadelis was a leader of the campus union and works as a supervisor in the boarding department.

[18] Dimitris Litsas, interview with author, Thessaloniki, March 1999, written, AFS Archives. Brought up on campus (son of Theodoros and Chrysanthi Litsas) Dimitris Litsas has played a major role in the school's land management plan.

government came to power, the school's salaries became subject to the national collective agreement.[19]

While the union at the farm school was not disruptive, members did strike, as a result of the fiscal disarray. In the 1970s, farm school workers took part in general strikes off-campus and just once they halted work on campus for a couple of hours and protested silently. At the time, the strikers handed out press sheets outlining their grievances to the journalists who came to cover the story and the event made the local papers. Apparently their complaints were directed at the government rather than at the school.

The formation of the union on campus caused a gulf between management and staff, a divide that became formal on 2 June 1976 when the executive committee of the board of trustees resolved "that with formation of a union at the American Farm School, the board of trustees hereby establishes the principle that all executive personnel are ineligible for membership in the union and that any member of the staff presently holding executive responsibility who desires to join the union must relinquish that responsibility."[20]

In the midst of this critical period in the farm school's history, a series of significant administrative changes were undertaken at the highest level. Philosophically, Bruce had always held that the American Farm School was only as good as its highest-placed Greek administrator. This piece of inherited wisdom he received from Charlie, who had relied heavily on the wise counsel and striking leadership qualities of Theodoros Litsas, then the assistant director. Litsas had remained on with Bruce as associate director until his untimely death in 1963. Nor was it unheard of either, as noted by a former ambassador to Greece (1964–1967) Phillips Talbot, for some schools founded by Americans throughout the world to turn over the top position to a citizen of the country. "During the past generation American-related schools in the Far East have found it both wise and necessary to move to indigenous

---

[19] Archimedes Koulaouzides, interview with author, Thessaloniki, 21 March 2000, written, AFS Archives. Koulaouzides, a graduate of the kidnapped class (1949) received a bachelor's degree from California Polytechnic University at San Luis Obispo. He served as an instructor and as head of recruitment before taking the position of personnel manager until retirement in 1989.

[20] "Minutes of the Executive Committee of the Board of Trustees," 2 June 1976, AFS Archives.

leadership, with new and more distant arrangements for American backup."[21]

With the intention of grooming the first Greek director in the farm school's history the board of trustees had approved Andonis Trimis's appointment as assistant director in July 1968, and in 1971 he was promoted to associate director to run the school's academic programs. Ultimately, in 1974, he assumed the duties of director. As noted earlier, he first had gained his reputation and work experience as farm school coordinator of the Community Development Program of Thessaloniki Prefecture in 1958. When the junta came to power, realizing the dictatorship would take control of the project, Trimis terminated his association with Community Development. As farm school director, Trimis ran the institution's day-to-day operations, reporting to Bruce, who assumed, for a time, the title of president. Referring to the junta, Ambassador Talbot wrote, "Even in the current national condition and climate in Greece, this would be a very poor moment to withdraw American overall leadership. Those Greeks who best understand American concepts of intellectual and social liberality and who would be the natural leaders of these institutions have less leverage than would be the case in normal times. So the idea of Trimis as director makes an important acknowledgement of the forces of the future while Bruce's continuation as president reflects the realities of the present."[22]

In his proposal to the chairman of the board of trustees, Bruce described his own position: "As president I would want to keep responsibility for policy through budget and finance as well as the coordination for activities between the United States and Greece. I would hope to keep my contact with the students both through my weekly chapel meeting with them as well as my seminars with the freshmen and senior classes. I would hope, however, to undertake long-term planning and to work with the board of trustees, with our Greek committees, and with the Greek government in the development of our new program."[23] Reporting to Trimis in this restructuring were the heads of each division

---

[21] Phillips Talbot to George Post, Darien CT, 10 June 1974, in Trimis, "My Relationship," enclosure 36.

[22] Ibid.

[23] Bruce Lansdale to George Post, Thessaloniki, 14 May 1974 in Trimis, "My Relationship," 16.

with the exception of the business manager and the personnel manager, who reported to Bruce.

Trimis, then, under this structure, was at Bruce's side during the junta period, fending off the dictators' intrusions, conversing with the students as they protested, helping with the implementation of the SEGE program, helping cultivate donors, aiding in the planning of the lyceum curriculum, and acting in Bruce's stead when the president was off campus. Additionally he did a stint in the New York office to get the feel of that indispensable operation, which supports trustee activities and acts as the official farm school representative in the United States. It seems that the diarchy, to an extent, was successful. But in June 1977 Trimis was informed that there would be another restructuring of the administration. The board of trustees with Bruce's concurrence concluded that during this period of financial stress and major curriculum change leading to a whole new educational status for the farm school, Bruce should return to manage all aspects of the school's administration.

Problems from 1967 through the 1970s abounded. It was a period of agitation with tough issues that beat their way through almost all significant aspects of the school's operation. Piled onto the difficulties already enumerated was a plunge in enrollment lasting until 1977, due in part to rural families' lack of enthusiasm for agricultural education and their migration to the cities. "The enrollment dropped by two-thirds, and the cost per student, owing to inflation, doubled. Staff members became demoralized, trustees grew disillusioned, and students worried about their futures," as Bruce characterized the predicament.[24] Trimis, too, reported the situation:

> The spring of 1975 during the Trustees' meeting the question "What is the mission of the School?" became the major issue. This was to be expected because...the gears were shifting but many people were unaware of this shift. Many people did not even know what the word SEGE was all about. Trustees who were used to charming little boys from villages with a thank you in their smiling faces and a "welcome, welcome" at the AFS gate, suddenly became shocked meeting young men with the question in their faces... "Why are you Americans helping me?"

---

[24] Bruce M. Lansdale, *Master Farmer: Teaching Small Farmers Management* (Boulder CO: Westview Press, 1986) 53.

There were other important questions in the Trustees' meeting in 1975, such as why these technologically advanced agricultural units? Why this big budget? Why so much staff for so few students? Why were customary accounting methods not used? Why? Why? The why's became like a torrential rain that swept away all good which was accomplished during all the past years.[25]

When SEGE was dropped and TEL and TES curricula were instituted in 1977, these new programs, promising as they were, nevertheless spread the institution's resources thin as the faculty scrambled to develop teaching packages, classroom aids, and all sorts of supplementary materials to enrich the lessons and adjust the practical program to the lyceum's more theoretical course of study. According to Bruce, during the last half of the 1970s the situation degenerated because of "the failure of the director [Bruce], the faculty, and trustees to clarify its objectives."[26]

The trustees, criticizing the management of the school, intruded into areas that were traditionally the purview of the director. One of the stress points in the late 1970s was the financial management, a matter that drew the board of trustees' attention to the school's management systems. The members complained that the accounting system needed to be updated. This concern occasioned another change at the top: the appointment of Homer Lackey, an executive who had retired from Hellenic Steel, in Thessaloniki. He was charged with overhauling the accounting system and modernizing the business practices, a complicated assignment when one considers that the farm component was in itself an impressively large business, never mind the problem of tracking funds given to the school by the American government, the Greek government, global institutions such as the World Bank, and individual and corporate donors in the United States and Greece. Add to that such factors as the dollar/drachma exchange, bank accounts in the United States and Greece, and salaries for over 100 employees in Greece and a handful in the New York office, it becomes clear that the farm school truly resembled an intricate, international corporation demanding, but never able to afford, high-level

---

[25] Trimis, "My Relationship," 29.
[26] Lansdale, *Master Farmer*, 53.

financial executives to aid the director in managing its various components.

Lackey, who stayed a year and a half, set up a system whereby each operation became a cost center that was analyzed according to its profitability. The school's self-service store, for instance, opened in 1966 to draw the public onto campus to buy the school's farm products and other groceries was judged unprofitable and eventually closed.

After consultation with staff and friends in 1978, the board decided the time had come to outline a new mission statement. That reflection led to the commissioning of Professor Irwin Sanders "to analyze the objectives of the School in the context of the monumental changes in rural Greece. Throughout the whole process, board members, staff, and volunteers cemented a positive relationship. What had originally been a near calamity for the School proved to be the focal point of a new spirit of cooperation for the faculty, trustees, and other friends."[27] Sanders' recommendations—simply stated and wise—served as guideposts through the next decade.

To characterize the 1970s at the farm school simply as a period of turbulence can be misleading. Despite the challenges, if we look beyond the gloom we can see the gleam of creative innovation. For example in 1970 a person-to-person venture, still cherished by its alumni as a defining life experience, was created: Greek Summer. The experience continues in this new century: former participants are now sitting on the board of trustees and sending their progeny on Greek Summer to undergo, as they themselves had, the "metamorphosis," as Bruce aptly named it.

> What is Metamorphosis?
> Is it a Greek word
> An Aegean village
> A natural phenomenon
> A change of heart
> Is it divinely inspired
> Is it a transfiguration
> Of time, space, spirit?[28]

---

[27] Ibid., 51.
[28] Lansdale, *Metamorphosis,* 3.

Greek Summer, envisioned as a kind of Peace Corps service for American high school students, though it attracted other nationalities, still captures the participants, heart and soul. The program offers some forty teenaged boys and girls a five-week adventure in international living and service, along with travel and fun. Greek Summer started long before adventure tourism became the commonplace it is today. At the heart of the program is a work project in a northern Greek village located near the farm school in Central Macedonia, where the youngsters live with Greek families while completing the project. For example, one summer they paved a street to the church and school in a small village; another summer they built a concrete village square and worked in the fields, helping the farmers. As a ritual, boys and girls would spend the night before they departed for the village camping at the Lansdales' country house high on a cliff above the magical seacoast in the Halkidiki village of Metamorphosis. There, as dusk cast a lavender glow over the Aegean, Bruce would amuse them with his inimitable Hodja stories and inspire them to become close to the Greek people and attend to their wisdom. At the end of the work project, their diligent labor is rewarded with a climb to the summit of Mount Olympus. But the youngsters seek no reward: they are all transformed—that metamorphosis is recompense enough.

Another innovation, Summer Work Activities Program (SWAP), instituted in 1979, the creation of Jeff Lansdale (Bruce and Tad's son who lived the first sixteen years of his life on the farm school campus, and works in the field of rural education in Central America.) and David Acker (director of International Agricultural Programs at Iowa State University and now a farm school trustee) was tailored to college students. University and post-university age international students from developing countries as well as Europe and the United States served on the farm when the regular workers went on vacation. The students were introduced to Greek culture and agriculture as well as the Mediterranean environment. In 1997 the program was redesigned and, under the new name of Summer International Internship Program, integrated into the curriculum of the Dimitris Perrotis College of Agricultural Studies.

In fall 1979, if the farm school felt any qualms about its worth to the Greek people, those uncertainties were set to rest by the multitude of Greek friends who came to campus to celebrate its seventy-fifth birthday. President of Greece Konstantinos Tsatsos, accompanied by his wife

Ioanna Tsatsos, sister of the Nobel Laureate George Seferis and poet in her own right, honored the school with his presence and in his impromptu speech—"I didn't come here with the intention of speaking, but this visit is so challenging I want to." He went on to emphasize the relationship between the American and Greek people. "The educational work that is being done here with the help of the United States is of great importance.... The Greeks do not forget, even during hard times, the many things that have been done with American help. The possibilities of educational cooperation are endless...."[29]

---

[29] As quoted in *The Sower*, no. 98, 1980.

# CHAPTER 11

# THE PAPANDREOU YEARS:
# A DIFFERENT DRUM

When measuring Greece's momentum decade by decade from the early 1950s to the end of the 1980s, we are struck by the country's transformation. A glance at the figures indicating the rise of per capita income—$300 in 1950 to $5,340 in 1989—suggests a straight line of economic advancement. But, as related in previous chapters, the forward motion resembled, rather, a jagged trajectory. Historian Mark Mazower writes, "From 1950 onward, Greece was at peace but it was a strange, strained peace guarded by what was formally a democratic order, but held in place by repression, persecution of the Left, and armed violence on the fringes of society. It was, arguably, not until the anticommunist Right was itself discredited with the fall of the junta in 1974 that the country could return to some semblance of tranquility."[1] Yet, for all the forward movement occurring in the 1980s, Greece was beset by continuing malaise: the decade was marred by a bundle of economic troubles, squabbles with other countries, and conflicting domestic and foreign policies.

The drumbeat to which the country and the farm school marched in the 1980s was orchestrated by Andreas Papandreou, who founded PASOK (the Panhellenic Socialist Movement); he set the rhythm for modernizing Greek politics and society. The first-ever leftist party to be voted into power, PASOK, through its philosophy and organization,

---

[1] Mark Mazower, ed., *After the War Was Over: Reconstructing the Family, Nation, and State in Greece, 1943–1960* (Princeton and Oxford: Princeton University Press, 2000) 7.

offered an outline for guiding Greek politics as the country approached
the twenty-first century. Papandreou dominated the decade, having
garnered majorities in the 1981 and 1985 elections, victories that
afforded him an unrivaled opportunity to drive home his profoundly held
antagonism toward the United States.

To the discomfort of some, particularly the American-chartered
institutions in Greece, the thrum of anti-Americanism was heard at
varying volume until 1989, although after 1985 Papandreou did soften
his rhetoric significantly. This antipathy especially affected the private
school sector. Papandreou's professed ideal of equal opportunity for all
youngsters was patently incompatible with the existence of private
schools, which in 1989 educated just under 10 percent of primary and
secondary students. And he regarded foreign-chartered institutions as
anathema. His aversion implied the shutting down of such American-
chartered schools as Pierce College in Athens, Anatolia College in
Thessaloniki, and of course the American Farm School, placing these
schools for a time under this sword of Damocles.[2] Although the threat of
closure set these excellent institutions trembling, once again the farm
school derived a measure of comfort from the high regard it enjoyed
among the Greek people and those stalwart officials who were still
proclaiming that the American Farm School was really *dikiá mas* (ours).
Despite its negative stance toward private schools, ultimately the
socialist government tolerated them.

A source of relief to the farm school was the government's priority
concerning education, which the socialists placed at the head of their
agenda: within the first four years of governance, they increased the
education budget by 50 percent. Their aim was to draw students away
from university and entice them into vocational technical junior colleges
to provide trained people for social services, technological fields, and
middle level management—personnel whom the country sorely lacked.

---

[2] The elite Athens College, located in the capital, founded along American lines with
an American president, and incidentally Andreas Papandreou's preparatory school, was
not actually chartered in the United States, but fell legally under the penumbra of a
private institution. Anatolia College, a farm school neighbor in Thessaloniki, once a
secondary school, has added a four-year liberal arts college, the American College of
Thessaloniki. Pierce College has evolved into the American College of Greece, an
institution comprising Pierce College, Junior College, and Deree College.

The socialists began by appraising existing facilities and curricula: they bemoaned the state of vocational education in Greece.

A most significant reform concerned the country's post-secondary technical schools such as the TEI (a replacement for the older KATEE); these junior colleges happily accepted graduates from the American Farm School's TEL. Between 1980 and 1989, encouraged by the ministry of education, the number of students matriculating in the TEI system tripled, as did government funds to support the larger enrollment. Among the most highly qualified applicants were optimistic farm school alumni, who entered the TEI with expectations of employment after graduation. An unfortunate outcome was that Greece's limited economy could not absorb the bulk of the TEI graduates, who in their frustration proceeded to compete with university graduates for positions in the civil service and to form groups to force the ministry of education to upgrade the TEI diploma to university level.[3]

Other reforms in the field of education also affected the school. To make high school education accessible to all, in fall 1981 the ministry of education abolished entrance exams to the lyceum, permitting all gymnasium graduates to continue on to lyceum. As a result of the new legislation, enrollment in the lyceum increased 30 percent throughout the country. Similarly, the number of applications to the farm school's TEL increased substantially, but at the same time the number decreased for those seeking entrance to the TES. The TES progam fit perfectly the school's basic mission of educating youngsters in practical ways, a curriculum and ethos bent on directing the graduates back to the land. Although the farm school took pride in the success of its lyceum, despite its variation on the mission theme, at the same time it strove to reinvigorate the TES in order to remain close to its raison d'etre. Plainly, the entrance for all into the TEL potentially opened the floodgates to university, the very current the government was trying to stem. This irony is neatly conveyed in a Hodja story. While Hodja faces one way

---

[3] Constantine Tsoucalas and Roy Panagiotopoulou, "Education in Socialist Greece: Between Modernization and Democratization," in Theodore C. Kariotis, ed., *The Greek Socialist Experiment: Papandreou's Greece, 1981–1989* (New York: Pella Publishing Company, 1992) 328–30. C. A. Karmas, A. G. Kostakis, and Thalia Dragonas, *Occupation and Educational Demands of Lyceum Students: Development Over Time* (Athens: Centre of Planning and Economic Research, 1990) offers a wealth of information on the gymnasium, lyceum, technical junior colleges and draws a connection from one to the other.

astride his donkey, the donkey faces the opposite way. Hodja explains the contradiction thus: "My friend here wanted to go one way and I wanted to go the other, so we are compromising."[4]

In the countryside, farm school students were witnessing subtle but nonetheless key changes in the general outlook. According to one observer, the rural areas had been held in "the tight-fisted control exercised by the local gendarmerie since the civil war."[5] The rural population's widespread support of PASOK indicated a loosening of the right-wing grip; in 1981, for the first time country people voted openly for the left. Once in power, PASOK continued to court the rural population by wiping out police files and doing away with the services of informers. The government saw to it that the farmers received credits, a large portion of which was allocated via cooperatives. Toward the end of the 1980s, 940,000 members belonged to 7,300 cooperatives.[6]

In a nation where parties (except for the communist party) had historically been formed in parliament under the patronage of stodgy political elites, PASOK's grass roots power represented modernity itself. Here were activists, often young graduates of American or Western European universities, moving out into the countryside and provincial towns to paste up their green and white posters, establish party branches, and enlist the participation of local people. The alluring media campaign, modern agenda, and energetic activists with their rallying cry, "*Allaghí! allaghí!*" (change) were a clarion call to broad elements of the population that had traditionally felt excluded from the centers of power. Suddenly all were called, an inclusion especially appreciated by people living in the provinces, 27 percent of whom in 1981 were still working in agriculture. The old Greece of antiquated political parties out of touch with their own people and the rest of Europe was being eclipsed. Among

---

[4] For this Hodja story, see Bruce M. Lansdale, *Master Farmer: Teaching Small Farmers Management* (Boulder CO: Westview Press, 1986) 80.

[5] James Petras, "Greek Socialism: Walking the Tightrope," *Journal of the Hellenic Diaspora* 9/1 (1982): 7.

[6] Foreign Area Studies, *Area Handbook for Greece* (Washington, DC: US Government Printing Office, 1994) 178–79. A comparison with other European Union countries is revealing. In Western Europe, the flow of products through cooperatives averaged 50 percent as compared to 10 percent in Greece in 1984. For more information on cooperatives, see *European Files* (Brussels: Commission of the European Union, 1992). In the case of Greece, most products passed through middlemen to the detriment of the farmer.

country folk a new sense of possibility and empowerment was afoot. The link to Western European countries with their tradition of real socialist parties (although the PASOK brand of socialism did not necessarily tally with mainstream European socialism) and full accession to the European Union in 1981 reinforced the mood of connection, a sentiment that came into full bloom at the end of the 1980s. In fact, three scholars remark, "it is fascinating to see the supportive attitudes of Greek youth towards European integration superseding those of their counterparts in other countries."[7] What future Greek farmers—garbed in blue jeans, humming the latest European or American hit songs, and living in a nascent consumer society—could possibly find their own images reflected any longer in those pastoral paintings of Millet.

In foreign relations, Papandreou's early premise was that the United States's rude intrusion into Greek government affairs after World War II, Greece's membership in NATO, and the installation of American military bases on Greek soil for use as a bridgehead against the Soviet bloc, the Middle East, and Africa made Greece a dependent country, its national sovereignty curtailed. He saw NATO forces as "an army of occupation."[8] Inflaming and exploiting the already strong anti-American sentiment in Greece, Papandreou kept American-chartered institutions on edge.

On the economic front, Papandreou was struggling with a full-blown crisis. Inflation had characterized the last two years of Karamanlis's government, but by 1985 it had skyrocketed to 25 percent while unemployment had reached between 9 and 10 percent. In 1983 the drachma was devalued more than 15 percent against the dollar. Adding to these woes, the service sector absorbed large amounts of investment and did not contribute concomitantly to the GDP. Moreover, exports receded while imports flooded the Greek market. When, owing to foreign

---

[7] Georgia Kontogiannopoulou-Polydorides, Mary Kotoulla, and Kelly Dimopoulou, "Citizen Education: Silencing Crucial Issues," *Journal of Modern Greek Studies* 18/2 (2000): 291.

[8] Angelos Elephantis, "PASOK and the Elections of 1977: The Rise of the Populist Movement," in Howard R. Penniman, ed., *Greece at the Polls: The National Elections of 1974 and 1977* (Washington, DC: American Enterprise Institute for Public Policy Research, 1981) 111. Also informative is Iatrides, "The United States and Greece" in *Greece and the New Balkans: Challenges and Opportunities*, ed. Van Coufoudakis et al (New York: Pella Publishing Company, 1999) 265–94, and the earlier *Greek American Relations: A Critical Review*.

loans and deficits, the debt burden increased precipitously in 1985 and the trade deficit reached a level of more than $3 billion, public sector borrowing leapt from 12.5 percent of the GDP in 1983 to 17.5 percent in 1985. The situation was truly cause for alarm.[9] With a deficit running at 8 percent of the GNP and a fall of real wages reaching 12 percent by 1987, Papandreou introduced the Stabilization Plan, masterminded by Kostas Simitis, the future prime minister. The Stabilization Plan was submitted to the European Union with a request for financial support. By the end of October 1985, the European Union agreed to the plan and granted a loan of nearly $1.75 billion. In the midst of this disordered economy, the farm school struggled to steady its budget.

During his first government (1981–1985), the prime minister worked ceaselessly to stir up agitation through confusing and antagonistic policies. He called for dismantling all United States military facilities in Greece and provocatively denounced NATO and the European Union. Despite his fuming against the United States, in 1983 his government did sign a new DECA (Defense and Economic Cooperation Agreement) to maintain the American military bases for another five years, a decision bound to confuse onlookers. Papandreou quarreled with NATO allies, insisting that the threat to Greece came from Turkey (a NATO member), not from the Soviet bloc on its northern borders, and pressed NATO for a security guarantee against future Turkish aggression, a demand that was ignored.

When PASOK came to power in October 1981, Greece's full accession to the European Union had occurred on 1 January of that year. When the accession treaty was ratified by the previous New Democracy government, PASOK and the communists had walked out in protest. Vociferous anti-Americanism was an old saga, and wrangling with NATO dated back to the Turkish invasion of Cyprus in 1974, but Papandreou's opposition to European Union membership, rooted deeply in the party's philosophy, injected troubling and serious contradictions into Greece's foreign policy. Membership in a powerful European organization ran counter to PASOK's fundamental principles. At its birth in September 1974, the party declared itself, "A movement of the working and underprivileged people" and set as its main target "the political and military independence of Greece and the creation of a state

---

[9] James Petras, "The Contradictions of Greek Socialism," 110.

free from control or influence by the economic oligarchy." The party saw Greece's European Union accession as a movement that would "consolidate the peripheral role of the country as a satellite in the capitalist system; ...render national planning impossible; ...seriously threaten Greek industry; and...lead to the extinction of the Greek farmers."[10]

No forecast could have been more inaccurate; Greek farmers would benefit enormously throughout the 1980s. After thirteen years in the European Union, Greece had received benefits worth $6 billion a year. By 1995, the per capita income had risen to $10,981, almost double the 1989 number.[11] Regarding European Union membership, the party failed to mention the prestige, stability, and monetary assistance to the country, all of which would lend the nation a stature that it had not previously enjoyed.

The year 1985, then, marked a turning point in the government's foreign policy. There is little doubt that the economic crisis was a catalyst for the rapprochement. The government ceased bickering with the European Union on such issues as changing Greece's terms of accession and demanding a wide range of special privileges and began to normalize relations with the United States. Suddenly, the leadership switched to a more expedient strategy of seeking advantages rather than continuing the unprofitable ploy of "bad boy," and, at least on a formal level, toned down the volume on its disputes with NATO. However cantankerous Papandreou's approach to the European Union, his handling of European Union negotiations was admired by farmers since it produced a generous price support system for Greek products and liberal funding for social and regional development subsidies.

For Greek agriculture, the advantages of cooperating with the European Union were colossal. By the mid-1990s, transfers exceeding $17 billion of the total $78 billion in benefits from the European Union went to price supports for agricultural products and to the modernization of the agricultural infrastructure. In particular, farmers in Northern Greece were big winners because generous price support was given for cotton, tobacco, and rice, crops grown widely in Eastern, Central and

---

[10] As quoted in Panos Kazakos, "Socialist Attitudes toward European Integration in the Eighties," in *Greek Socialism*, 261.

[11] P. C. Ioakimidis, "Greece, the European Union and Southeastern Europe: Past Failures and Future Prospects," in *New Balkans*, 171.

Western Macedonia. The benefits were observable. According to the *Financial Times Survey*, annual growth there reached 2.3 percent, as compared to an average 1.7 percent throughout the rest of the country.[12] Another indicator of the prosperity in Northern Greece in the 1980s was the unemployment estimate: the national average was 9 percent while in Northern Greece it was only 6 or 7 percent. If anything possibly could, these results augured well for slowing the rate of migration from the countryside to congested Athens and burgeoning Thessaloniki. While the Greek standard of living did not bear comparison to that of the rest of Europe, it was elevated to a height that meant that farm school students, especially those from Northern Greece were no longer plodding to campus from utterly impoverished circumstances.

The European Union adopted the Delors I package in 1988; it remained in effect until 1992. Through Delors I, structural funds were transferred to Greece in the amount of $8 billion. To understand how important the agricultural sector was to the Greek economy, it is worthwhile to note that in 1992, for example, support funds granted by the European Union for Greek agriculture amounted to 62 percent of all funds received by Greece. Another program, CAP (Common Agricultural Policy), aimed at maintaining an equitable income for farmers and keeping agricultural prices at reasonable levels by controlling the prices of agricultural goods, limiting production, and giving subsidies (cash grants) to farmers.[13]

Although improvements in infrastructure did indeed ensue, critics claim that a preponderance of the money went to price supports rather than to infrastructure much to the detriment of the development of Greece's agriculture.[14] Delors II (1993–1999) the subsequent grant package, emphasizing education and training, transferred over $30 billion to Greece. During those years the European Union was taking cognizance of Greece's economic plight and studying ways to reform its system of transfers to help people in the poorest regions, areas where the population had a per capita income below 70 percent of the rest of the

---

[12] Karen Hope, "Survey of Northern Greece," *Financial Times*, 4 November 1992, 3.

[13] Roy C. Macrides, ed., *Modern Political Systems Europe* (Englewood Cliffs, New Jersey: Prentice Hall, 1987) 360. By 1994, income and price supports were almost all eliminated by restrictions enacted by CAP according to *Area Handbook, 1994*, 178. In the late 1990s, the total European Union budget was $100 billion.

[14] George Daoutopoulos to the author, Thessaloniki, 28 August 2001, AFS Archives.

member states of the European Union. A report stated that the "ten most prosperous regions within the organization were three times as rich and invested three times as much in their basic economic fabric as the ten poorest regions in Greece and Portugal."[15]

With these disadvantages in mind, the European Union in 1986 adopted the Single European Act. The Single Market, as it was called, committed its members to a series of social, technological, and environment obligations. The expectation was to establish by 1992 a single market. It is a measure of the European Union's success that by 1989 the organization's aid had stimulated a 2.6 percent increase in Greece's GDP.

As a recipient of these funds, it behooved Greece to modernize its policies to insure efficient use of these transfers and to comply with the European Union's regulations. For instance, certain requirements such as the strengthening of regional areas, making them more independent of the central government, represented a radical modification for Greece, which ranked among the most highly centralized countries in Europe.

In the agricultural domain, Greece rose to a new level of sophistication, planning what and how much to produce, working within the framework of the Single Market. As for individual farmers, plainly they were pressed to synchronize their enterprises with the total plan for the good of the country, to say nothing of their own self-interest, since their cooperation meant substantial benefits in subsidies. Such coordination presupposed an educated, informed rural population; farmers would succeed or fail on how they gathered and interpreted information and applied management principles. Greece, it was clear, would need to push forward into the information age. To contribute to this effort, as stated earlier, the farm school was among the first secondary schools, if not the very first, to introduce computers in its classrooms and on the farm, and thus gained distinction as a pilot institution for harnessing information technology to the field of agriculture. In addition, management principles were being taught at the school in both classrooms and in the fields.

---

[15] *European Files* (Brussels: Commission of the European Union, 1992). Under the Third Community Support Framework (CSF) 2002–2006 Greece is slated, according to *Kathimerini*, to receive an average per year of some $5 billion, net of its own payments into the European Union budget. *Kathimerini*, 17 June 2002, 5.

To help farmers cope in the global business environment, the European Union created two programs. Regulation 159 called upon farmers to join with extension agents to draft detailed studies of each farming unit. The presentation of these studies entitled farmers to obtain long-term subsidized loans that they could use for land improvement, investment in fixed assets, and the purchase of livestock. Also, to streamline their operations, farmers received subsidies for keeping accounts, a management tool that was not commonly employed, especially by older farmers. The other, regulation 160, provided advisers to farmers for socioeconomic affairs. The advisers regularly counseled the unproductive, usually older farmers to retire or find another means of livelihood, the purpose being to shift their land to more productive, vigorous people and to enlarge holdings, a strategy promoted by the European Union in the interest of efficiency. Sensitive to the psychology of farmers, Bruce points out that "this is one of the most controversial aspects of the farm modernization scheme. People question the right of outsiders to advise any farmer to relinquish the joy and stimulation of farming for less satisfying work elsewhere."[16]

The country was modernizing on many fronts. In the area of media, the arrival of private radio stations and television channels after the 1986 municipal elections, first in Athens, Thessaloniki, and Piraeus, and later in other cities, brought the fresh sound of non-regulated news, social commentary, and entertainment to an elated public; the deregulation was due, in part, to European Union policies and regulations. The Greek people, historically curious, took readily to television. Commercial television regarded audiences as consumers who would be drawn to lively subjects rather than tedious announcements dwelling on the machinations of party politics. At the same time, newspapers continued to proliferate, the Greek people being avid readers of dailies.[17] Now, students arriving from the villages to attend the farm school were savvy, well aware of the bustle outside Greece and were taking their first steps, via the enlivening media, into the information age.

---

[16] Lansdale, *Master Farmer*, 106.

[17] Ministry of Press and Mass Media, Secretariat General Information, *About Greece* (Athens: n.p., 1999) 342; Manolis Paraschos, "Is Greek Media at the Threshold of an Historical Change?" *Emphasis: A Journal of Hellenic Issues* 1/1 (1995): 94–100 discusses the development of the media in Greece.

At the dawn of the socialist culture in the early 1980s, labor unions were working in harmony with the government, but that spirit of comity proved short-lived as the country slid into hard times, especially during the Stabilization Plan. By 1985, PASOK had begun to pack the unions with its own people, essentially bringing labor organizations to heel. Try as it would, though, the government could not prevent a strike supported by all major unions in 1987. Throughout this whole period, the school's labor organization, formed in that season of discontent in 1976, remained active, but preferred negotiation to striking.

If PASOK's political philosophy and election campaign of earlier years represented genuinely modern approaches to life, the beginning of the second Papandreou government in 1985 showed unmistakable signs of regression. The civil service became bloated and scores of debt-ridden companies, supported by the government, drained the country's coffers. Seeking popularity in anticipation of the 1989 election, Papandreou had abandoned the Stability Program in 1987; public borrowing surged from 13.5 percent of the GNP to 21.8 percent by 1989. In that year, Papandreou, tainted by personal scandal and charges of public corruption, saw his party voted out of office. As the result of elections, the government reverted to the New Democracy Party, with Konstantinos Mitsotakis as prime minister. Konstantinos Karamanlis (who served as president from 1990 to 1995) was succeeded by Kostis Stephanopoulos.

With Andreas Papandreou again as prime minister, PASOK returned to power in the fall of 1993. As tempestuous as the party had been in earlier years, with time it would mellow until finally in the hands of Prime Minister Kostas Simitis, who became prime minister in 1996, the Greek people, their country at peace, their democracy secure, their economy improving, confidently crossed the threshold into yet another millennium.

# CHAPTER 12

# THE FARM SCHOOL IN THE EIGHTIES: A FASTER BEAT

The miasma that had engulfed the school in the 1970s, by 1980 had lifted. Beside student strikes, labor unrest, and friction between administration and board, all of which had clouded the latter half of the 1970s, the school had also reached an uncharacteristic stasis. First, the Sanders Report, a document evaluating the school and suggesting a set of actions, was in progress until fall 1977, hampering the director from mapping a long-term strategy until its completion. Crucial to planning were the government's imminent educational reforms: the TEL, the first lyceum in the school's history implemented in the fall of 1977, promised a radical departure from its traditional curriculum. No wonder the administration and faculty paused, perplexed as to how their rural clientele would adjust to the new program of study. Another cause for the stasis was the government's reorganizing agricultural price structure and policy, as well as examining labor costs, which caused the farm school to rethink agricultural strategy concerning its own farm. Adding one more cause for paralysis, the land management program was put on ice pending details of the new educational program, since the land plan had to be handled within the educational context. In 1977, the farm school's usual dynamism had been immobilized by all of these factors.[1]

---

[1] Bruce Lansdale, "An Introduction to the Briefing Papers," trustees meeting, May 1977, AFS Archives. For the management of its properties, the school began a series of studies in 1975. For a summary of the school's land development process, see "Report on the Development Potential of the Core Property Holdings of the American Farm School,"

By 1981 these major internal issues had been at least clarified and to a large extent resolved. The school then moved forward at a speed reflecting the compelling demands placed on Greece to adapt to a global environment, mindful of the competition, political, legal, and financial organization, and advanced technology that membership in the European Union implied.

The Sanders Report appeared in October 1977, laying out specific step-by-step recommendations that all parties welcomed and used for guidelines.[2] Professor Irwin Sanders' advice was always highly respected. Most importantly, he championed the advantages of coeducation, which led directly to the acceptance of girls in 1978, and guided the school toward implementing the TEL (technical vocational lyceum) and TES (technical school) programs. He sorted out the conundrum of using the farm for both production and education, urging a balance, clarified the benefit of teaching English language in a Greek agricultural school, and underlined the need for audiovisual teaching techniques, all burning subjects of the moment. Moreover, Sanders, envisioning a Greece with a global perspective as the country moved toward full European Union integration, suggested that the staff create a plan for making the school a more truly international center.

As for the educational reforms, the TEL, a success by any measure, worked as a magnet for motivated boys and girls, girls accounting for a third of the enrollment by 1990. The issue of the farm's financial status began to improve, thanks in part to the on-the-spot presence of Homer Lackey, who successfully applied his organizational and financial acumen for eighteen months before his departure in February 1979. Farm productivity was substantially aided by the production-minded veterinarian, Alexandros Michaelides, who was promoted from acting farm manager to farm manager in 1979. Finally, taking their cue from the TEL and TES curricula, members of the land management committee returned to planning, fixing their eyes on booming Thessaloniki, its supermarkets and roads extending around the school grounds. Pushing

---

1993, AFS Archives, and "A Proposal for the Effective Management of the A.F.S. Real Estate Assets," September 1998, AFS Archives.

[2] Irwin T. Sanders with the assistance of Basil Moussouros, "Future Directions for the American Farm School: A Report Prepared for the Board of Trustees," October 1977, AFS Archives. The report is of lasting historical value for the sociological context that Professor Sanders included to anchor his guidelines.

the metropolis inexorably outward, especially in the direction of the airport just past the farm school's doorstep, the city's population surged to 858,661 by 1981, up from 710,352 in 1971 (and to over a million in 1990). Hefty increases in land values were an inevitable outcome of this expansion. The school's real estate, redundant pieces of which the trustees began selling, substantially aided the school's finances. According to trustee Pavlos Condellis, "The sale of the land (the old Girls School site and building) to the German School and the Children's Home has provided the cash for the purchase of the 67.5 acres of land next to the 125 acres which Dimitris Zannas is giving the School. We paid for the 67.5 acres with the sale of the approximately five acres."[3]

In a spirit of euphoria, Bruce reported to the board of trustees in 1981, "Four years ago, we had an enrollment of 79 students compared to today's 200. The staff was demoralized to a point of wondering if there was a future for the School compared to the enthusiastic, optimistic teachers of today. At that time, production-demonstration units were barely breaking even and many were losing money. This year we anticipate a return in excess of $30,000."[4]

During Bruce's last ten years as director of the American Farm School (he retired in 1990) the institution was indeed flourishing. "Last year," he reported, "90 percent of our students who took exams for higher schools, passed. Eight-three percent of the graduating class were accepted into TEI compared to 30 percent of vocational graduates across the country."[5] Pavlos Condellis concluded that the school was in "the best shape it has been for many years."[6]

As usual, at least since 1967, there were stiff challenges issuing from the government. Rapport existed with PASOK, certainly, but on a lesser scale than with earlier governments; generally, the socialists were reluctant to offer wholehearted support as the following statement by

---

[3] "Report of the Resident Trustee Committee," 12 February 1981, AFS Archives. Pavlos Condellis, longtime head of the resident trustee committee and general manager of Pavlos J. Condellis, S.A., has been a generous donor of agricultural machinery, vehicles and other items over decades. Trustees who reside in Greece constitute the resident trustee committee and have the power to act in behalf of the board for some operational issues affecting the school directly. In 1981 the resident trustees were Pavlos Condellis, Konstantinos Kapsakis, George Legakis, Stavros Constantinidis, and Dimitris Zannas.

[4] "Director's Report to the Board of Trustees," June 1981, AFS Archives.

[5] "Director's Report to the Board of Trustees," October 1983, AFS Archives.

[6] "Director's Report, Board of Trustees Meeting," June 1982, AFS Archives.

Bruce reveals: "Although we have not had visits from top officials to the School, we maintain a good working relationship with the ministries. At graduation and at recent meetings in Athens, the Ministry of Agriculture attended in full force, although no party officials were present."[7] American-born Margaret Papandreou, Andreas Papandreou's influential former wife, who was intimately concerned with social issues, did not visit the campus until 21 May 2001. Papandreou, although he served a total of eleven years in Greece's top political office, was educated in the United States and held American citizenship for some twenty years, never visited the campus. Tad described the fanfare when the school had celebrated its seventy-fifth anniversary in 1979: "Students, staff, and friends gathered to greet President and Mrs.Tsatsos. Greek flags lined the road, which was solid with clapping students. A succession of black limousines disgorged their passengers at the church: ministers, members of parliament, nomarchs, the American ambassador, the agricultural attaché, slews of top-ranking military, legions of secret service men, hordes of frenetic photographers and TV cameramen."[8] The contrasting inattention in 1981 just two years later was not born solely of anti-Americanism. The brusque winds of modernization were starting to blow away the sense of obligation; the provincial family-like rituals and displays of gratitude the small Balkan state had historically displayed toward nationals of the "protecting power" were evaporating. Greece was gaining a sense of itself and no longer a client, rid at last of foreign patrons.

Some of the older sentiments and customs would linger. A particular instance: in 1987, the Greek government decided to demonstrate its gratitude afresh by awarding Bruce the Commander of the Order of Honour, through the good offices of the minister of Northern Greece, Ioannis Papadopoulos. The award, the highest given by Greece to a private foreign citizen, recognized Bruce's contribution to Greek agriculture and agricultural education during the span of his directorship. The medal was a sign, remarked Papadopoulos, "of the love that is here for Bruce Lansdale."[9] Bruce responded with his customary blend of eloquence and humility: he had come to Greece to be a teacher, but "finally I realized that it was I who was being taught. I came to help

---

[7] "Director's Report to the Board of Trustees," October 1983, AFS Archives.
[8] Tad Lansdale to Ann Kellogg House, quoted in *The Sower*, no 98, 1980.
[9] Ioannis Papadopoulos, as quoted in *The Sower*, no. 122, 1987.

others, but it was I who was helped. I came to share my skills but others taught me theirs."[10] Present at the ceremony were the American ambassador to Greece, the deputy minister of agriculture, the commanding general of the Third Army Corps, the local bishop, and several members of parliament.

Bruce, his sensors out not only to official relationships but to every nuance in Greek-American relations, wrote in 1983, "Reports of anti-Americanism in Greece must be of concern to both Greek and American trustees. It is heartening to have reassurances from so many that none of the feeling is against the School, which represents the American people, but rather against the American government's policy primarily because of the general support of Turkey by the State Department, particularly on the Cyprus issue."[11]

If the heads of ministries were now and then dismissive and a handful of prominent government officials were keeping their distance, nevertheless financial support from the ministry of agriculture was still forthcoming: in 1983, the ministry paid for 50 percent of the boarding expenses and 25 percent of training expenses. In citing good omens, Bruce noted that the government aired a half-hour TV program focusing on the farm school.

Another positive note: at one point in PASOK's first months in office in 1981, the ministry of agriculture had decided to run the on-campus KEGE independently of the farm school. However, in fall 1983, the government reversed its decision and requested the farm school to manage the center, feed its trainees, oversee its extracurricular activities, and supplement the ministry of agriculture's instructors, all functions the farm school had handled according to an agreement made in 1949 and enduring until 1981.The school rejoiced at the government's turnabout since the on-campus KEGE served a particular purpose for the school by opening a window on the problems and aspirations of some 3,500 women and men farmers. The KEGE also played a role in student recruitment by introducing the farm school to parents, village leaders, and extension agents who attended the courses. The renewing of the agreement was emblematic of the country's need for the school's continuing assistance, and for the breadth of practical education that only

---

[10] "American Farm School Honor," *Athens News*, 14 January 1987, 5.

[11] "Director's Report to the Board of Trustees," February 1983, AFS Archives.

the farm school was able to deliver, so essential since Greece's accession in the European Union.

In 1982, the new KEGE building on the farm school campus, built by the Greek government with a loan from the World Bank, was completed, rendering architectural definition to the lower section of the campus. The structure had a capacity for sixty overnight guests, as well as dining and classrooms facilities. Keeping men and women farmers up-to-date on agricultural developments, KEGE offered courses on dozens of different subjects, including dairy technology, irrigation, management, environmental protection, machine embroidery, and canning.

Unsettling, though, was the government's continued demonizing of private education. "The private school teachers have already started scattered strikes and our information is that there will be more," warned Andonis Stambolides in 1983. "The private school teachers as well as private school parents and students are all uneasy because of all the uncertainty about our future."[12] This uncertainty notwithstanding, the applications to the farm school poured in at an historic high. It would take another two years and a national financial crisis for those officials who controlled the levers of government to temper their rhetoric against private schools.

When PASOK came into office, the party hustled to flush out officials from the previous government and replace them with PASOK faithful. For the farm school, once again the exercise in diplomacy became a primary focus. Many officials were standoffish; even if they were not anti-American by conviction, they would give the wrong impression, they reasoned, if they treated visiting administrators from the school with cordiality. There were others less frosty, casting aside the American identity of the school in their insistence that the American Farm School was not in any full sense American, but really *dikiá mas*. In view of that sentiment, some supportive officials in Athens advised Bruce to drop the popular name American Farm School and use the formal Thessalonica Agricultural and Industrial Institute when conducting business with the government. Bruce readily ordered a large supply of formal letterhead stationery.

In the early-to-mid-1980s the school was stressed by economic hardship owing to the recession and inflation that also troubled the

---

[12] Andonis Stambolides to Irwin Sanders, Thessaloniki, 7 February 1983, AFS Archives.

United States and walloped Greece with special vengeance. The school's revenue sources could not keep pace with rising costs. The board of trustees and director watched helplessly as the endowment fund shrank—it lost $25,000 in 1982–1984.[13] That the drachma went through a series of devaluations did not help the situation; on the other hand the dollar was worth more when spent in Greece, except for a period in 1987 when it lost value against all European currencies. To make matters worse the Greek government, which in 1985 had submitted its Stabilization Plan, cut back its support to the school to 90 percent of the agreed amount, a sum that represented a loss in revenue of $20,000. It seemed apparent that contributions gathered from supporters in Greece would not be sufficient to offset the inflation that was climbing to 20 percent.

Faced with this worrisome economy, the school sought cost-cutting measures. On principle, the leadership vowed not to cut back any aspect of the TEL or TES programs, widely considered as the best agricultural curricula in the country. Likewise, the demonstration farm, with its outstanding dairy herd whose offspring could be found in twenty-one counties throughout the country, 20,000 turkeys, 25,000 hens, and the clean but voracious pigs, was not to be scaled back. Turning to personnel, the school granted its employees a cost of living allowance in line with government policy, but the administration decided in 1981 not to replace some people who left and to cut cleaning staff as well as others in non-essential positions. The school increased student fees for room and board by 20 percent. Then, thankfully, the Greek government agreed to a 20 percent increase in its annual grant, but based on 90 percent of the amount promised earlier. Moreover, the farm, as a rule dependable for its bounty, proved a lifesaver in its financial contribution to the school.[14] Improved management, elimination of costly units, and upgrading of facilities and equipment led to an eleven-fold increase in the farm's contribution between 1978 and 1982. To cite an instance of improved

---

[13] "Director's Report to the Board of Trustees," May 1984, AFS Archives.

[14] The farm had withheld its bounty in the period just after the student and staff strikes and the painful disharmony among the school's various constituencies. It might have struck some people that the school's problems were assuming Biblical proportions in 1977 when the farm yielded the worst crop in over twenty years, resulting in deficits of over $38,000.

output, milk production went from 442 metric tons in 1974–1975 to 823 metric tons in 1981–1982.

In 1983 the government froze salaries. The employees grew restive and the on-campus union members agitated over their straitened financial condition. School contacts with the local inspector of labor indicated that the union had lost credibility with local government agencies because of the extreme position taken by union officers, a stance that did not apparently reflect the feelings of the school staff as a whole.[15]

Undismayed by the worldwide recession and inflation, in 1983 the board of trustees launched a capital campaign, the first since 1931, on the occasion of the school's approaching eightieth birthday. Spearheaded by campaign chairman and trustee Mimi (Mrs. Charles F.) Lowrey, a person of uncommon vigor and confidence, who remained unawed by the Herculean labor set before her of raising $4.5 million, the campaign accomplished much more than raising a striking sum of money. If the school yearned for vindication that it was still capable of capturing the public's heart both in Greece and the United States, and of bonding people from two cultures in a common philanthropic enterprise, here was proof. "To suggest that a Farm School should even consider raising $4,500,000 at a time when relations between our two countries are strained would be absurd in the minds of many people," commented Bruce. Greece "is going through a period of economic depression, the worst since World War II. This makes it even more impressive that the School development committees in Athens and Thessaloniki have exceeded their goals this year by more than twenty-five percent."[16] In fact, donations from Greek sources—individual, corporate, and gifts in kind—increased perceptibly throughout the 1980s.

The lead support in the campaign had come from four trustees: ever-faithful Stavros Constantinides, who had first referred to Princeton Hall as "the Parthenon," George Livanos, and Denison B. Hull. By 30 September 1984, more than $3 million had been received in gifts and pledges.[17] The campaign ended in 1985.

---

[15] "Board of Trustees Report," October 1983, AFS Archives.

[16] "Director's Report to the Board of Trustees," May 1984, AFS Archives.

[17] Mimi (Mrs. Charles F.) Lowrey, interview with author, Thessaloniki, 19 October 2000, written, AFS Archives. Mimi Lowrey joined the board of trustees in 1975. She served as vice chairman (1980–1985) and as chairman (1985–1989, 1990–1992), while continuously playing a vital role in fund raising for the school.

The board of trustees was faced with a costly project, the reconstruction of Princeton Hall, which had suffered structural damage in the 1978 earthquake. This building stood as the emotional center of the campaign. Not only was its function indispensable for a fully lived life on campus, but its form evoked for many the permanence and architectural grace of the Parthenon. Trustee Jenny Fenton articulated this sentiment, describing her reaction to the building when she had worked at the school in the 1960s: "People stayed on campus in those days. The School was still isolated, no one had cars, transportation was scarce. The road in front of Princeton Hall was not yet paved. Faculty and students, staff and visitors, members of resident families strolled back and forth in front of the building, discussing this and that, and I thought then, yes, this is what it must have been like in ancient days, the citizens, in exactly this manner, walking to and fro under the Parthenon, talking about great affairs or exchanging the day's news."[18]

As part of the capital campaign, USAID gave two grants for Princeton Hall: $295,000 for structural repairs and $350,000 for the library and information center. Each of these donations, as with all funds from USAID, had to be matched by private contributions. The renovated building was designed to accommodate a crafts center for weaving and ceramics, dining hall, computer room, library, meeting room, playroom, student laundry, staff apartments, and space for other functions.

The capital campaign answered another need for the school: amassing an endowment fund of $1,500,000 to put toward scholarships, faculty salaries, staff training, and a salary for a resident priest. An endowment fund had been established many decades earlier by trustee Leander T. Chamberlain, but it was added to only sporadically. By 1967, overseen by Craig R. Smith, who served for twenty-two years as chairman of the board of trustees, the endowment had reached a market value of $500,000; by 1983 it was worth approximately $2.8 million; by 1991 it was worth $8 million. Weighing on everyone's mind was the decline of faculty salaries owing to inflation. "The cost of living adjustments are straining the School to the limit, while the standard of living and pension prospects of staff are declining."[19]

---

[18] Jenny Fenton, interview with author, Thessaloniki, 20 October 2000, written, AFS Archives.

[19] "Director's Report to the Board of Trustees," May 1984, AFS Archives.

Plainly in need of enhancement was the curriculum: computer purchases and instruction for students in technology on how to solve farm management problems loomed as an imperative. Also, the English department looked forward to acquiring instructional aids and books, and renovation of the language laboratory. Gratified by the contributions from Greece, Lowrey, who had become chairman of the board in 1985 when Ambassador Labouisse resigned, noted, "Gifts from Greek companies prove the level of faith the Greek people have toward the training given young students and their adult counterparts."[20]

Picking up pace in all its activities, the school had grown in the 1980s into a true outreach center, spreading far beyond the Greek-American axis. Outreach was historically a feature of farm school life, but the board of trustees and the administration were cautious for fear of overextending. For instance, as early as 1954, a six-nation conference, sponsored by the Food and Agricultural Organization of the United Nations in the Near East, with top executives from Greece, Turkey, Yugoslavia, Cyprus, Israel, and the United States, met at the school. Now, consistent with the Sanders Report, the administration and board kept on the lookout for such opportunities. The board was keen on serving people from developing countries as a way to continue attracting support from the United States; some trustees feared that Greece's entry into the European Union, represented a sign of prosperity and might cause American philanthropists to cut back their donations to Greece and to the school in particular.

In that regard, a cooperative venture in partnership with the Aristotle University of Thessaloniki's department of agricultural extension and rural sociology and the University of Reading in England, which had sent graduate students from developing nations in Africa to the school for courses each year for eleven days starting in 1982, proved a rewarding experience for the three institutions. Based on that success, the school joined with the Aristotle University again in 1986 to secure initial funding from the European Union in the amount of $30,000 for a series of seminars known as the Thessaloniki International Training Program (TITP). The first initiative of its kind in Greece, TITP ran from 1988 to 1991. The seminars welcomed on campus teachers, extension officers, farm program coordinators, rural administrators, and

---

[20] As quoted in *The Sower*, no 16, 1984.

horticultural specialists from developing countries in Africa and the Middle East. These professionals attended classes designed to help them improve their skills in planning and running educational programs for young people and adults. TITP drew largely on the experience of Professor John Crunkilton from the department of agricultural education at Virginia Polytechnic Institute and State University, a longtime advisor to the farm school and later trustee, and his colleague Professor James Clouse.[21]

What a commentary that this small institution on the fringe of Europe could extend a hand to assist agriculturalists from other continents. The rewards for both the visitors and students were immense, but with this caveat: "The Farm School is embarking cautiously, and with some hesitation, on its international program, aware that nothing must detract from its primary mission of educating the rural youth of Greece." The justification here was that "the greater interaction with agricultural leaders abroad will strengthen the School's own programs while giving it an opportunity to help others who are still struggling with domestic agricultural problems to produce enough food to feed their people."[22]

Younger students from abroad flocked to the farm school. As described in the previous chapter, the Greek Summer Program, initiated in 1970, was an immediate success. The program continued, keeping its original pattern: traveling; a work project in a village and establishing a kinship with the host family; an ascent to the summit of Mount Olympus; getting acquainted with the farm school and in some inexplicable way in such a short time, grasping its special ethos.

An enduring program that continues to enrich student life is the Student Exchange Program. Farm school students travel to the United States and Europe to visit other agricultural schools and farming families, while European and American students are received at the farm school. For the Greek students this program accomplishes two aims at

---

[21] John Crunkilton, "Analysis of the 1988, 1990 and 1991 Seminars on Methods of Practical Programs: The American Farm School," AFS Archives. Trustee John Crunkilton frequently offers his services to the school on matters of agricultural education. He was instrumental in writing the curriculum for TITP and later for the Dimitris Perrotis College of Agricultural Studies. Sperry Lea, longtime member of the board of trustees, endorsed TITP heartily and undertook to keep the trustees involved and supportive.

[22] *The Sower*, no 124, 1998.

once. Since English is the program's language of communication, the youngsters are motivated to work hard in the classroom and in their spare time to improve their English language proficiency in order to qualify. As a second benefit, their experience abroad, according to English teacher and Student Exchange Program Coordinator Mary Chism, "makes them realize what sorts of improvements they can make at home. What they always have taken for granted as the 'way it's done' is shaken up and makes them open and more willing to try something new."[23] Speaking of outreach, it was no small matter that by 1982, twenty-four members of the faculty and staff had studied to some extent in the United States, and had visited seventeen countries to work in or observe agricultural education and development.

The thumping of hammers and clang of steel vibrating across the campus announced yet another form of activity. In the 1980s, the farm school received the largest number of grants from USAID since the American organization had started awarding them in the early 1960s. Through this assistance the school underwent the most accelerated and substantial building and renovation phase since that time. In addition to two grants earmarked for the renovation of Princeton Hall, USAID gave from 1980 to 1989 a total of $ 2,913,000 for renovation of James Hall and Rochester Hall and construction of a biogas system (1980); extension to the biological sewage treatment plant, dormitory furnishings, improvements to the utility systems, improvements to poultry facilities (1982); construction of Henry R. Labouisse Youth Center and improvements to the biogas system (1985); extension to Massachusetts Hall, and improvements in water supply system (1988); improvements to biogas system (1989). The biogas system that converted farm animal manure to methane to be used for energy proved uneconomical and technically flawed. After more than a decade of experimentation it was replaced by a manure composting system.[24] Serving as an athletic gymnasium and general assembly building capable of seating 1000 spectators is the Henry R. Labouisse Center. Dedicated on graduation day, May 1988, this functional, multipurpose structure

---

[23] Mary Chism, interview with author, Thessaloniki, 18 March 2000, written, AFS Archives.

[24] George Strikos, interview with author, Thessaloniki, 2 November 2001, written, AFS Archives. George Strikos, a farm school graduate with an associate degree from the American College of Thessaloniki, is head of the maintenance department.

lends completeness to the campus, especially to the athletic life of the school.

One of the most successful stories beginning in the 1980s was the implementation of the many uses of information technology on campus; the technology was not confined to the student laboratory or administration. For the graduate follow-up program, Anastasios Pougouras compiled a database of 500 graduates who had stated their specific area of interest. This enabled the school to send them information from magazines, books, and other printed materials, and most importantly communications from the ministry of agriculture and the European Union concerning rules, regulations, and guidelines.[25]

However, more work needed to be done. As late as 1993, a special evaluating committee complained that schoolwork was not assigned to students to encourage computer use.[26] With the opening of the Dimitris & Aliki Perrotis Library in 2001, computers became available to students throughout the secondary and post-secondary programs and even to the general public.

Money from the European Union began filtering into the farm school as early as 1982, when the institution received a grant of $18,000 to develop a dairy management course for farmers in the region. By the end of the 1980s, European Union funds were becoming an ever-increasing source of farm school revenues.[27] A graph, entitled "EU Inflows" charts the farm school receipt of European Union funds from 1987 through 2001, showing total receipts of approximately $20 million calculated according to the average dollar drachma rate. The peak year was 1993 when just under $3.5 million was granted. By 2001, the amount received slumped to approximately $180,000.[28] Although the monies were regarded as a substantial benefit, delays in payment caused the school to confront a cash-flow problem The cumbersome procedure required the farm school to present its requests to the Greek government, which in turn submitted them to the European Union. If approved by the

---

[25] Victor Walker, "The American Farm School in Thessaloniki: An 86-Year-Old Institution Still Filling a Vital Need," *Greek-American Trade* (November-December 1990): 42.

[26] "Report of the Academic Study Team," 26–30 April 1993, AFS Archives.

[27] Ann Elder, "Farming the Future, Class of 1993," *Athenian Magazine* (July 1993): 16–19.

[28] "EU Inflows," 2001, AFS Archives.

European Union, the funds were paid to the government, which in turn remitted them to the school.

To support Greece's role in the European Union the farm school pinpointed agricultural specialties to emphasize in the practical program. The horticultural department was tagged as one focal point since Greece was considered to have a competitive advantage over other European Union members thanks to its climate: Greek farmers could market many crops at least a month before their northern neighbors could harvest them, especially if greenhouse use was intensified and irrigation systems perfected. Providing students with a venue to try their hand at practical horticultural applications, greenhouses were added on campus. Also, the faculty worked on teaching packages for greenhouses, irrigation systems, and crop production. Moreover, three new laboratories were set up where students could work on problems concerning fruit tree, flowers, and vegetable cultivation, and could experiment with soils and fertilizers. In April 1984 work was completed on a 10,764 square-foot, double-skin, insulated greenhouse, the first of its kind in Greece. Also in 1984, the American government made it possible for the school to acquire fourteen pieces of state-of-the art farm machinery and to lay down the required equipment in order to demonstrate the latest drip systems for irrigation.

The pressure exerted by European Union membership on a small country like Greece, with its poverty of resources, is extreme. This Mediterranean country, spreading over a total land area of only 50,942 square miles of which only 30 percent is cultivable land, presents a monumental challenge to agriculture. Characteristically, Bruce looked beyond the competitive edge to steady the staff and to remind the students of the essentials of life: "As Greece begins its first year in the European Union, we continue to place major emphasis on the quality of life, that which will give it meaning in the years ahead—the sounds, the smells, the touch, the feel of life, which give young people a sense of pride as well as satisfaction in being close to the soil and an agricultural existence."[29]

While the European Union provided an undeniable cornucopia, its policies presented an ironic side: they required the agricultural sector in Greece, as well as in other countries, to become more productive, the plan being to reduce Greece's farming population to eventually 5 or 6

---

[29] As quoted in *The Sower*, no. 103, 1981.

percent from the 25 percent or so in the 1980s, thus making it possible for the European Union to scale back its massive agricultural funding programs to all its members, which were reported to absorb half its budget.[30] Certainly, this policy contradicted the American Farm School's once cherished but now anachronistic ideal: keeping its graduates on the land. As a practical matter, this impending plan could have held negative ramifications for the school's recruitment strategy if the future meant a paring down of the farming population.

What could not be foreseen in the 1980s was that, nearing the end of the century, the Greek government would replace the TEL (technical vocational lyceum) curriculum. The replacement, EL (comprehensive lyceum) offers little by way of theoretical or practical agricultural content. However, the new TEE (technical school) curriculum, as we learned earlier, retains its agricultural concentration. In keeping with the practical approach to agricultural education, the Dimitris Perrotis College of Agricultural Studies offers advanced courses in applied agricultural science and presents management principles for experimentation on the perfect learning medium: the Zannas Farm. If the number of Greek farmers were to be drastically reduced, what was called for was an efficient, intense, and highly scientific approach to farming and sustainability: the school is more than ever in a position to educate modern stewards of the land.

Visitors to the school in the 1980s—some 10,000 of them annually from around the world—would observe the 250 or so students studying in the morning in accordance with the curriculum set up by the ministry of education. In the afternoons the youngsters would pull on their dark blue coveralls and dusty boots to participate in the practical program authorized by the ministry of agriculture. Exhibiting the unstoppable energy and infectious humor that possess adolescents during the high school years, the boys and girls would work away at one of the specialties: horticulture, crop production, mechanized agriculture, animal husbandry, or farm machinery. Friendly and welcoming, they would wave a *káli méra* (good day) or the less formal *yiá sou* (a greeting that conveys the informal flavor of "hi") to the attentive onlookers.

School life in the 1980s took place on grounds spreading over 375 acres, by now some of the most productive land in Northern Greece.

---

[30] "Germans Offer Plan to Remake and Centralize European Union," *International Herald Tribune*, 1 May 2001, 1.

Eighty-nine buildings, all of them built or refurbished since 1950, housed the staff, the students at work or play, not to mention the sheltering of cows, pigs and chickens. A dedicated, well-educated staff of 110, 40 of them directly engaged in teaching the students, inspired the learning process. The school was run on a $4 million annual budget.

As it entered the last decade of the twentieth century, the American Farm School was the only coeducational agricultural school in Greece with a residential life, invigorated by athletics, cultural activities, international components, and guidance for spiritual growth. Student life was overseen for many years by Nikolaos Papaconstantinou, director of residential life, and animated by Dimitris Pantazis, head of recreation, two men who were key in carrying on the original spirit of the school; both these men attained legendary status on the campus. Nikolaos Papaconstantinou, whose biography reads like an emblem of earlier times, came to the farm school in 1954 as a poverty-stricken orphan, his childhood ruined by the murder of his father in the civil war. Felled by tuberculosis in 1957, he regained his health in a sanitarium where he resided for five months. He returned to the school to graduate in 1959. The school sent him to Chapman University in California, where he received a bachelor's degree in mathematics and a master's degree in education. He was responsible for student life from 1972 until his retirement in 2001. As a student under Charlie House and Theodoros Litsas, and a top staff member under Bruce, he seemed to embody the very essence of the school. Dimitris Pantazis, also an alumnus, attended the farm school from 1952 through 1956. After working at the school starting from 1964, he moved with his family to Sweden from 1971 to 1977. From 1977 to his retirement in 2002, he served as head of recreation and dormitory and as dining room supervisor. Under his spirited leadership, the American Farm School Dance Team won three major prizes in Los Angeles, Strasbourg, and Greece.

The extensive plant, still up to date in the 1980s, facilitated the agricultural and industrial hands-on experience provided by no other institution in the country—unique, right down to its distinctive returnable glass bottles filled to the tinfoil cap, the milk unadulterated with preservatives, delivered fresh each day to the stores.

# EPILOGUE

# WHY AN AMERICAN
# FARM SCHOOL IN GREECE?

While the epilogue, focusing on the 1990s, is meant to bring the history of the American Farm School close to its centennial celebration in 2004, it also serves a second purpose. As a synopsis, it outlines how the school validates its continuing presence in this remarkable land. If the fundamental question is "Why an American Farm School in Greece?" the answer would turn on the purpose the school serves and the Greek people's assessment of its services.

## *Greece in the 1990s*

The twentieth century in the Balkans died in much the way it was born: tormented by ethnic strife. However, in the 1990s the points of contention were not in Greece but to the north in disintegrating Yugoslavia. Extraordinary geopolitical conditions had resulted from the demise of the Soviet bloc in 1989. In the absence of the superpower struggle for hegemony, which had held the belligerent ethnicities in check, came a long season of grisly fighting in the region. The Greek nation, divided from its Balkan neighbors for some forty years during the Cold War, was compelled to define itself anew in a reconfigured world.

The internationalizing pressures of the European Union and the pull of globalization impelled the American Farm School to discover fresh ways to serve the rural population and to help improve Greece's agricultural position. Toward that end, ideas were germinating and plans were gelling at the farm school to establish an agricultural college. Since the concept included accepting students from outside of Greece and outside of member states of the European Union, to include the Balkan countries, it was audacious, but replete with pitfalls.

Greece had been alarmed by the multiple convulsions in the Balkans—the violence in Croatia and Bosnia from 1991 to 1995 and, before long, the tragedy in Kosovo. But the worst foreboding emanated from the disintegration of Yugoslavia in 1991, when the Republic of Macedonia seceded to form a nation calling itself Macedonia. However, it was called FYROM (the Former Yugoslav Republic of Macedonia) by Greece and by some international bodies such as the UN, or occasionally called Skopje by Greece. Contiguous to Northern Greece, the breakaway republic in its name alone articulated an irredentist ideology. Symbols, too, such as its flag, a sunburst of red and gold appropriated from the ancient Macedonian kings of the Argead dynasty that included Philip II and Alexander the Great, unnerved the Greeks. Unless the country changed its name and unless it struck from its constitution articles that expressed a provocative agenda, the Greek government refused categorically to grant diplomatic recognition to FYROM. Over this issue, the Greek government stood at loggerheads with European friends and NATO allies, who failed to grasp the depth of the social psychology and historical realities underpinning Greece's apprehension. When in February 1994, the Greek government closed the frontier with FYROM and banned its use of the port of Thessaloniki, which handled 70 percent of FYROM's imports and exports, the European Union threatened legal action against Greece. Domestically, polls showed support for the embargo at around 80 percent.[1] In Thessaloniki more than one million "demonstrators, festively clad in traditional costumes from Greece's Macedonia region, waved a sea of blue and white flags and carried banners proclaiming that 'Macedonia is not for sale" in the largest rally ever staged in the city," led by "black-clad Orthodox priests carrying religious icons."[2] Adding to the strain between Greece on the one side, and NATO and the European Union on the other, Greece remained the only member of those organizations to voice sympathy for the Serbs of Yugoslavia. Axel Sotiris Wallden, explaining the political motives that prompted Greece's policy toward Yugoslavia, writes that

[1] "Support for Embargo," *Athens News*, 29 March 1994, 3.
[2] "Macedonia Is Not for Sale," *Hellenic Chronicle*, 7 April 1994, 1.

"Athens...hoped that Belgrade would solve the 'Macedonian' problem, initially by not allowing Skopje to become independent, and later by reintegrating it into a new federation as part of a solution to the Yugoslav crisis."[3]

Simultaneously, a row broke out in another Balkan quarter: the Albanians, Greece charged, were persecuting the Greek Orthodox Church and the Greek minority living in Northern Epiros, a section of Albania that has a sizable Greek minority. Meanwhile Greece endured the influx of hundreds of thousands of illegal Albanian immigrants fleeing their impoverished and destabilized homeland.

As luck would have it in 1995, as President George Draper and the board of trustees were formulating their plans for inaugurating a college the next year, Greece softened its policy toward FYROM, about the time that the Dayton Peace Accords were signed.[4] Soon after, Kostas Simitis, having succeeded Andreas Papandreou as prime minister in January 1996, concluded that so securely was his country nested in the stability and prosperity of Western Europe that he could aspire to a policy of reconciliation within the Balkans and a cooperative stance toward NATO and the European Union. According to observer James Pettifer, "in Simitis' world-view the era of Greek nationalism was over, and a very substantial degree of surrender of economic and political sovereignty was required for the needs of Greek security in the modern world.... Papandreouist themes were to be abandoned in favor of accelerated European integration and accompanying defense cooperation, the latter as a counterweight to the prominence of Turkey in NATO affairs."[5] No foreign policy decision could have been more propitious in encouraging the founding of an American Farm School post-secondary program in keeping with the student demographic pattern that the president and trustees had envisioned. Despite the quarrels with FYROM and Albania, and the tumult directed from Belgrade, George Draper is of the opinion

---

[3] Axel Sotiris Wallden, "Greece and the Balkans: Economic Relations," in *Greece and the New Balkans: Challenges and Opportunities*, ed. Van Coufoudakis et al (New York: Pella Publishing Company, 1999) 80. For an analytical view supportive of Greece's position, see C. M. Woodhouse, "Recognizing 'Macedonia' Defies History," in *Christian Science Monitor*, 28 October 1992, 19.

[4] P. C. Ioakimidis, "Greece, the European Union and Southeastern Europe," in *New Balkans*, 180.

[5] James Pettifer, *The Greeks: The Land and People Since the War* (London: Penguin, 2000) 63.

that "the Greek business community saw great potential for leadership in
the Balkans, and...most reasonable Greeks saw and still see the need for
common ground and productive joint efforts of all kinds as the most
likely road to peaceful coexistence in that troubled region."[6]

In continuing the policy of reconciliation, the Simitis government
adroitly joined NATO in certain important ways in an effort to contain
the Serbs during the Kosovo crisis. Although Greece had voted against
the NATO bombing of Yugoslavia and refused use of its airbase at
Preveza for that purpose, it did allow NATO troops and materiel to be
unloaded in Thessaloniki to be transported north, and did cooperate in
other NATO operations. Moreover, assuming a peacemaking posture,
Greek officials engaged in a series of diplomatic contacts spearheaded by
Greece's foreign minister, George Papandreou, in an effort to deter
Belgrade from its exploits in Kosovo. The bombing became a
contentious issue dividing Greek and American friends of the farm
school; the board of trustees split along national lines over the
issue—some Americans defending the bombing and the Greeks
denouncing it—when they gathered at a board meeting in Greece in
1999. But each faction, bound together in a common philanthropic
enterprise, for the sake of the school and for the sake of their affection
for one another, made a strenuous effort at least to understand the
position of the other.[7]

Coming into play in 1990, the European Union's regional strategy
for all its member states unquestionably influenced farm school plans
regarding its own social and educational role among its neighbors. The
European Union formulated INTERREG, an innovative program directed
at stimulating cross-border cooperation throughout Europe. As pertains
to Greece, INTERREG involved Albania and Bulgaria as well as border
regions within Greece itself in an effort to stimulate social and
commercial development in such categories of infrastructure as
transportation, tourism, education, and rural development.[8] As might be

---

[6] George Draper to the author, Sorrento ME, 17 July 2001, AFS Archives.

[7] For an outline of Greece's position regarding the bombing of Yugoslavia and other
NATO operations, see *The Strategic Regional Report: A Publication of the Western
Policy Center*, 4/2 (1999): 7.

[8] INTERREG I (1991–1993) allocated some $270 million of European Union funds
to Greek border regions while INTERREG II (1994–1999) allocated $370 million.
INTERREG III 2000–2006 is its most recent phase. For information about PHARE, see

expected, "Cooperation was strained in INTERREG I due to Greek hesitation at the time. Cross-border cooperation was much more prominent in INTERREG II," comments Wallden.[9] In cooperation with the Greek ministry of agriculture, the farm school wholeheartedly participated in those INTERREG activities that corresponded to its mission. For example, during 1993 with INTERREG I, farm school staff under the direction of Dimitris Michaelides, head of the school's Office of European Union Affairs, conducted a total of 142 separate events (seminars, conferences, workshops on agricultural and other topics) reaching from Kastoria in northern Greece to Orestiada on the Turkish and Bulgarian border in the east.[10]

In the 1990s not only foreign relations but also other issues turned critical. The destruction of the Greek environment on a monumental scale had been in process since the 1960s. The visual and aural evidence—unplanned urban growth and sprawl throughout the countryside, pollution smudging the once-fabled blue sky, the roar of machinery, reek of diesel fuel, smoke plumes rising from burning farm fields—caused alarm. Greek scientists and rural sociologists, among them George Daoutopoulos and Myrto Pyrovetsi, have conducted intensive studies of the environment, especially in Northern Greece, and conclude that in the post-war period the constant growth of total agricultural output has "accelerated soil erosion, accumulation of toxic chemicals, and the Greek farmers' reliance on diminishing and unpredictably priced petroleum resources." These and other abuses to the land, they claim, have pushed Greek agriculture into crisis.[11] The lack of water exemplifies the crisis. Rivers have been diverted to help farmers improve production. If we add the prolonged drought of 2000–2001, the population increase due to booming tourist traffic, the effluence from augmented industrialization, development of infrastructure for the 2004 Olympic Games, the problem looms as a genuine calamity. The farm

---

www.europa.eu.int/scadplus/leg/en/vb/e5004.htm; for INTERREG see www.europa.eu.int/comm./regional_policy/index_en.htm.

[9] Wallden, "Economic Relations," in *New Balkans*, 116.

[10] Dimitris Michaelides served as personnel manager from 1989 to 1993 and as head of the Office of European Union Affairs from 1993 to 1999.

[11] George Daoutopoulos, Myrto Pyrovesti, E. Petropoulou, "Greek Rural Society and Sustainable Development," in K. Eder, K. and M. Kousis, eds., *Environmental Politics in Southern Europe: Actors, Institutions, and Discourses in a Europeanizing Society* (Dordecht: Kluwer Academic Publishers, 2001) 151.

school, sensitive to the deterioration of the environment, has participated in varied efforts to educate students and the public on this subject.

Among the positive outcomes for Greece in this period, the end of the Cold War heralded a meaningful economic advantage: "After 1989, the downward trend in Greece's Balkan trade was reversed.... Greece's trade deficit of $150 million in 1989 was transformed into a $520 million surplus by 1996, with the export-import ratio rising from 62 percent to 175 percent," notes economic historian Wallden.[12] To the extent that the European Union influenced Greece to shape its economy and to craft its foreign policy, the multinational organization's very nature obliged it to promote a spirit of community. Its regulations and guidance, underscored by substantial funding drawn from its members, made it an irresistible force throughout Europe. We should remember that in its first fifteen years of membership, Greece had received $6 billion a year from the European Union. In this same period, the per capita income had risen to $10,981.

Since a major effort during the last decade of the twentieth century at the farm school was the planning for and establishing of the Dimitris Perrotis College of Agricultural Studies, it is useful to touch on the Greek ministry of education's many projects for higher education: these projects bear heavily on the future of Perrotis College. Much has changed in the years since the planning stage in the early 1990s. By 2001 the ministry had doubled the number of university positions for lyceum graduates. In fact, at this writing most candidates holding a lyceum diploma are guaranteed entry into a Greek university or TEI. Furthermore, the ministry has upgraded the system of TEIs (polytechnic institutes), giving the degree earned by those graduates the same recognition as that earned by graduates of the public universities. Additionally, two state universities are now offering bachelor of science degrees in areas similar in course content to offerings at the Dimitris Perrotis College of Agricultural Studies: the University of Thrace (since 1999) and the University of Ioannina (since 1998). These additional state programs draw students from the same pool as does Perrotis College, obviously presenting stiff competition. How the college can attract Greek students in this competitive situation arises as a key question to be

---

[12]Wallden, "Economic Relations," in *New Balkans*, 71–121. For more economic analysis, see Press and Information Office, *Greece: A News Review from the Embassy of Greece*, 4/10, 1998.

answered in the future according to how and if Greek laws change to allow equivalent status for private universities.

## The Farm School during the 1990s

Taken together with Bruce Lansdale's retirement in 1990, how would the institution respond to all the external pressures also placed upon it? Was it even possible for the institution to act dynamically and innovatively situated just south from this Balkan mayhem? How would Bruce's successor, in comparison a novice, cut from different cloth, measure up to the many challenges?

George Draper, raised in an atmosphere different from his predecessors, was not steeped in religious tradition. Rather, possessing a secular outlook, he never conveyed the notion that he was counting on divine intervention. What he did possess in keeping with the school's ethos, was a heightened sense of social commitment and service, a diligent work ethic, a special aptitude for identifying opportunities, and gumption to see his plans through.

Boston bred, he and his wife, Charlotte, both went to Harvard, where they met. Like the husband and wife teams before them, they always worked in unity to lead the school. Charlotte had received a master's degree in library science and had worked as a school librarian. The couple collaborated on works of fiction as well as a series of books for teaching English as a second language, first published in 1984, and now in its third edition. (The titles of the ESL books are, *Great American Stories I, II,* and *III*). While at the school, Charlotte reestablished the farm school's historical archives and prepared an informative photo history entitled, *The American Farm School of Thessaloniki: A Family Album,* published in 1994.

Unlike Charlie and Bruce, both engineers, George had earned a bachelor's degree with a major in government and a master's degree in American and English literature. "He had an intellectual turn of mind, a quality that distinguished him from other candidates. That, I liked," remembers Edmund Keeley, a member of Draper's search committee.[13]

---

[13] Edmund Keeley, telephone interview with author, 29 June 2001, written, AFS Archives. Edmund Keeley taught English and athletics at the school under one of the first Fulbright fellowships to Greece. He was a chaired professor at Princeton University, a

Whereas Bruce was loquacious, prone to belabor a matter, he was also amusing before the public; George tended to cut to the point, on occasion could even be jaunty. Bruce had a whole team of farm school "titans" serving on his top staff during his first two decades, men who had once served under Charlie and possessed an institutional memory and kept devotedly to the farm school tradition. Likewise, George was also fortunate. He had on his top staff farm school veterans: Nikolaos Papaconstantinou, head of residential life and special programs; Alexandros Michaelides followed by Konstantinos Evangelou, farm managers; and Dimitris Pantazis, head of recreation, the dormitory, and the dining room, all men who had witnessed the old days as students (except for Michaelides) and had served under Bruce. Andonis Stambolides, associate director for education, and David Willis, associate director for administration, had also served under Bruce, as had Panayiotis Rotsios, dean of studies. Dimitris Litsas, head of land development, completed the group.

A strong distinction should be drawn between the two men's sense of mission to the school. Bruce embarked on a passionate lifetime commitment to serve, while George pledged only a ten-year stint. Even so, some day when we look back, George's stewardship in the 1990s may well stand out as among the most dynamic in the life of the American Farm School. And how could it have been otherwise? Not to be outpaced in the new global order, the institution was obliged to speed ahead or stall completely.

Positioned on the threshold of its centennial, the school was described in a press interview by its fifth president, David Buck (who succeeded George Draper in 1999), as standing neatly on five pillars. These pillars also provide a tidy organizational scheme for our discussion here. The first pillar, the secondary school, which accommodates 250 students, receives applications from all over Greece and is, significantly, the only all-boarding school in the country. The second pillar, the Dimitris Perrotis College of Agricultural Studies, founded in 1996 under George Draper's presidency and drawing students from an international pool, inaugurated for the American Farm School an unprecedented venture. The Lifelong Learning Program, the third pillar, was the work of

---

founding member of the Modern Greek Studies Association (which he served as its first president), and president of the PEN American Center. He is a novelist and noted translator of Greek poets and a farm school trustee.

George late in his tenure, as the successor to two significant postwar programs for training adults: short courses for farmers and community development programs. In keeping with European Union trends, Lifelong Learning carries agricultural and environmental programs into the villages and conducts workshops and seminars for farmers and community leaders on the school grounds. The fourth pillar encompasses the two educational and demonstration farms (one situated on campus, the other, the Zannas gift, by the Axios River), providing practical training venues for the Perrotis College students, as well as adult trainees. Giving significance to the total structure is the fifth pillar, the spirit of the farm school.[14] Each of these pillars holds the answer to the fundamental question: "Why an American Farm School in Greece?"

The secondary school remains at the very core of the farm school's mission. This entity, founded a century ago, has been revered through the generations by all those connected with it: students, alumni (in 2002 living graduates numbered some 3,000,) friends in the United States and Greece, faculty, and board of trustees. Sentimental reasons aside, there exists a cogent reason for the school's centrality. The institution's legal base is derived from its secondary school. Recognition from the Greek government is of overriding importance to all operations at the farm school. Parents, too, are particularly attached to the secondary school; they admire the way the boarding component teaches their youngsters the delicate art of community living, and the way it builds character. "With new influences penetrating even the remote villages," George told interviewer Ann Elder, "parental anxiety is most evident." The school" believes in trying to protect and strengthen village mores, which we value greatly."[15] Student life offers extracurricular enhancements, reinforcing strong traditions in Greek life including dancing, singing, weaving, ceramics, athletics, and theater presentations, along with such activities as student government that are generally offered in American high schools.

Challenging to the president, staff, and board members, however, was the transformation in the curriculum in 1998. When the technical vocational lyceum (TEL) was replaced by the comprehensive lyceum (EL), as noted in chapter 3, the switch brought to the farm school a disconcerting proposition. The 1998 curriculum mandated by the Greek

---

[14] David Buck, "The American Farm School," *Hellenic Voice*, 13 June 2001, 5.

[15] Ann Elder, "The American Farm School," *Athens News*, 17 September 1999, 12.

government set the school's lyceum in lockstep with all other lyceums throughout the country. Since it no longer offers a heavily vocational or agricultural component, EL appears to be at a remove from the farm school's core purpose. Germane to its ethos, however, is the three-year TEE program (instituted at the same time as EL and replacing the previous TES) but with a pronounced agricultural concentration that fits squarely with the mission. EL and TEE still offer tuition scholarships to all secondary school students, while families cover part of the boarding costs. This philanthropic aspect is admired throughout the country.

For a number of reasons, the founding of the Dimitris Perrotis College for Agricultural Studies in 1996 seems astonishing. The college is, along with all other private institutions on the tertiary level in Greece, necessarily categorized as a Laboratory of Free Studies, meaning that it is not recognized by the government, a classification that places Greek students at a serious disadvantage: they receive no deferments for military service and no possibility to transfer credits toward a degree at a Greek university, to mention only a few drawbacks.[16] The plan constituted an act of daring, an initiative taken by George Draper, who had never before presided over an institution as complex as the American Farm School. Being so close to President McGrew's deliberations for founding a college at Anatolia no doubt influenced his conviction that something similar could be achieved at the farm school. At Anatolia, as teacher and later as vice-president in the late 1960s and 1970s, he developed indispensable expertise: fluency in the Greek language; affection for and knowledge of the country's customs; and, most importantly, insight into the educational environment in Greece. Two points should be emphasized here. Beyond fitting smoothly into Greek culture, George was in tune with the spirit of the European Union, recognizing early that much could be gained by setting the farm school into this wider context. Also, given his natural inclination to mine the creativity of his staff, he arranged the reporting structure to allow for fresh ideas and initiative. Staff members reacted readily, initiating highly innovative programs.

When George envisioned the possibility of constructing this pillar of advanced education, his tenure at the farm school had hardly begun. A

---

[16] For an evaluation of the Greek university system at the time the farm school was planning its college see Dusko Doder, "Chaos in Greece," *Chronicle of Higher Education*, 4 May 1994, 45–46.

crucial fact to remember is that the founding of the Dimitris Perrotis College of Agricultural Studies departed radically in two essentials from the farm school's pattern: the catering to a university-age clientele and, at least for the last seventy years, to non-Greek students. First, never before were students above the secondary level enrolled for long-term residential education. The administration had feared that older students would engage in behavior inappropriate for younger people, and lead the innocents astray. In the world of the 1990s, that notion seemed outmoded. Second, the plan to accept students from European Union member states made good sense since Greece was a member benefiting in many ways from the organization, and desiring to move closer to Western Europe. But it was another matter entirely to recruit a multinational student body from implacable Balkan states, former and possibly future enemies, during a period when the region was seething with ethnic mistrust and violence.

In fact, the miracle occurred: students from Greece, Albania, Bulgaria, and FYROM (the first students from FYROM entered in 2000) study together, eat together, and sleep in the same dormitory in harmony. As they stroll across campus, they are "cool," like college students all over the world, belting out snatches of the latest hits, bantering, laden with books, and dressed as nearly as possible in the latest American fashion. They relish any greeting from a native English-speaker who is passing by, giving them a chance to exercise their language skills.

True, the school had hosted hundreds of non-Greeks in a variety of short programs over the years, but rarely and not since the decade following World War I, in secondary schooling. For this precedent-shattering plan, George had proceeded judiciously: "In seeking advice from the Greek community the School consulted with the private sector, especially the School's trustees, and the business communities and educators in Athens and Thessaloniki. All sources, unanimously and without exception, urged us to make the college Balkan in scope and Greek in focus and personality."[17] These radical changes act as a barometer gauging how Greek sensibilities were broadening and how the farm school was redefining its mission to keep apace.

In the Balkans, the European Union's INTERREG program—in addition to the PHARE program, its major development effort in Eastern

---

[17] George Draper to the author, Sorrento ME, 10 April 2002, AFS Archives.

Europe—was taking an active role in helping to stabilize the region by aiding nations that were not even European Union members. Programs such as these, as well as the European Union's TEMPUS program for cooperation with former Eastern Bloc states in the area of higher education, brought the farm school into close contact on many levels with neighboring countries.

Funded by the European Union's INTERREG, the farm school in 1992 studied the feasibility of transferring its know how to Bulgarian and Albanian agricultural schools and universities. Contacts with government and academic leaders in the two countries laid the foundation for future cooperative efforts, including relationships between academic institutions in those countries and Perrotis College. Interestingly, George and Charlotte and trustee John Crunkilton and his wife Sherry drove to Bulgaria to visit agricultural schools there. In that neighboring country they were hosted at a school founded in the early twentieth century by Reverend Edward Haskell, who, it will be remembered, with John Henry House, was cofounder of the American Farm School. Some of the same customs and ethics—the humanistic tradition—instilled by the pair of missionaries at the American Farm School in Thessaloniki were also a vital force in Pordim, Bulgaria, the location of the Edward B. Haskell Agricultural School.[18]

In addition to extending its reach, the farm school's involvement in INTERREG had more than one facet. The school could apply 10 to 15 percent of INTERREG grants to capital improvements on campus, maintenance, and administrative costs. Since the activities involved support for farmers, they fell neatly within the farm school's mission. Of course, what was truly extraordinary about INTERREG was the extension of programs across borders, inspiring people of one hostile nationality to put aside grievances and join in rewarding endeavors.

George recalls that "the wonderful thing about the work in Albania was that the key players from the start were American Farm School 'graduates,'" as he fondly calls them all: the three real graduates were Panayiotis Rotsios and Archimedes Koulaouzides (both retired from the school by then), and Evangelos Vergos. Rounding out the group were Bruce Lansdale in the first years of his retirement, Randall Warner, who came to the school in 1989 to administer the TITP program and would

[18] *The Sower*, no. 131, 1991.

stay on as head of the school's publications office, and George Koulaouzides who was to become assistant to the dean of Perrotis College.[19]

INTERREG offered another advantage. The program drew the school closer to the Greek ministries, promoting substantive dialogue with Athens as the institution explored options and submitted proposals to the government. Cooperating in this manner in European Union affairs with Greek government agencies meant that the school served as a useful partner by providing additional channels through which the government could distribute European Union funds to the rural population. That the European Union funds continued to be dispensed through Greek ministries to the farm school signaled that Greece was still recognizing the institution as *dikiá mas.*

Noting the workings of the European Union, especially with respect to the INTERREG program, puts a sharper perspective on the farm school's recruitment of Balkan nationals for the college. Even if Greece from 1990 to1995 were pursuing a policy in the Balkans that diverged from its allies, the country, nonetheless, remained responsive to European Union efforts to stabilize the region, agreeing that education would serve as an excellent pacifying tool. From the European Union the school received a real honor when the European Commission President Jacques Santer, the highest-ranking official in the European Union, visited the campus in May 1998. At that time the farm school was the only educational institution to have been visited by Santer.

George, along with the board of trustees (some members needing convincing, others, most importantly resident trustee Yiannis Vezyrogou, were plainly enthusiastic), proceeded with plans to establish a post-secondary component, another unique institution, this one to prepare students for careers in middle-level management, modeled on American

---

[19] George Draper to the author, 10 April 2002, AFS Archives. "The key person in Albania, especially at the start, was Bruce. It is impossible to overstate his influence in getting things going. He alone had the knack of, for example, getting the Albanian Ministry of Education to talk to the Albanian Ministry of Agriculture! He knew how to deal and inspire at that cabinet level, and he knew how to bring USAID and even EU agencies into the mix in a way that kept both ministries interested. I also remember talking to him on his return from his first trip there. 'Wonderful!' he said, beaming. 'It's just like Greece fifty years ago!'" George Draper to the author, Sorrento ME, 12 September 2002, AFS Archives.

lines, using English as the language of instruction. In a clear-headed statement of purpose, George later summed up the rationale:

> In an era when political and economic conditions are changing significantly, when information technology is developing with phenomenal speed, and at a time when the role of rural society is being defined, the need for applied higher agricultural education is taking on new urgency. Moreover, the rapid changes in the structure of Greek and European agricultural economy and the dynamic and qualitative development of the food processing industry indicate the need for an education that will prepare students for new professions in the broader agricultural sector.[20]

He pointed to the significance of European Union leadership and the power of globalization:

> Greece's membership in the European Union as well as the general globalization of standards and regulation will continue to exert a strong influence on the Greek agricultural and food industry. The influence is characterized not only by the imposition of standards, regulations, ceilings on production, and incentives, but almost equally by policies that encourage and support professional re-training, cooperation between Balkan countries and the "animation" *(re-development)* of rural areas.[21]

A study conducted in 1994, and recommended to the board of trustees in 1995, advised that

> the American Farm School establish on its campus a College of Agricultural Studies whose mission would be to produce outstanding managers, technicians, and leaders in the

---

[20] George Draper, "Report from the Dimitris Perrotis College of Agricultural Studies," American Farm School Meeting of the Board of Trustees, 18–20 February 1999, AFS Archives.

[21] "President's Report to the Board of Trustees," 9–11 November 1994, AFS Archives. Development of the rural areas is better known as "animation" in Europe and in the United States as "agricultural extension."

new era of agricultural industry. The College would be built on the ninety-year old American Farm School tradition of mixing theoretical education with practical, hands-on training and would emphasize professional orientation and guidance, establishing formal and informal relationships in the private sector of the agricultural and food industry in Greece and the Balkans, including internships and industrial placement.[22]

At this meeting George underlined certain points as core assumptions:

> The need for high quality secondary education will continue
> The college must not channel resources away from the secondary school. The college will be administered separately, keep separate books and accounting. The college will be self-sustaining
> The college will focus exclusively on agricultural studies in a broad sense
> The curriculum will emphasize practical management and business skills. Scientific and academic studies will be taught on applied basis
> The application pool will be of high quality and geographic and socio-economic diversity, and will seek to be attractive to other Balkan as well as Greek students.[23]

To ensure that the new institution would not draw funds away from the secondary school, a capital campaign, led once again by trustee Mimi Lowrey in the United States and Pavlos Condellis in Greece, was launched in 1995. Joann Ryding, director of development and public relations since 1989, coordinated the campaign in Greece. Benefactor Aliki Perrotis, an Athenian and long-time donor to the farm school scholarship fund, stepped forward with a donation of $4.9 million, the largest ever received by the school, to endow the college in her late husband Dimitris' memory. The campaign exceeded its goal of $6,650,000. Although every college student pays tuition, 40 percent of the student body does receive partial scholarships.

---

[22] Ibid.
[23] Ibid.

Based on the American academic model, the two-year curriculum at Perrotis College offers options in tune with worldwide demand for trained professionals in the agriculture, food, and natural resource sectors, and focuses on the overall goal of sustainability in agriculture. Students specialize in such disciplines as management for agribusiness, marketing for the food industry, or management of agricultural production systems, leading to an associate degree. Knitting together Perrotis College and the Greek business community are eight-week paid internships carried out by students in agricultural enterprises and research laboratories. When graduates face the world, diplomas in hand, they are much sought after owing to their practical skills, management training, fluency in English, and experience in information technology. Beginning with the first graduating class in 1998, Perrotis alumni have gone on to complete their undergraduate degrees at American and European universities, some ranking first in their class and obtaining fellowships for graduate study or career placements of their choice in the agricultural industry.

Although it was on the rise, familiarity with information technology was, in 2001, still not prevalent in Greece. One observer stated that "Greece started late and has been kept back by bad technology and inadequate infrastructure." "Getting people wired has become an important goal for Greece, which lags behind the other 14 members of the European Union." The explanation offered for the tardiness is that "Computer use has been slow to spread through an economy rooted in agriculture and small, family-owned businesses."[24] At the farm school, a fiber optic network linking 8 major campus and farm buildings provides high-speed internet access to more than 150 computers used by students and staff. And software for farm business, record keeping, livestock management, and field mapping was already in place to support all components of the college teaching program. In the secondary school, the student to computer ratio in 2000 was 5:1 as compared to an average for state schools of 26:1. Unlike at other schools, all the farm school computers for students' use have internet access.[25]

The role of the European Union in the school's financial stability and in its ability to broaden its scope cannot be overstated. Speaking frankly about the European Union's powerful role, George admits, "We

[24] "Survey of Technology," *International Herald Tribune*, 23 March 2001.
[25] *The Sower*, no. 149, 2000.

may look back on this decade as the decade of European Union funding. We have survived because of that funding." In fact, from 1995 through 1999 the school received from Delors II Social Fund $1.6 million per year, and from Mediterranean Overall Program (MOP), from approximately 1992 through 1999, some $2 million for capital projects.[26] Thanks principally to this support, which the school sought aggressively, the endowment tripled in the course of George's tenure from $8 million to $24 million; in fact, it became unnecessary to draw on the endowment until his last two years. Another helpful feature of European Union funds was that the farmers received subsidies enabling them to pay their children's fees toward room and board, unlike earlier years when they simply could not scare up a drachma to fulfill their obligations.

Under another European Union umbrella came the European Center for Agricultural Information, a venue where discussions on rural development take place in the form of workshops, seminars, and conferences that function in tandem with the department of Lifelong Learning. Mindful of the wide variety of programs in adult education and rural development the school has offered since the end of World War II, the Greek government in 1997 designated the institution as a center for adult education. Since then, the school's Lifelong Learning department has brought seminars and workshops out to the villages, and welcomed to campus those in need of retraining, especially the unemployed. The importance of this service becomes evident when we learn that according to a 1999 report released by the European Commission, Greece demonstrated one of the highest rates of jobless people who remained unemployed more than twelve months before finding another job. Greece, with 11 percent unemployment, a scourge that affects especially young people, shared first place with Italy and Spain.[27]

Another farm school innovation came to the fore when in 1991, a pilot computer network was developed, linking geographically dispersed Greek farmers to databases located on the school's campus. In 1996, the school, in cooperation with the Young Farmers of the Argolis

---

[26] George Draper, interview with author, Wayland MA, 5 February 2000, written, AFS Archives; George Draper, "Report to the Board of Trustees," 19–20 May 1999, AFS Archives.

[27] For an excellent article on the link between unemployment and education in Greece, see Apostolos Lakassas, "What Isn't Taught: The Degrees of Unemployment," *Kathimerini*, 22 March 2000, 3.

(Peloponnesos), the Greek Telecommunications Organization and other partners, created a system of more than twenty-five clusters of Web-accessible services for farmers and others, including commodity prices, new research findings, and information on subsidies, training, and employment opportunities.

The fourth pillar, comprising the school's two farms, facilities unrivaled in Greece or the Balkans, are used for demonstration and practical training. The campus farm is divided into horticulture, minor livestock, and dairy departments. Horticulture includes field crop, vineyard, nursery and greenhouse production. Eggs, chicks, broilers, and turkeys are raised by the poultry unit. The purebred Holstein dairy herd is an industry leader in productivity and quality of milk. At the Zannas Farm by 1999, projects such as soil reclamation, drainage, and irrigation systems were completed and the next year the buildings for classrooms and lodging were refurbished. A total of $1 million had been invested, providing a singular venue for Perrotis College to conduct practical applications, demonstrate its accomplishments, and undertake practical training in agricultural production, agribusiness, and natural resource management. The farm school is still the only place in Greece where students at all levels can receive practical agricultural training in such breadth.

In view of the country's environmental problems, the farms provide a welcome setting for study and application of sustainable agricultural projects, a major interest for the secondary school and Perrotis College. For instance, in just one of several field studies during 2002, "Sweet Corn Production Using Low Input," seven students under the supervision of Associate Professor of Agricultural Production Athanasios Gertsis, applied 50 percent fewer nutrients than the average for Greek farmers, avoided all spray with fungicides or insecticides, and applied a drip irrigation system that reduced water use by nearly 50 percent.[28]

In George's first year, he dedicated his stewardship decade, 1990–2000, to the environment with the goal of fostering agriculture that was sustainable and environmentally safe. To that end, a composting system replaced the failed biogas system. Acreage was set aside as chemical-free areas where student learned to grow various crops using sustainable agricultural methods. To aid in water conservation, field

---

[28] *The Sower*, no. 149, 2000.

crops are irrigated using mainly recycled wastewater. In an intriguing project initiated by George in 1994, plots of campus land attract some 200 avid gardeners of all ages from greater Thessaloniki, who cultivate mainly local and heritage varieties of vegetables and flowers using no chemicals and no mechanization. They are guided by Perrotis College instructor Beatrice Winterstein, who is also a pioneer in the process of organic certification in Greece.

The total farm school plant appeals to numerous communities. In 1995, for example, CEDEFOP (European Centre for the Development of Vocational Training), a European Union agency, relocated from Berlin to Thessaloniki. In searching for temporary quarters, the organization was attracted to the school. With upgrading, the old AMAG building would serve as office space; the farm and campus environment seemed the perfect setting for a group dedicated to comparative studies in the field of vocational studies; other facilities for occasional conferences could be rented from the school; infrastructure for computer installation was in place. All in all, the inviting prospect proved to be exactly the site for CEDEFOP'S first home in Greece.[29]

Four years later when CEDEFOP moved to its new installation up the hill behind the farm school, it vacated a renovated, handsome structure to which the organization had added a large three-story building. This complex houses both the academic facilities and dormitory for the Dimitris Perrotis College of Agricultural Studies. Every building on campus is a work in progress, but this one, originally the barracks built by the German occupying forces, has lost its original identity. In its latest permutation its corridors reverberate with the voices of students and faculty. Only those who knew the original German barracks in the dark days of the war, or a decade or so after, grow pensive as they contemplate it in its new guise, with its cheery pastel façade and large windows.

## 2000 and Beyond

One of Greece's highest honors was bestowed on the American Farm School at the 27 December 2001 awards ceremony of the Academy of

---

[29] Johan van Rens, director of CEDEFOP, interview by author, Thessaloniki, 19 March 2000, written, AFS Archives.

Athens. Established in 1926 as a continuation of Plato's Academy, the Academy of Athens is the country's nonpareil scientific and scholarly body, whose annual awards—usually to individuals but rarely to institutions—make up a national honor roll of distinguished achievements in every aspect of science and the humanities. The academy singled out the American Farm School of Thessaloniki for "the education it provides to rural youth and for its contribution to the rural development of Greece since 1904." Accepting the award on behalf of the school was Pavlos Condellis, vice chairman of the board of trustees.[30] The school was first honored by the academy in 1932 with a silver medal "for the education of 80 to 100 rural boys per year."

On hand along with Pavlos Condellis to accept the award was the president of the school since 1999, David Buck. David assumed the presidency of the American Farm School after a career in international education on four continents in such places as Prague, Vienna, Kuala Lumpur, the Dutch Antilles, and Cyprus. No stranger to Greece, he had come with his wife, Patti, to Thessaloniki in 1977 to work at the Pinewood School, he as a math teacher and Patti as a middle school teacher. At the farm school, David and Patti have immersed themselves in environmental issues, and both are devoted to supporting English language teaching to insure that students reach their potential. An experienced administrator, David uses his expertise to maintain the institution on firm financial footing while providing for the growth of both endowment and campus infrastructure.

The fifth pillar, which infuses the entire structure with meaning, is the spirit of the school—the sensibility that sets it apart from all other schools in Greece and probably in the United States. It is the spirit that has historically engaged the thirty-six members of the board of trustees, a band of devoted people who give their time, dollars, and euros to insure that the school endures, as well as enthusiastic volunteers and friends in the United States and Greece. The original spiritual infusion communicated by the Protestant missionaries still maintains if one can say that respect for the individual, moral rigor, generosity to one's community, heartfelt hospitality to the stranger, hard work ("to work is to pray"), a balanced personality ("the head, the hands, and the heart"), a striving for excellence, and a spiritual inclination are strictly missionary

---

[30] "Extract from the Report of the General Secretary of the Academy of Athens, Nikolaos Matsaniotis," 27 December 2001, Academy of Athens Archives.

values. Of course they are not. But the school's constant striving for these principles ties it to the original ethic. True, these principles have been secularized, worked into the boarding experience, worked into school celebrations, worked into daily student life, expressed in the long line of capable and successful graduates, and for all we know, even worked into the soil, as Dimitris Zannas has suggested. The religious underpinning has for decades been supplied by the Greek Orthodox faith, the heritage of most of the students. And that, concludes every one concerned, is exactly how it should be.

Looking toward Mount Olympus from the steps of Princton Hall, 1927.

Bruce McKay Lansdale and Elizabeth [Tad} Krihak Lansdale

Queen Frederica and King Paul, center, with Bruce Lansdale, left, and Charlie House, right. Ann House is in background. 50th Anniversary Celebration, 1955.

Queen Frederica, King Paul, Crown Prince Constantine with Charlie and Ann visit Haskell Cottage, the first Farm School building, constructed in 1903. 50th Anniversary Celebration, 1955.

50<sup>th</sup> Anniversary Parade.

The Lansdale family: Christine, Michael, Jeffrey and David with Bruce and Tad.

Bruce Lansdale ensured that a practical, hands-on approach would be the hallmark of an American Farm School education.

The visit of Queen Anna Maria of Greece in 1967.

"Then it really doesn't matter which road you take."

One day Hodja was resting at a crossroads on the edge of the village. A stranger stopped to ask for directions. When Hodja asked him which village he was heading for, the stranger looked hesitant and said he was not really sure. "Then it really doesn't matter which road you take," said Hodja, with a trace of a smile on his face.

Bruce used parables of Middle Eastern sage Nasredin Hodja to enlighten and guide audiences around the world in *Cultivating Inspired Leaders*, the title of a book he published in 2000.

Lyndon Baines Johnson visited the School as U.S. Vice President in 1963, accompanied by Lady Bird and Linda Bird. Students presented the visitors with a donkey and the vice presidential family reciprocated with the gift of a tractor.

Tad and Bruce accompanying Greek leader Konstantinos Karamanlis on a visit to the U.S. Trade Pavilion at the Thessalonki International Fair, early 1960s.

U.S. Ambassador to Greece Henry R. Labouisse with his wife, Eve Curie Labouisse. Bruce Lansdale is left and Theodoros Litsas right, early 1960s.

His Holiness Athenagoras I, Patriarch of the Greek Orthodox Church, greatly honored the School with his visit in 1963.

Celebration of the American holiday of Thanksgiving is a time-honored tradition at the Farm School.

Farm School students today.

The Farm School has been vital to the dairy industry of Greece throughout the century. Greek war relief, June 1948: calves born to Elsadora and Elsalita, heifers donated to Greece by the Borden Company, with two students from the Girls School.

A member of the
Farm School's
famous Holstein
Friesian herd.

Farm School ceremonies provide a stage to promote the latest farming practices.

Short Courses for adult farmers spread Farm School know-how throughout the rural areas.

Theodoros Litsas was a towering figure in the School's history, from his arrival as an Asia Minor refugee born in Smyrna to his tragic death in 1963. Four decades of students and adult trainees were touched by his charismatic spirit and love of life.

Licensed to drive a tractor, early 1950s.

Women poultry Farm School graduates with their own poultry units.

Farm School students learned to fabricate hot water showers and other modern fixtures in their village homes.

Norman Gilbertson, center, circa 1946, leading a team of relief workers in Northern Greece.

Joice Nankivell Loch and Sydney Loch.

The Tower in Ouranoupolis.

Activities of the earliest days of the Girls School, documented by photojournalist Nancy Crawshaw, whose papers and photographs are housed in the Manuscript Division of the Department of Rare Books and Special Collections of the Princeton University Library. Copyright © Nancy Crawshaw.

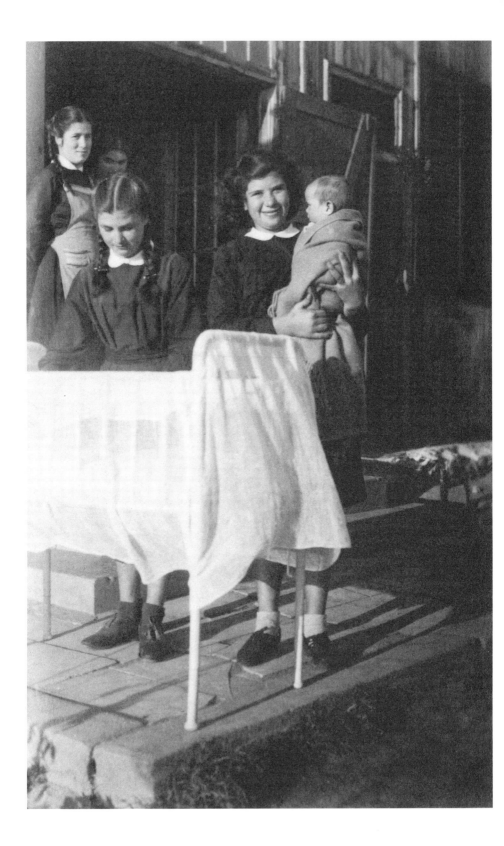

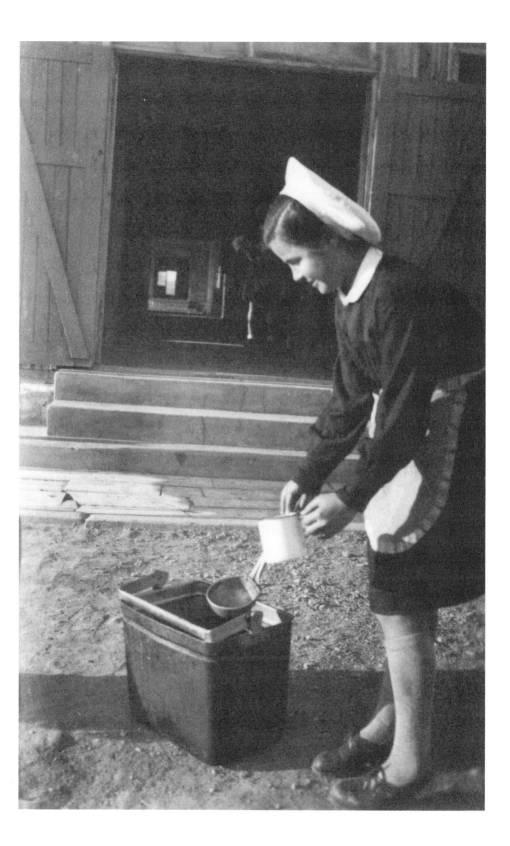

Girls School students marching with their American Farm School male counterparts, 1949, probably in the Saint Dimitrios Day Parade on 26 October.

Pegasus design by Joice Loch.

"Manos the Music Man Casts Spell Over the Students" reads the headline from the Farm School magazine, *The Sower*. Staff and students showed their deepest respectfor the legendary Greek composer during his 2 November 1973 campus visit by making him an honorary graduate of the American Farm School.

Princeton Hall, condemned after the earthquake of 1978 and reconstructed with trustee leadership and support. Reopened in 1986.

The Farm School founded Greek Summer, a community service program for U.S. teenagers based in Greek villages, in 1970. It is among the most successful and longstanding programs for youth in international living.

George Draper and Charlotte Whitney Draper

Mrs. Aliki Perrotis cutting the ribbon to inaugurate the Dimitris Perrotis College of Agricultural Studies, 7 October 1996.

European Commission President Jacques Santer visited the School 11 May 1997; to the right, George Draper.

David Wesley Buck and Patricia Hoag Buck

Inauguration of the Dimitris and Aliki Perrotis Library, 7 October, 2001. David Buck to the left of Mrs. Perrotis, Chairman of the Board of Trustees Alexander Drapos to right, with portrait of Dimitris Perrotis.

27 December 2001. Vice Chairman of the Board of Trustees Pavlos J. Condellis, left, accepting the Award of the Academy of Athens commending the American Farm School for "the education it provides to rural youth and for its contribution to the rural development of Greece since 1904."

American Farm School campus circa 2000.

American Farm School
Campus circa 2000

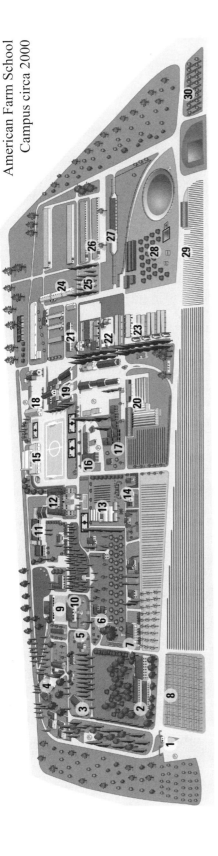

1. Entrance
2. AMAG Buildings
3. Chapel of St. John Chrysostomos
4. Haskell Cottage
5. Metcalf House
6. Hastings House
7. Cincinnati Hall
8. Yiannis Boutaris Demonstration Vineyard
9. Massachusetts Hall
10. James Hall
11. Charles and Ann House Dormitories
12. Princeton Hall. The Dimitris and Aliki Perrotis Library
13. Student Greenhouses
14. Garden Center and Winery
15. Labouisse Youth Center
16. Rochester Hall
17. Short Course Center
18. Constantinides Farm Machinery Laboratory
19. Sherrill Quadrangle. The Dimitris Perrotis College of Agricultural Studies
20. Demonstration Greenhouse
21. Piggery
22. Dairy
23. Research Quadrangle
24. Incubator/Hatchery
25. Cemetery
26. Poultry
27. Cow Shed
28. Waste Management Facility
29. DPCAS Experimental Plots
30. Community Gardens

# Selected Bibliography for Book 2

## I Primary Sources

A. The history of the American Farm School has been drawn extensively from documents in the American Farm School Archives. Some of these items lack dates, titles, page numbers, or authorship, including newspaper clippings and magazine articles. Items that lack complete citations are identified in the endnotes and here as AFS Archives to indicate that they are filed in the American Farm School Archives. Materials from the archives used in research for this book include:

Articles in Greek and American newspapers and magazines
Minutes of the American Farm School board of trustees
Reports by director or president to the board of trustees
Farm School media such as the *Newsletter, The Sower, O sporéas*
Special Reports written by outside evaluators or Farm School evaluating committees
Bruce and Tad Lansdale's correspondence
Unpublished reports that bear on education, academic and agricultural matters
Unpublished memoirs and documentary photos with text

B. Interviews conducted through correspondence or in person gave a valuable dimension to understanding the school. Those interviewed are cited in the endnotes and include:

Government officials
American Farm School alumni
Members of the American Farm School board of trustees
American Farm School faculty and administration

Educators from various other Greek and American universities and
schools
Professionals in the field of agriculture

C. To develop the Greek historical context some primary sources
were consulted such as:
Documentary photos/ texts
Reports from the European Union and other agencies
Greek government reports.
Research by American, British, and Greek scholars

A selected list appears below:

Crawshaw, Nancy. "Quaker Experiment in Peasant Education: The
Domestic Training School in Salonika." *Sunday Times* (London),
17 August 1951, education supplement, n.p.
Cultural Affairs Office, American Embassy, Athens, Greece.
"Education in Greece." Unpublished, 1965. AFS Archives.
Ministry of Foreign Affairs of Greece, Department of Political
Science and Public Administration. *Documents of the History of
the Greek Jews: Records from the Historical Archives of the
Ministry of Foreign Affairs*. Researched and edited by Photini
Constantopoulou and Thanos Veremis. Athens: Kastaniotis
Editions, 1998.
European Commission. *The Agricultural Situation in the Community,
1993*. Brussels, 1994.
Elder, Ann. "Farming the Future." *The Athenian Magazine* (July
1993): 16–19.
Food and Agriculture Organization of the United Nations. *The
Management of Agricultural Schools and College: A Manual for
Practical Use*. Rome: 1985.
Fourtouni, Eleni. *Greek Women in the Resistance: Journals, Oral
Histories*. Selected, translated, and introduced by Eleni Fourtuni.
New Haven: Thelphini Press, 1986.
Gage, Nicholas. *Eleni*. New York: Random House, 1983.
Gibertson, Norman. "Greece 1945–1953." Unpublished, 1997. AFS
Archives.

Hadjimankos, Nicholas. "An Incredible Adventure: The True Story of My Capture and Escape From the Communists." Unpublished, 1986. AFS Archives.

Karmas, C. A. et al. *Occupation and Educational Demand of Lyceum Students: Development Over Time.* Athens: Center of Planning and Economic Research, 1990.

Karmas, C. A., et al, *Prosdokíes kai thesis ton spoudastón tou technoloikoú ekpaideftikou idrímatos Athinón (Expectations and Outcomes of Students in the Technological Institutes of Athens.* Athens: Ministry of Education and Religion, 1986.

Klavdianou, Papadaki, et al. "La Pluriactivité et les activités agrotouristiques-agroartisanals, une solution pour le mantien des femmes-agricultrices dans l'éspace rural grec: la zone montagneuse de Pindos." Unpublished paper, n.d. Author's possession.

Klavdianou Papadaki A."Agrotikí ekpaídefsi: geotechnikí paidía kai érevna" ("Agricultural Training: Vocational Education and Research"). Unpublished paper, n.d. Author's possession.

Kontos, C. William. "Address to the Meeting of the Farm School Board of Trustees." 4 February 1967. AFS Archives.

Koulaouzides, George Archimedes. "Product Evaluation of the Curriculum of the American Farm School and a Proposal for the Development of a Distance Learning System for Agricultural Adult Education 1998." M.S. thesis, University of Surrey, 1998.

Koutsouris, Alexandros. "Dierévnisi ton krísimon paragónton pou sinthéondai me tin ekpaídefsi ton neoeiseroménon sti yeoryía" ("Identification of the Critical Factors That Relate to the Education of New Professionals in Agriculture"). Ph.D. dissertation, Agricultural University of Athens, 1994.

Linn, Alan. "Thessaloniki Farm School: The American Farm School Is Having a Valuable Impact on Greek Agriculture." *The Farm Quarterly* (Summer 1966): 74–149.

Littell, Robert. "They're Helping the Greeks to Help Themselves." *Readers' Digest* (September 1960): 129–32.

Loch, Joice. *A Fringe of Blue: An Autobiography.* London: John Murray, 1968.

———. *Prosforion Ouranopolis: Rugs and Dyes.* Istanbul: American Board Publication Department, 1964 .

McGrew, William. "Report to the Board of Trustees of Anatolia College, 1974–1999." 1999. AFS Archives.

Ministry of Press and Mass Media, Secretariat General of Information. *About Greece*. Athens: 1999.

Papadopoulos, George. *To pistévo mas* (*Our Creed*). Volume 4. Athens: 1969.

Post, George. "Eulogy for Alexander Allport." 20 November 2000. AFS Archives.

Sanger, Clyde. *The Unitarian Service Committee Story*. Toronto: Stoddart Publishing Co., 1986.

Prosser, Margaret. "A Macedonian Oasis." *National Plant, Flower and Fruit Guild Magazine* (1934): n.p.. AFS Archives.

Terzis, Nikos. "Continuing Education, Initial and Continuing Training and Adult Training as a Continuum: The Case of Greece." 1 November 1994. Author's possession.

Trimis, Andonis. "My Relationship with the American Farm School of Thessaloniki, Greece Since 1954." Unpublished, 1977. AFS Archives.

"The American Farm School in Thessaloniki: An 86–Year Institution Still Filling a Vital Need." *Greek-American Trade* (November/December 1990): 36–42.

## II Secondary Sources

### A. Monographs, General Histories

Allen, Harold. *Come Over into Macedonia*. New Brunswick NJ: Rutgers University Press, 1943.

Bita, Lili. *The Scorpion and Other Stories*. New York: Pella Publishing Co., 1998.

Bien, Peter, *Kazantzakis and the Linguistic Revolution in Greek Literature*. Princeton NJ: Princeton University Press, 1972.

Blum, Richard and Eva. *Health and Healing in Rural Greece*. Stanford: Stanford University Press, 1965.

Campbell, John and Philip Sherrard. *Modern Greece*. London: Ernest Benn Limited, 1969.

Clogg, Richard, and George Yannopoulos, editors. *Greece Under Military Rule*. London: Secker & Warburg, 1972.

Clogg, Richard. *A Short History of Modern Greece*. Cambridge: Cambridge University Press, 1980.

*Close, David, H.* The Origins of the Greek Civil War. *London: Longman, 1995.*

Coufoudakis, Van et al. *Greece and the New Balkans: Challenges and Opportunities*. New York, Pella Publishing Co., 1999.

Couloumbis, Theodore A. et al. *Foreign Interference in Greek Politics: An Historical Perspective*. New York: Pella Publishing Co., 1976.

Couloumbis Theodore A., and John O. Iatrides, editors. *Greek American Relations: A Critical Review*. New York: Pella Publishing Co., 1980.

Dicks, T. B. *The Greeks: How They Live and Work*. New York and Washington: Praeger Publishers, 1971.

Eder, K., and K. and M. Kousis, editors. *Environmental Politics in Southern Europe: Actors, Institutions and Discourses in a Europeanizing Society*. Dordecht: Kluwer Academic Publishers, 2001.

*Area Handbook for Greece, 1970*. DA PAM 550. Washington, DC, 1970.

*Area Handbook for Greece, 1977*. DA PAM 550-87. Washington, DC, 1977.

*Area Handbook for Greece*. Washington, DC, 1994.

Hanson, Victor Davis. *The Other Greeks: The Family Farm and the Agrarian Roots of Western Civilization*. Berkeley: University of California Press, 1999.

Holden, David. *Greece Without Columns: The Making of the Modern Greeks*. London: Farber and Farber, 1972.

Jelavich, Barbara. *History of the Balkans: Twentieth Century*. Volume 2. Cambridge: Cambridge University Press, 1983.

Kariotis, Theodore, editor. *The Greek Socialist Experiment: Papandreou's Greece, 1981–1989*. New York: Pella Publishing Co., 1992.

Kassimeris, George. *Europe's Last Red Terrorists: The Revolutionary Organization 17 November*. London: Hurst & Company, 2001.

Keeley, Edmund. *Inventing Paradise: The Greek Journey, 1937–47*. New York: Farrar, Straus and Giroux, 1999.

Lansdale, Bruce M. *Master Farmer: Teaching Small Farmers Management*. Boulder CO: Westview Press, 1986.

———, and Robert A. McCabe (photographs). *Metamorphosis Or, Why I Love Greece*. New Rochelle NY: Caratzas Brothers, Publishers, 1979.

———. *Cultivating Inspired Leaders: Making Participatory Management Work*. West Hartford CT: Kumarian Press, 2000.

Legg, Keith, and John Roberts. *Modern Greece: A Civilization on the Periphery*. Boulder CO: Westview Press, 1997.

Macrides Roy, editor. *Modern Political Systems: Europe*. Englewood Cliffs, NJ: Prentice-Hall, Inc. 1987.

Mazower, Mark, editor. *After the War Was Over: Reconstructing the Family, Nation, and State in Greece, 1943–1960*. Princeton and Oxford: Princeton University Press, 2000.

———. *Inside Hitler's Greece: The Experience of Occupation, 1941–1944*. New Haven and London: Yale University Press, 1993.

———. *The Balkans: A Short History*. New York: Random House, 2000.

McNeill, J. R. *The Mountains of the Mediterranean World: An Environmental History*. Cambridge: Cambridge University Press, 1992.

McNeill, William H. *American Aid in Action: 1947–1956*. New York: Twentieth Century Fund, 1957.

McNeill, William H. *The Metamorphosis of Greece Since World War II*. Chicago: University of Chicago Press, 1978.

Pettifer, James. *The Greeks: The Land and People Since the War*. London: Penguin Books, 2000.

Reynolds, David. *One World Indivisible: A Global History Since 1945*. New York: Viking Press, 2000.

Sakellariou, M. B., editor. *Macedonia: 4000 Years of Greek History and Civilization*. Athens: Ekdotike Athinon, 1983.

Sanders, Irwin, T. *Rainbow in the Rock: The People of Rural Greece*. Cambridge MA: Harvard University Press, 1962.

Stavrianos, Lefteris. *The Balkans Since 1453*. Hinesdale IL: Dryden Press, 1958.

Stevens, Everett, *Survival Against All Odds: The First 100 Years of Anatolia College*. New Rochelle: Arisitide D. Caratzas, 1986.

Tsoucalas, Constantine, *The Greek Tragedy*. Baltimore: Penguin Books, 1969.

Vouras, Paul. *The Changing Economy of Northern Greece Since World War II*. Thessaloniki: Institute for Balkan Studies, 1962.

Woodhouse, C. M. *The Restorer of Greek Democracy*. New York: Oxford University Press, 1982.

Journal Articles

Avdela, Efi. "The Teaching of History in Greece." *Journal of Modern Greek Studies* 18/2 (October 2000): 239–49.

Cavounidis, Jennifer. "Capitalist Development and Women's Work in Greece." *Journal of Modern Greek Studies* 1/2 (October 1983): 321–38.

Close, David, H. "Environmental Crisis in Greece and Recent Challenges to Centralized State Authority." *Journal of Modern Greek Studies* 17/2 (October 1999): 325–52.

Daoutopoulos, George. "The Prospects for Community Development in Greece." *Annals of Public and Cooperative Economics* 67/2 (1996): 281–90.

Dragonas, Thalia and Anna Frangoudaki. "Introduction." *Journal of Modern Greek Studies* 18/2 (October 2000): 229–38.

Goss, Peter, editor. "The Marshall Plan and its Legacy." *Foreign Affairs* 76/3 (May–June 1997): 157–221.

Halkias, Alexandra. "Give Birth for Greece! Abortion and Nation in Letters to the Editor of the Mainstream Greek Press." *Journal of Modern Greek Studies* 16/1 (May 1998): 111–38.

Jackson, Richard. "The Role of U. S. Colleges Abroad: Anatolia as a Case Study." *Mediterranean Quarterly* 10/4 (Fall 1999): 24–43.

Kassotakis, Michael. "Technical and Vocational Education in Greece and the Attitudes of Greek Youngsters Toward It." *Journal of the Hellenic Diaspora* 8/1–2 (Spring–Summer 1981): 81–93.

Kontogiannopoulou-Polydorides et al. "Citizen Education: Silencing Crucial Issues." *Journal of Modern Greek Studies* 18/2 (October 2000): 287–303.

Nassiakou, Maria. "The Tendency toward Learning in the Greek Countryside." *Journal of the Hellenic Diaspora* 8/1–2 (Spring–Summer 1981): 63–69.

Noutsos, Babis, "Change and Ideology in the General Lyceum Program: Two Examples." *Journal of the Hellenic Diaspora* 8/1–2 (Spring–Summer 1981): 53.

Papacosma, S. Victor. "The Jews of Salonika." *Midstream* (December 1978): 10–14.

Paraschos, Manolis. "Is Greek Media at the Threshold of a Historic Change." *Emphasis: A Journal of Hellenic Issues* 1/1 (April–June 1995): 94–100.

Petras, James. "Greek Socialism, Walking the Tightrope." *Journal of the Hellenic Diaspora* 9/1 (Spring 1982): 7–15.

Sideris, Aloe. "Some Information About Private Schools." *Journal of Hellenic Diaspora* 8/1–2 (Spring–Summer 1981): 55–61.

Voulgaris, Yannis, "The Political Attitudes of Greek Students." *Journal of Modern Greek Studies* 18/2 (October 2000): 269–85.

Wassar, Henry, "A Survey of Recent Trends in Greek Higher Education." *Journal of the Hellenic Diaspora* 6/1 (Spring 1979): 85–95

# APPENDIX:

# SECONDARY LEVEL CURRICULUM: AMERICAN FARM SCHOOL 1904–2003

**1904–1929**
Five-year curriculum
Students enroll after elementary school
Language of instruction: English

**1929–1965**
Four-year curriculum
Students enroll after elementary school
Language of Instruction: Greek

**1965–1971**
Four-year curriculum
Students enroll after elementary school
Graduates receive certificate equivalent to gymnasium (junior high school)

**1971–1977**
SEGE Three-year curriculum
Students enroll after completing nine years of education
Graduates receive a Foreman's Certificate
Graduates eligible to enter junior technical colleges and later to enter university

## 1977–1998

TEL Three-year curriculum
Students enroll after completing nine years of education
Graduates receive a lyceum (senior high school) certificate
Graduates eligible to enter university
and
TES Two-year curriculum
Students enroll after completing nine years of education
Graduates receive a vocational school certificate
or
Graduates eligible to enter TEL

## 1998–

EL Three-year curriculum
Students enroll after completing nine years of education
Graduates receive a comprehensive certificate
Graduates eligible for university
and
TEE Two- or Three-year curriculum
Graduates receive a vocational school certificate or can enter EL

# INDEX